新能源发电

英汉—汉英技术词典

主　编　陈铁华　李　岩
副主编　李树刚　范永玉
陈晓玉

中国水利水电出版社
www.waterpub.com.cn

内 容 提 要

以英汉、汉英两种方式收录了风能、太阳能、核能、水能等新能源发电常用专业技术词汇，并对风能发电和太阳能发电专业词汇词义进行解释和说明。以汉英方式收录的相关、相近专业常用词汇包括电力继电保护、电气工程、机械工程、监控与通信、工程力学、流体力学等。

本词典可供从事新能源发电工程的科研人员、工程技术人员、高等院校师生以及相关专业的技术人员参考与使用。

图书在版编目（CIP）数据

新能源发电英汉-汉英技术词典 / 陈铁华，李岩主编. -- 北京 : 中国水利水电出版社，2014.3
ISBN 978-7-5170-1814-8

Ⅰ. ①新… Ⅱ. ①陈… ②李… Ⅲ. ①新能源－发电－词典－英、汉 Ⅳ. ①TM61-61

中国版本图书馆CIP数据核字(2014)第049142号

书　名	新能源发电英汉—汉英技术词典
作　者	主　编　陈铁华　李岩 副主编　李树刚　范永玉　陈晓玉
出版发行	中国水利水电出版社 （北京市海淀区玉渊潭南路1号D座　100038） 网址：www.waterpub.com.cn E-mail：sales@waterpub.com.cn 电话：(010) 68367658（发行部）
经　售	北京科水图书销售中心（零售） 电话：(010) 88383994、63202643、68545874 全国各地新华书店和相关出版物销售网点
排　版	中国水利水电出版社微机排版中心
印　刷	北京瑞斯通印务发展有限公司
规　格	140mm×203mm　32开本　15.5印张　599千字
版　次	2014年3月第1版　2014年3月第1次印刷
印　数	0001—3000册
定　价	**48.00**元

前言

近年来，能源气候变化成为各国关注的主要议题。在国家政策支持下，我国以风能发电、太阳能发电、核能发电为主的新能源发电规模迅速攀升，装机容量占全国电力总装机容量的比例不断提高。

目前，风力发电在全世界已处于一个蓬勃发展的时期。在我国风力发电成本已接近常规能源发电成本，风电规模也受国家政策及能源发展趋势的影响迅速扩大，风电技术得到明显提高。

太阳能更是一种清洁、高效和永不衰竭的新能源。在新世纪中，各国政府都将太阳能资源利用作为国家可持续发展战略的重要内容。太阳能发电具有安全可靠、无噪声、无污染、制约少、故障率低、布置简便以及维护方便等特点，应用面较广，现在全球装机总容量已经开始追赶风力发电。

科技发展与技术进步，逐渐提高新能源发电技术水平，同时也吸引了大量科技工作者投身到新能源发电领域，从事科研、设计、建设、运行与维护工作。越来越多的科技工作者需要阅读新能源发电技术英文文献或编写英文资料，与国外技术交流的形式也日益增多。为了满足新能源发电领域科技工作者阅读专业文献和编写英文资料的需求，编写了这部《新能源发电英汉—汉英技术词典》，添补了新能源发电领域科技英语工具书的空白。

本词典收录了风能发电、太阳能发电、核能发电和水能发电等专业词汇、均按英汉、汉英两种方式编辑。在附录中对部分风能发电和太阳能发电专业术语及词汇定义进行了解释。词典还收集了机械工程、电气工程、流体工程、工程力学、继电保护、监控与通信等相关专业常用词汇。

本词典由长春工程学院陈铁华、李岩，中水东北勘测设计研究有限责任公司李树刚，松辽水利水电开发有限责任公司范永玉、陈晓玉共同编写。编写人员在编译时结合相关标准和规范，力求达到专业、简明、实用、现代的特点。本词典在编辑过程中，参阅了大量国内外专业资料，并有所引用，在此向相关作者及出版单位表示感谢。

由于时间仓促以及主编人员水平有限，本词典还有许多不足之处，希望读者能在使用过程中，提出宝贵意见，以便使本词典更加完善。

作者

2013 年 8 月

使用说明

1. 英汉词汇按字母顺序排列，汉英词汇按拼音顺序排列。
2. 圆括弧内的英文或汉字，表示可以省略，并且排序时不考虑括弧内的英文或汉字。圆括弧内的英文，是缩写词汇的完整表达或是英文词汇的缩写形式。圆括弧内的中文，是对释义的进一步解释或应用范围所作的说明。

   如：airfoil chord 翼（型）弦（线），即“翼弦”。

   LWECS (Large scale wind energy conversion system) 大型风能转换系统

   fixed-speed fixed-pitch (FS-FP) 恒速定桨距

   array spacing 阵内间距（风轮机的）
3. 汉语相同范畴释义间采用逗号“,”；不同范畴或不同词性的释义间采用分号“;”。
4. “/”前后的英文单词可以相互代换。如：air inlet /intake 等同于 air inlet 或 air intake，其含义都是“通风口，进气口”。
5. [ ] 表示专业范围或领域。如：reduced frequency [ˈfrɪkwənsi] [力] 简约频率，换算频率；约化频率；斯德鲁哈尔数。[力] 表示该词属于力学范畴。
6. 手册只对较难拼读的英文单词标注了音标。

目　录

英 汉 词 汇

风 能 发 电

1-D momentum theory [məʊ'mentəm] ['θɪərɪ] 一维动量矩理论

2-D height alimeter 二维测高仪

2-DOF system 二自由度系统

3-D aerodynamics 三维空气动力学

3-dimensional 3-component model 三维三分量模型

4-legged steel truss tower 四腿钢桁架式塔架

50 year return gust 50 年一遇的阵风

A

A weighted or C weighted sound pressure level A 计权或 C 计权声压级

absolute humidity 绝对湿度

ac synchronous converter ['sɪŋkrənəs] 交流同步转换器

AC transmission system [trænz'mɪʃən] 交流输电系统

accelerated test 加速试验

accelerating 加速

acceleration amplitude ['æmplɪtju:d] 加速度幅值

acceleration potential 加速势

accelerometer [əkˌselə'rɒmɪtə] 加速表，加速计

acceptance test 验收试验

accumulator [ə'kju:mjʊleɪtə] 储压罐

accuracy ['ækjʊrəsɪ] 精度

ACE (Aylesbury Collaborative Experiment) 埃尔兹伯里（风荷载）台做实验

acid fog 酸雾

acid gas 酸气

acid rain 酸雨

acknowledgement [ək'nɒlɪdʒmənt] 确认

acorn ['eɪkɔ:n] 整流罩

acoustic damping effect [ə'ku:stɪk] 声阻尼效应

acoustic environment 声环境

acoustic pollution 声响污染

acoustic radar 声雷达

acoustic reference wind speed [ə'ku:stɪk] 声的基准风速

acoustics [ə'ku:stɪks] 声学

across-flow/cross-flow 横流向的

across-wind correlation 横风向相关

across-wind cross correlation 横风向互相关

across-wind response 横风响应

across-wind test 横风响应试验

across-wind/cross-wind 横风向的

actinography [ˌæktɪ'nɒgrəfɪ] 日射计

actinometer [ˌæktɪ'nɒmɪtə] 日射表

activation power [ˌæktɪ'veɪʃən] 临界

功率

activation rotational speed 临界转速

active appendage [ə'pendɪdʒ] 主动附件

active control 主动控制

active current 有功电流

active pitch control 变桨距控制

active power 有功功率

active stall control 主动失速控制

active yaw/yawing [jɔː] 主动偏航

actual extracted power 实际输出功率

actual load 实际载荷；有效载荷

acute angle [ə'kjuːt] 锐角

added mass 附加质量

addendum modification on gears [ə'dendəm] 齿轮的变位

adhesive [əd'hiːsɪv] 带黏性的；胶黏，黏合剂

adiabatic boundary layer [ˌeɪdaɪə'bætɪk] 绝热边界层

adiabatic change 绝热变化

adiabatic chart 绝热变化图

adiabatic compression 绝热压缩

adiabatic condensation [kɒnden'seɪʃ(ə)n] 绝热凝结

adiabatic cooling 绝热冷却

adiabatic curve 绝热曲线

adiabatic degassing [diː'gæsɪŋ] 绝热放气

adiabatic energy storage 绝热贮能

adiabatic equation 绝热方程式

adiabatic equilibrium [ˌiːkwɪ'lɪbrɪəm] 绝热平衡

adiabatic evaporation [ɪˌvæpə'reɪʃən] 绝热蒸发

adiabatic expansion [ɪk'spænʃ(ə)n] 绝热膨胀

adiabatic flow 绝热流

adiabatic gradient ['ggreɪdɪənt] 绝热梯度

adiabatic heating/warming 绝热增温

adiabatic index 绝热指数

adiabatic lapse rate 绝热递增率

adiabatic layer 地热层

adiabatic line 绝热线

adiabatic process 绝热过程

adiabatic region 绝热区

adjustable pliers ['plɑɪəz] 可调钳

adjustable spanner 活动扳手

adjustable wall 可调（风洞）壁

adjusting plate 调整板

admissible concentration 容许浓度

admittance 导纳

admittance function 导纳函数

admittance parameter 导纳参数

adrift 飘浮

adsorption [æd'səːpʃən] 吸附

adsorption column 吸附柱

adsorption effect 吸附效应

advance ratio 前进比；进距比

advanced and innovative wind system ['ɪnəvetɪv] 革新型风能系统

advection [əd'vekʃ(ə)n] 平流

advection fog 平流雾

advection layer 平流层

advection region 平流区

advection scale 平流尺度

advective term 平流项
adverse pressure gradient [ˈgreɪdɪənt] 逆压梯度
adverse wind 逆风
aeolian deposit [iːˈəʊlɪən] 风积物
aeolian erosion [ɪˈrəʊʒ(ə)n] 风蚀
aeolian excitation [ˌeksɪˈteɪʃ(ə)n] 风成脱涡激励
aeolian feature 风成地貌
aeolian soil 风积土
aeolian tone 风声调
aeolian vibration 脱漏风振
aeragronomy 航空农业
aeration [eɪəˈreɪʃən] 换气
aerial [ˈeərɪəl] 航空的；天线
aerial cable 架空缆索；架空电缆
aerial contaminant [kənˈtæmɪnənt] 空气污染物
aerial contamination [kənˌtæmɪˈneɪʃən] 空气污染
aerial current 气流
aerial radiation thermometer [θəˈmɒmɪtə] 空气辐射温度计
aerodromometer 气流流速表
aerodynamic admittance [ˌeərəʊdaɪˈnæmɪk] [ədˈmɪt(ə)ns] 气动导纳
aerodynamic airfoil [ˈeəfɒɪl] 气动翼型
aerodynamic appendage [əˈpendɪdʒ] 气动力附件
aerodynamic area 气动力面积
aerodynamic attachment [əˈtætʃm(ə)nt] 气动力附件
aerodynamic augmentation device [ɔːgmenˈteɪʃ(ə)n] 气动增强装置
aerodynamic balance 气动天平；气动力平衡
aerodynamic behavior 气动性能
aerodynamic boundary layer 气动力边界层
aerodynamic brake 空气动力制动器
aerodynamic brake [ˌeərəʊdaɪˈnæmɪk] 空气动力刹车，气动刹车
aerodynamic braking system 气动制动系统
aerodynamic centre 气动中心
aerodynamic characteristics 气动特性
aerodynamic characteristics of rotor 风轮空气动力特性
aerodynamic chord of airfoil [kɔːd] [ˈeəfɒɪl] 翼型的气动弦线
aerodynamic damping 气动阻尼
aerodynamic data 气动力数据
aerodynamic derivative [dɪˈrɪvətɪv] 气动导数
aerodynamic derivative coefficient 气动导数系数
aerodynamic design 空气动力学设计
aerodynamic destabilizing 气动力失稳
aerodynamic device 气动装置
aerodynamic dilution [daɪˈluːʃn] 气动力稀释
aerodynamic directional instability 气动方向不稳定性
aerodynamic dissipation

[dɪsɪˈpeɪʃ(ə)n] 气动力耗散
aerodynamic disturbance [dɪˈstɜːb(ə)ns] 气动力扰动
aerodynamic effect 气动力效应
aerodynamic efficiency 气动效率
aerodynamic end effect 气动力端部效应
aerodynamic environment 气动环境
aerodynamic excitation [ˌeksɪˈteɪʃ(ə)n] 气动力激发
aerodynamic field 气动力场
aerodynamic force（空）气动力
aerodynamic force coefficient 气动力系数
aerodynamic force derivative [dɪˈrɪvətɪv] 气动力导数
aerodynamic loads 气动载荷
aerodynamic noise 气动噪声
aerodynamic power 气动功率
aerodynamic radius [ˈreɪdɪəs] 空气动力学半径
aerodynamic rotor 气动转子
aerodynamic roughness 气动粗糙度
aerodynamic roughness length 气动粗糙度长度
aerodynamic roughness parameter 气动粗糙度参数
aerodynamic self excitation 气动力自激
aerodynamic self-starter 气动自动启动装置
aerodynamic shadow 气动阴影，气动尾迹
aerodynamic shape 气动外形
aerodynamic shape correction 气动外形修改
aerodynamic shape optimization [ˌɒptəmɪˈzeʃən] 气动外形最优化
aerodynamic similarity [sɪməˈlærətɪ] 气动相似性
aerodynamic smoothness 气动光滑度
aerodynamic solidity [səˈlɪdɪtɪ] 气动实度
aerodynamic sound 气动声音
aerodynamic source of sound 气动声源
aerodynamic spoiler [ˈspɒɪlə] 气动扰流板
aerodynamic stability 气动稳定性
aerodynamic staring 气动力启动
aerodynamic stiffness 气动刚度
aerodynamic surface 空气动力表面；气动力作用面
aerodynamic test 气动力试验
aerodynamic torque [tɔːk] 气动力矩
aerodynamic trail 气动尾迹；航迹云
aerodynamic transfer function 气动力传递函数
aerodynamic twist 气动扭转
aerodynamic wake 气动尾流
aerodynamically fully rough [ˌeərəʊdaɪˈnæmɪkəlɪ] 气动充分粗糙的
aerodynamically induced vibration 气动力致振

aerodynamically smooth 气动光滑的
aerodynamically stable 气动力稳定的
aerodynamics [ˌeərəudaɪˈnæmɪks] 空气动力学
aeroelastic characteristics [ˌeərəuɪˈlæstɪk] 气动弹性特性
aeroelastic derivative [dɪˈrɪvətɪv] 气动弹性导数
aeroelastic effect 气动弹性效应
aeroelastic equation of motion [ɪˈkweɪʒ(ə)n] 气动弹性运动方程
aeroelastic feedback 气动弹性反馈
aeroelastic instability [ɪnstəˈbɪlɪtɪ] 气动弹性不稳定性
aeroelastic load 气动弹性载荷
aeroelastic mode 气动弹性模态
aeroelastic model 气动弹性模型
aeroelastic modeling 气动弹性模拟
aeroelastic oscillation [ˌɒsɪˈleɪʃən] 气动弹性振动
aeroelastic response 气动弹性响应；气动弹性反应
aeroelastic stability 气动弹性稳定性
aeroelastic stiffness 气动弹性刚度
aeroelastic structure 气动弹性结构物
aeroelastic transfer function 气动弹性传递函数
aeroelastic vibration 气动弹性振动
aeroelasticity 气动弹性
aeroelasticity [ˈeərəuˌelæsˈtɪsətɪ] 气动弹性，气弹性
aeroelastics 气动弹性力学
aerofoil [ˈeərəfɒɪl] 机翼
aerofoil data 翼型数据
aerofoil family 翼型族
aerofoil (airfoil) [ˈeərəfɒɪl] 翼型；[车] 导风板
aerogenerator [ˌeərəuˈdʒenəreɪtə] 风力发电机
aerological ascent [ˌeərəuˈlɒdʒɪkəl] 高空观测
aerological station 高空观测站
aerology [eəˈrɒlədʒɪ] 高空气象学
aerometric network [ˌeərəˈmetrɪk] 高空气象网
aeromotor [ˈeərəməutə] 风力发动机；航空发动机
aeronautical type wind tunnel [ˌeərəˈnɔːtɪkl] 航空风洞
aeronautics [eərəˈnɔːtɪks] 航空学
aeropause [ˈeərəpɔːz] 大气上界
aeroradioactivity 大气放射性（强度）
aero-sensitive 气动力敏感的
aerosol [ˈeərəusɒl] 气溶胶；气雾剂；烟雾剂
aerosphere [ˈeərəusfɪə] 大气层；空气圈
aerostatics [ˌeərəuˈstætɪks] 空气静力学
aeroturbine 风轮机
aerovane [ˈeərəˌveɪn] 风向风速仪
after sales service 售后服务
afterbody effect 后体效应
against wind 逆风，迎风

ageing test 老化试验
ageostrophic wind 非地转风
aileron [ˈeɪlərɒn] 副翼
aileron control 副翼控制
aiming 对风，调向（风轮的）
air at rest 静止空气
air bearing 空气轴承
air boundary layer 空气边界层
air braking system 空气制动系统
air change rate 换气率
air circulation 空气环流；空气循环
air cleaning 空气净化
air conditioning 空（气）调（节）
air contaminant [kənˈtæmɪnənt] 空气污染物
air cooler 空气冲却器；风冷装置
air cooling 风冷
air current 气流
air curtain 气幕
air cushion [ˈkʊʃ(ə)n] 气垫
air dam [车] 前扰流板
air damping 空气阻尼
air deflector 导风板
air density 空气密度
air diffusion [dɪˈfjuːʒ(ə)n] 空气扩散
air discharge 排气；空中放电
air draft 气流；空气通风
air drag 空气阻力，气动阻力
air drain 空气流泄；通风管
air dry weight 风干重
air duct 风道
air eddy 空气涡旋
air exhaust [ɪgˈzɔːst] 排气
air flow 气流；空气流量
air flow guide 导流罩，导流板
air flow meter 空气流量计
air flow noise 气流噪声
air flow rate 气流流速；气流流量
air flow structure 气流结构
air flue 风道；烟道
air friction [ˈfrɪkʃ(ə)n] 空气摩擦阻力
air gap 气隙
air gauge [geɪdʒ] 气压表
air hatch [hætʃ] 通风口
air humidity 空气湿度
air inclusion 空气杂质
air infiltration [ˌɪnfɪlˈtreɪʃən] 空气渗透，漏风
air injection [ɪnˈdʒekʃ(ə)n] 空气引射
air inlet/intake 通风口，进气口
air jet 空气喷射；喷气口
air load 气动载荷；风载
air management [ˈmænɪdʒm(ə)nt] 大气管理
air mass 气团
air meter 风速计；气流表
air mixture [ˈmɪkstʃə] 空气混合物
air monitor 大气（污染）监测器
air monitoring station 大气监测站
air outlet 空气出口；放气口
air parameter [pəˈræmɪtə] 空气环境参数
air permeability [ˌpɜːmɪəˈbɪlɪtɪ] 透气性
air pressure 空气压力
air pressure measurement 空气压

力测量
air resistance [rɪˈzɪst(ə)ns] 空气阻力；气动阻力
air resistance brake 气动阻力制动器
air resonance [ˈrez(ə)nəns] 空气共振
air scoop [skuːp] 进气喇叭口；进风口
air seal 气密
air sizing 风力分级
air spoiler [ˈspɒɪlə] 扰流板
air stream 气流
air temperature [ˈtemp(ə)rətʃə] 气温
air temperature measurement 空气温度测量
air tight 气密的
air to air heat exchanger 空气-空气热交换器
air torrent 空气急流
air turbulence [ˈtɜːbjʊl(ə)ns] 大气湍流；大气湍流度
air vane 导风板；风标
air vent 排气口；通风孔
air ventilation [ˌventɪˈleɪʃ(ə)n] 通风，换气
air ventilation window 通风窗
air-water interface [ˈɪntəfeɪs] 气水界面
airbrake 空气动力制动器
airflow line（气流）流线
airflow pattern 流谱
airflow pulsation [pʌlˈseɪʃən] 气流脉动
airfoil [ˈeəfɒɪl] 翼型
airfoil camber 翼型弯度
airfoil characteristics 翼型特性
airfoil chord 翼（型）弦（线）
airfoil contour 翼型廓线
airfoil drag 翼型阻力
airfoil profile [ˈprəʊfaɪl] 翼型；翼型廓线
airfoil section 翼型，翼剖面
airfoil shaped blade 翼型叶片
airfoil shield [ʃiːld] [车] 顶装导风罩
airfoil theory 翼型理论
airfoil wind machine 翼型板风力机
airfoil with reverse camber [rɪˈvɜːs] 反弯度翼型
airiness 通风
air-mass fog 气团雾
air-screw 空气螺旋桨
airscrew slip 螺旋桨滑流
air-shaft 通风井
air-spring 气垫；空气弹簧
airstream deflector 导风板
air-termination system [ˌtɜːmɪˈneɪʃən] 接闪器
align [əˈlaɪn] 对准；校直；定位
allen key [ˈælən] 六方
allen wrench 六方扳手
allowable limit [əˈlaʊəbl] 容许极限
allowable stress 容许应力
alloy [ˈælɒɪ,əˈlɔɪ] 合金
all-steel disc couplings 全钢膜片
along wind deflection [dɪˈflekʃ(ə)n] 顺风向挠度
along-flow 顺流向的

along-wind 顺风向的
along-wind acceleration [əkseləˈreɪʃ(ə)n] 顺风向加速度
along-wind buffeting 顺风向抖振
along-wind correlation 顺风向相关
along-wind cross-correlation 顺风向互相关
along-wind fluctuation [ˌflʌktjʊˈeɪʃən] 顺风向脉动
along-wind galloping [ˈgæləpɪŋ] 顺风向驰振
along-wind oscillation [ˌɒsɪˈleɪʃən] 顺风向振动
along-wind response 顺应，顺风向反应
alteration [ˌɔːltəˈreɪʃən] 变更；改造
alternately spaced vortex [ˈvɔːteks] 交替分布涡
alternating current (AC) 交流电
alternating current machine 交流电机
alternating load 交变荷载
alternating pressures 交变压力
alternating stress 交变应力
alternating voltage 交流电压
alternating vortex shedding 交替旋涡脱落
alternative energy source 替代能源
alternative energy tax law 替代能源税收法
alternative machine/alternator 交流发电机
altimeter [ˈæltɪmiːtə] 高度表
altitude [ˈæltɪtjuːd] 海拔
aluminum bars [əˈljuːmɪnəm] 铝棒
aluminum vertical ladder 铝爬梯
ambient humidity [ˈæmbɪənt] [hjʊˈmɪdɪtɪ] 环境湿度
ambient noise 环境噪声
ambient pollution burden 环境污染负荷
ambient power density 周围风能密度
ambient temperature [ˈæmbɪənt] [ˈtemp(ə)rətʃə] 环境温度，周围温度
ambient turbulence [ˈtɜːbjʊl(ə)ns] 周围湍流（度）
ambient vibration 环境振动
ambient wind angle 周围迎风角
ambient wind 周围风
ambient windspeed 周围风速
ambient windstream 周围气流
America windmill; Western windmill 美国风车；西部风车
ammeter [ˈæmɪtə] 安培计；电流表
amplifier [ˈæmplɪfaɪə] 放大器
amplitude [ˈæmplɪtjuːd] 幅值
amplitude histogram of wind-speed [ˈæmplɪtjuːd] 风速较差直方图
amplitude modulation (AM) 调幅
amps 安培数
anabatic flow [ˌænəˈbætɪk] 上升气流
anabatic wind 上坡风；山谷风
anaflow 上升气流
analog signal [ˈænəlɒg] 模拟信号
analogue board 模拟盘
analogue control 模拟控制
anchor [ˈæŋkə] 锚；抛锚；锚定

anemocinemograph 电动风速计
anemogram [ə'neməgræm] 风力自记曲线
anemograph [ə'neməgrɑːf] 风速计，风力记录仪
anemology [ˌænɪ'mɒlədʒɪ] 测风学
anemometer [ˌænɪ'mɒmɪtə] 风力计；风速仪
anemometer factor 风速计系数
anemometer mast 测风杆
anemometer tower 测风塔
anemometer with stop watch 停表风速计
anemometrograph [ˌænɪmə'metrəˌgrɑːf] 自计风速计
anemometry [ˌænɪ'mɒmɪtrɪ] 风速测定法
anemoscope [ə'neməskəʊp] 风向仪
aneroid barometer ['ænərɒɪd] [bə'rɒmɪtə] 无液气压表；无液晴雨表；空盒气压计
angle design 倾角设计
angle grinder 角锉
angle of approach 接近角
angle of attack of maximum lift 最大升力攻角，临界攻角
angle of attack (of blade) (叶片)几何攻角
angle of declination [ˌdeklɪ'neɪʃ(ə)n] 偏斜角
angle of deflection [dɪ'flekʃ(ə)n] 偏移角，挠度
angle of depression [dɪ'preʃ(ə)n] 俯角
angle of inclination [ɪnklɪ'neɪʃ(ə)n] 倾角
angle of lag 滞后角，移后角
angle of oscillation [ˌɒsɪ'leɪʃən] 摆动角
angle of roll [航] 滚转角；[车] 侧倾角；[船] 横遥角
angle of scattering ['skætərɪŋ] 散射角；散布角
angle of shear [ʃɪə] 剪切角
angle of spread 展开角；散布角
angle of twist 扭转角
angle of wind approach 迎风角，来流迎角
angle of yaw 偏航角
angle plate 角盘
angle position encoder 角度位置编码器
angular connection 弯接头
angular contact ball bearing 角接触球轴承
angular correlation ['æŋgjʊlə] 角相关
angular deflection [dɪ'flekʃ(ə)n] 角偏转
angular displacement [dɪs'pleɪsm(ə)nt] 角位移
angular distribution [dɪstrɪ'bjuːʃ(ə)n] 角分布
angular momentum [mə'mentəm] 角动量；动量矩
angular velocity [vɪ'lɒsɪtɪ] 角速度
anisotropic turbulence [ˌænaɪsə(ʊ)'trɒpɪk] ['tɜːbjʊl(ə)ns] 各向异性湍流
annealing plate [ə'niːlɪŋ] 退火

annual available wind energy [ˈænjʊəl] [əˈveɪləb(ə)l] 年有效风能
annual average 年平均
annual average wind speed 年平均风速
annual average wind-power density 年平均风能密度
annual Energy Calculation 年度能源计算
annual energy production 年发电量
annual extreme mean daily temperature 年最高日平均温度
annual extreme-mile wind speed 年极端里程风速
annual flow 环状流
annual growth rate 年增长速度
annual maximum 年最高
annual mean wind speed 年平均风速
annual variation 年变化
annual wind direction diagram 年风向图
annual wind regime 年风况
annulus [ˈænjʊləs] 环面
annulus gear 内齿圈
antenna [ænˈtenə] 天线；触角
anti condensation heater 防冷凝加热器
antifriction [ˌæntɪˈfrɪkʃən] 润滑剂
anti-resonance frequency 反谐振频率
anti-rolling device 消摇装置
antitriptic wind 摩擦风
apparent angle of attack [əˈpær(ə)nt] 视迎角
apparent gravity 视重力
apparent mass 表观质量
apparent power [əˈpærənt] 视在功率
apparent sound power level 视在声功率级
apparent specific gravity 表观比重
apparent velocity 视速度
apparent wind 视风
applied aerodynamics [ˌeərə(ʊ)daɪˈnæmɪks] 应用空气动力学
applied fluid dynamics 应用流体动力学
applied force 外加力
approach velocity 来流速度
approaching wind [əˈprəʊtʃɪŋ] 迎面风；来流
approximate similarity [əˈprɒksɪmət] [sɪməˈlærətɪ] 近似相似型
arbor [ˈɑːbə] 柄轴；心轴
arc discharge 电弧放电
arc profile 圆弧翼型
arc reignition 电弧重燃
arc suppression coil [səˈpreʃən] 消弧线圈
arc welding 电弧焊
arc-control device 灭弧装置
arc-extinguishing chamber 灭弧室
Archimedean screw [ˌɑːkɪˈmiːdɪən] 阿基米德螺线
area density [ˈdensɪtɪ] 面密度
area moment 面积（一次）矩
area moment of inertia [ɪˈnɜːʃə] 面

积惯性矩；面积二次矩
area velocity 掠面速度
area-mean pressure 面积平均压力
armature [ˈɑːmətjʊə] 电枢
array losses 数组损失
array of wind turbines [əˈreɪ] 风轮机阵
array pattern [ˈpæt(ə)n] 阵列型式
array spacing 阵内间距（风轮机的）
articulated [ɑːˈtɪkjuːleɪtɪd] 接合；链接；有关节的
articulated blade [ɑrˈtɪkjʊleɪtɪd] 铰接叶片
articulation [ɑːˌtɪkjʊˈleɪʃ(ə)n] 铰接
artificial climate [ɑːtɪˈfɪʃ(ə)l] [ˈklɑɪmət] 人工气候
artificial cloud 人造云
artificial disturbance [dɪˈstɜːb(ə)ns] 人为扰动
artificial draft [drɑːft] 人工通风
artificial precipitation [prɪˌsɪpɪˈteɪʃ(ə)n] 人工降雨
artificial precipitation stimulation [ˌstɪmjəˈleʃən] 人工催雨
artificial rainfall 人工降雨
artificial velocity gradient [vɪˈlɒsɪtɪ] [ˈgreɪdɪənt] 人工（形成）速度梯度
artificial ventilation [ˌventɪˈleɪʃ(ə)n] 人工通风
artificial wind 人造风
ascendant current [əˈsend(ə)nt] 上升气流
ascending air [əˈsendɪŋ] 上升空气
ash air 含灰空气
ash content 含灰量
ash fall 灰尘沉降；烟尘阵落
ash-laden gas 含尘（烟）气
aspect ratio 纵横比；叶片展弦比
aspirated psychrometer [ˈæspəreɪtɪd] [psɑɪˈkrɒmɪtə] 通风式干湿表
aspirated thermometer [θəˈmɒmɪtə] 通风式温度计
assisted stall 辅助失速
assorted files 分类排列
asymmetric damping [ˌesɪˈmətrɪk] 非对称阻尼
asymmetric force 非对称力
asymmetric mode 非对称模态
asymptotic distribution [ˌæsɪmpˈtɒtɪk] [dɪstrɪˈbjuːʃ(ə)n] 渐近分布
asymptotic method 渐近法
asymptotic stability 渐近稳定性
asynchronous generator [eɪˈsɪŋkrənəs] [ˈdʒenəreɪtə] 异步发电机
asynchronous machine 异步电机
atmosphere 大气压
atmosphere moisture [ˈætməsfɪə] [ˈmɒɪstʃə] 大气湿度；大气水分
atmosphere stability 大气稳定性
atmospheric advection [ætməsˈferɪk] [ədˈvekʃ(ə)n] 大气平流
atmospheric boundary layer [ˈbaʊnd(ə)rɪ] 大气边界层
atmospheric circulation [sɜːkjʊˈleɪʃ(ə)n] 大气环流
atmospheric convection [kənˈvekʃ(ə)n] 大气对流
atmospheric damping 大气阻尼

atmospheric density [ˈdensɪtɪ] 大气密度
atmospheric depth 大气深度；大气厚度
atmospheric diffusion [dɪˈfjuːʒ(ə)n] 大气扩散
atmospheric diffusion equation 大气扩散方程
atmospheric pressure 大气压力；大气压强
atop (mountain) 在（山）上
attachment coefficient [əˈtætʃmənt] [ˌkəʊɪˈfɪʃənt] 附着系数
attenuation [əˌtenjʊˈeɪʃən] 变薄；稀薄化；变细；衰减
attenuation factor 衰减系数
audibility criterion 可听度判定标准
auger [ˈɔːgə] 打孔钻；螺丝钻
auto correlation function 自动相关函数
automated polishing (of blade surface) 自动抛光（叶面）
automated trimming 自切割成型（叶片）
automatic control 自动控制
automatic generation 自动生成
automatic meter reading 自动抄表
automatic oscillograph [ɒˈsɪləʊgrɑːf] 自动录波仪
automatic voltage controller (AVC) 自动电压控制器
auto-reclosing cycle 自动接通周期
autotransformer [ˌɔːtəʊtrænsˈfɔːmə] 自耦变压器
auxiliary circuit [ɔːgˈzɪljərɪ] 辅助电路
auxiliary device 辅助装置
auxiliary electrical equipment 配套电气设备
availability [əˌveɪləˈbɪlətɪ] 可利用率；有效率
availability factor 可用系数
available data 现有数据；可用数据
available power 可用功率；有效功率
average life 平均寿命
average noise level 平均噪声
average wind speed 平均风速
A-weighted filter A 加权滤波器
A-weighted noise measurements A 加权的噪声测量
axial flow 轴流
axial flow induction factor 轴流诱导因子
axial flow interference factor [ˌɪntəˈfɪərəns] 轴流干扰因素
axial induced velocity 轴向诱导速度
axial induction factor 轴向诱导因子
axial momentum equation 轴向动量方程
axial pitch 轴向齿距
axial thrust 轴向推力；风轮推力
axial thrust force 轴向推力
axial velocity 向速度
axle 轴心
axle box 轴柜

azimuth angle [ˈæzɪməθ] 方位角
azimuth flange [flændʒ] 方位角法兰
azimuthal binning 方位角分级
azimuthal position 方位角位置

B

back saw 背锯
background noise 背景噪声
background response 背景响应
backlash [ˈbækˌlæʃ] 后座；反斜线；后冲
backup power 备用电源
back-up system 备用系统
ball bearing 滚珠轴
ball peen hammer 圆头手锤
ball saddle 滚珠支撑
ball-eye 球头挂环
ball-hook 球头挂钩
ball-nut 球形螺母
ball-screw 球形螺钉
band-pass filter 带通滤波器
bar chart 柱状图
bar magnet 磁条
bare conductor 裸导线
barometer [bəˈrɒmɪtə] 气压计；晴雨表
base（风力电机）底座
base pad 基座
basic error 基准误差
basic regional wind velocity [ˈriːdʒənl] 地区基本风速
battery backup 备用电池
battery charger 电池充电器
battery module 蓄电池模块
baud 波特
baud rate 波特速率
Baufort scale 蒲田风级表
beam trammel [ˈtræməl] 骨架
beam wind 横风；侧风
bearer [ˈbeərə] 支架；托架；支座；载体
bearer rope 承载钢缆
bearing 轴承
bearing block 轴承座
bearing distortion 轴承失真
bearing friction 轴承摩擦
bearing housings 轴承箱
bearing life 轴承寿命
bearing seats 轴承座
bearing skidding 轴承打滑
bearing without outer ring 无外圈轴承
Beaufort force 蒲福风力
Beaufort number 蒲福（级）数
Beaufort scale 蒲福风级
bedplate 底座（机舱）
before the wind 顺风
bell crank 曲柄
belt drive 皮带传动
BEM method [ˈmeθəd] 边界元法；叶素动量方法
BEM model 边界元模型
bend natural frequency [ˈfrɪkwənsɪ] 弯曲固有频率

bending deflection 弯曲挠度
bending displacement 弯曲位移
bending frequency 弯曲频率
bending machine 折弯机
bending mode 弯曲模态
bending moment [ˈbendɪŋ] 弯曲力矩；挠矩
bending oscillation [ˌɒsɪˈleɪʃən] 弯曲振动
bending response 弯曲响应；弯曲反应
bending stress 弯曲应力
beneficial wind effect 有利风效应
benefit/cost ratio [ˈreɪʃɪəʊ] 性价比
Bernoulli binomial distribution [baɪˈnəʊmɪəl] 伯努利二项分布
Bernoulli constant 伯努利常数
Bernoulli surface 伯努利面
Bernoulli trail 伯努利尾迹
Bernoulli vector [ˈvektə] 伯努利矢量
Bernoulli's equation [ɪˈkweɪʒən] 伯努利方程
Bernoulli's law 伯努利定律
Bessel functions 贝塞尔函数
Betz 贝兹
Betz' law 贝兹定律
Betz limit 贝兹极限
bevel gear [ˈbevəl] 斜角；斜齿轮
bevel-edge shisel 斜缘薄钢板
bidimensional flow [ˌbaɪdɪˈmenʃənl] 二维流动
bidirectional and reversible 4 quadrant operation 双向可逆四象限运行
bi-directional vane 双向风标
bifurcated rivet [ˈbaɪfəːkeɪtɪd] 开口铆钉；分叉的铆钉；开叉钉
bifurcation plume [baɪfəˈkeɪʃ(ə)n] 分岔型羽流
bilinear approximation [əˌprɒksɪˈmeɪʃən] 双线性近似
bilinear or 'Tustin' approximation 双线性或"塔斯廷"逼近
bimetallic [ˌbaɪmɪˈtælɪk] 双金属的
binomial distribution [baɪˈnəʊmɪəl] 二项分布
binomial expansion [ɪkˈspænʃ(ə)n] 二项展开式
binormal distribution 双正态分布
Biot-Savart law 毕奥—萨伐尔定律
bivane [ˈbaɪveɪn] 双向风标
black body 黑体
black ice 雨凇
blackout [ˈblækaʊt] 断电；停电
blade [bleɪd] 叶片
blade aerodynamics [ˌeərə(ʊ)daɪˈnæmɪks] 桨叶空气动力（学）
blade angle 叶片角，桨叶角
blade angle of attack 叶片迎角
blade assembly [əˈsemblɪ] 叶片组装
blade axis [ˈæksɪs] 叶片轴
blade azimuth angle [ˈæzɪməθ] 叶片方位
blade bearing 变桨轴承
blade bonding 叶片焊接
blade calculation 叶片计量
blade chord [kɔːd] 叶弦

blade coning angle 桨叶锥角
blade connection 叶片连接
blade deflections 叶片偏转；叶片挠度
Blade Element - Momentum Theory [ˈθɪərɪ] 叶素—动量理论
Blade Element - Momentum(BEM) method [ˈmeθəd] 叶素—动量方法
blade element theory 叶素理论
Blade Elements Method 叶素方法
blade geometric twist [ˌdʒɪəˈmetrɪk] 叶片几何扭转
blade geometry [dʒɪˈɒmɪtrɪ] 叶片几何
blade loss 叶片损失
blade material [məˈtɪərɪəl] 叶片材料
blade of constant chord width 等截面叶片
blade orbital angle [ˈɔːbɪt(ə)l] 叶片运行角
blade pitch 桨距，叶片节距
blade pitch (angle) 桨距角
blade pitch control 桨距控制
blade pitch set angle 叶片安装角
blade planform taper ratio 片平面形状尖削比
blade profile [ˈprəufaɪl] 叶片翼型；叶片轮廓
blade repair 叶片修补
blade resonance 叶片共振
blade root [ruːt] 叶根
blade root area 叶根面积
blade root load 叶根载荷
blade section 叶片截面
blade section pitch 叶片剖面桨距
blade segment 叶段
blade setting angle 叶片安装角
blade shape 叶片形状
blade shedding of ice 叶片甩冰
blade solidity [səˈlɪdɪtɪ] 叶片实度
blade spar 叶梁
blade stall 叶片失速
blade structural design [ˈstrʌktʃərəl] 叶片结构设计
blade structure 叶片结构
blade strut 叶片撑杆
blade test bed/bench 叶片试验台
blade tip speed 叶尖速度
blade tip-root loss factor 叶端损失系数
blade twist [twɪst] 叶片扭曲
blade vibration [vaɪˈbreɪʃən] 叶片振动
blade weight 叶片重量
bladed cuff 桨叶根套
blade-passing frequency [ˈfrɪkwənsɪ] 叶片穿越频率
blades bearing 叶片轴承
blade-tip vortex 叶尖涡
blade-tip-speed 叶尖速度
blade-tower clearance [ˈklɪərəns] 叶片—塔架间隙
blankcted area 遮蔽面积；气动阴影
blanketing effect 气动阴影效应
blast 强风；过载
bleeder valve [ˈbliːdə] [vælv] 溢流阀

block and tackle 滑轮组
blocking 锁定
blocking control 堵塞（油路）控制
blowpipe [ˈbləʊpaɪp] 吹风管
bluff body 钝体
bluff body aerodynamics [ˌeərə(ʊ)daɪˈnæmɪks] 钝体空气动力学
bolt [bəʊlt] 螺栓；螺钉；支持；维持
bolt assembly 螺栓组装
bolt fatigue stresses [fəˈtiːg] [stres] 螺栓疲劳应力
bolt load increment [ˈɪnkrɪmənt] 螺栓载荷增量
bolted flange joint 螺栓法兰接头
bonding bar 等电位连接带
bonding conductor 等电位连接导体
bore [bɔː] 钻孔；钻
bound circulation 附着环量
bound vortex [ˈvɔːteks] 附着涡；约束涡
bound vortex system 附着涡系
bound vorticity 附着涡量；束缚涡
boundary conditions 边界条件
boundary constraint 边界约束；（风洞）洞壁约束
boundary effect 边界效应；（风洞）洞壁效应
boundary layer 边界层
boundary layer accumulation [əkjuːmjʊˈleɪʃ(ə)n] 边界层堆积；边界层增厚
boundary layer profile 边界层（速度）廓线
boundary layer reversal 边界层回流
boundary layer separation [sepəˈreɪʃ(ə)n] 边界层分离
boundary layer structure 边界层结构
boundary layer vorticity [vɔːˈtɪsətɪ] 边界层涡量
box transformer substation 箱式变电站
bracket 托架；支架
brake（for wind turbines）刹车（风力机）；制动器
brake block 刹车片
brake design 制动器设计
brake disc 刹车盘
brake duty 刹车
brake flap [flæp] 制动盘；减速板
brake fluid [ˈflu(ː)ɪd] 刹车油
brake lining [ˈlaɪnɪŋ] 闸；刹车的衬里；闸衬片
brake mechanism 制动机构
brake pad 刹车片
brake position 刹车位置
brake releasing 制动器释放
brake setting 制动器设定；闭合
brake shoe [ʃuː] 闸瓦；刹车片
brake torque 制动转矩
brake-caliper [ˈkælɪpə] 制动钳
braking [ˈbreɪkɪŋ] 制动系统
braking disk 制动盘
braking loads 制动载荷
braking releasing 制动器释放

braking vane 刹车尾舵
break wind 防风林；风障
breakaway starting current 最初起动电流
breakaway torque 最初起动转矩
breakdown（电）击穿
breather（风洞）换气装置
breeder reactor 核能反应堆
broad range variable speed [ˈveərɪəbl] 广泛变速
broken line graph 折线图
bubble breakdown 气泡击穿
built-in coning 内置锥形
bulk modulus of elasticity [ˈmɒdjʊləs] [elæˈstɪsɪtɪ] 体积弹性模量
bulkhead 隔壁；防水壁
bulky cargo 超大货物；大件运输
bull gear 从动齿轮
buoyancy（lift）[ˈbɒɪənsɪ] 浮力；浮升
burble 涡流；湍流
burble angle 失速角
burnish 抛光
bus 总线
bus tie breaker 母联断路器
busbar 母线
bus-bar expansion 母线伸缩节
bus-bar separator 母线间隔垫
bus-bar support 硬母线固定金具
bushing 轴衬；套管
bushing tap grounding wire 套管末屏接地线
by-pass switch 旁路开关
byte 字节

C

C weighted sound pressure level C计权声压级
cabinet screw driver [ˈkæbɪtɪt] [skruː] 内阁螺丝刀
cable cutter/shear 电缆剪
cable reel 电缆盘
cable shoes 电缆靴
cable tie 尼龙扎带；束线带；绑线
cage rotor 笼型转子
calibrate 校准
caliper [ˈkælɪpə] 测径器；卡钳；弯脚器
callipers 卡钳
cam [kæm] 凸轮
camber line 弧线
Campbell diagram 坎贝尔图
camshaft [ˈkæmʃɑːft] 凸轮轴
cantilever [ˈkæntɪliːvə] 伸臂；悬臂；悬臂梁
cantilever beams 悬臂梁
cap nut 螺帽
capacitance [kəˈpæsɪtəns] 电容；电容量
capacitive components 电容元件
capacitor [kəˈpæsɪtə] 电容器
capacitor bank 电容器组
capacitor box 电容柜
capacitor for voltage protection

[ˈvəʊltɪdʒ] 保护电容器
capacity factor 利用率；功率；能力系数
carbon brush [ˈkɑːbən] 碳刷
cardan joint 万向接头
cardan shaft 万向轴
carrier to interference ratio (C/I) [ˌɪntəˈfɪərəns] 载波干扰比
Cartesian coordinate system 笛卡儿坐标系
cascade transformer 级联变压器
cast iron [ˈɑɪən] 铸铁
cast steel hub 铸钢轮毂
castellated coupling [ˈkæstəleɪtɪd] 牙嵌式连接
casting 铸件；铸造
castle nut 开槽螺母
catastrophic failure (for wind turbines) [ˌkætəˈstrɒfɪk] 严重故障(风力机)
cavity [ˈkævətɪ] 空穴；腔
cement [sɪˈment] 水泥；接合剂；粘牢
center distance 中心距
center gear [gɪə] 中心轮
center of buoyancy 浮力中心
center of disturbance 扰动中心
center of gravity [ˈgrævɪtɪ] 重心
center of lift 升力中心
center of mass 质量中心
center of oscillation [ˌɒsɪˈleɪʃən] 振动中心；摆动中心
center of pressure 压力中心
center of resistance [rɪˈzɪst(ə)ns] 阻力中心
center of rotation 转动中心
center of span 跨度中点；桥跨中点
center of stiffness 刚度中心
center of turn [车] 转向中心
central/centralized lubricating system 集中润滑系统
central calm 中心无风区
central limit theorem 中心极限定理
central span 中跨
centralized control 集中控制
centralized wind energy systems [ˈsentrəlɑɪzd] 集中式风能转换系统
centre bit 中心位
centrifugal fan [senˈtrɪfjʊgəl] 离心式通风机
centrifugal forces [senˈtrɪfjʊgəl] 离心力
centrifugal governor [ˈgʌv(ə)nə] 离心式限速器
centrifugal loads 离心载荷
centrifugal release units 离心释放装置
centrifugal stresses 离心应力
ceramic media 陶瓷研磨膏；陶瓷介质
ceramics [sɪˈræmɪks] 陶瓷；陶瓷技术
certification 认证
certified product 通过认证的产品
CFRP 碳纤维复合材料
chain wheel 滑轮
change-over circuit 换接回路；转

换回路

channeling effect 夹道效应

characteristic area 特征面积

characteristic correlation length 特征相关长度

characteristic curve [ˌkærəktəˈrɪstɪk] 特性曲线

characteristic dimension [dɪˈmenʃ(ə)n] 特征尺寸

characteristic equation [ɪˈkweɪʒ(ə)n] 特征方程

characteristic frequency 特征频率

characteristic function 特征函数

characteristic length 特征长度

characteristic life 特征寿命

characteristic parameter [pəˈræmɪtə] 特征参数

characteristic temperature 特征温度

characteristic velocity 特征速度

characteristic wavelength 特征波长

characteristics 特性；特性曲线

charge and discharge test 充/放电试验

charging reactive power 充电无功功率

charging (damping) resistor 充电（阻尼）电阻

cheese-head screw [skruː] 开槽圆柱头螺钉

chemical corrosion [kəˈrəʊʒən] 化学腐蚀

chisel [ˈtʃɪzəl] 凿子；砍凿

chord 弦长

chuck [tʃʌk] 用卡盘夹住

circuit breaker (CB) [ˈsɜːkɪt] 断路器

circular pitch 齿距

circulatory flow 环流

circumferential backlash 圆周侧隙

clamping screw 固定螺钉

clamping unit 夹具

clearance [ˈklɪərəns] 排除故障

clevis 挂板

climate 气候

climate zones 气候带

climatic chamber; climatizer 气候实验室

climatic chart 气候图

climatic classification 气候分类

climatic control 气候控制

climatic data 气候资料

climatic effect 气候效应；气候影响

climatic element 气候要素

climatic environment 气候环境

climatic fluctuation [ˌflʌktjʊˈeɪʃən] 气候变动

climatic noise 气候噪声

climatic stability 气候稳定度

climatic wind speed 气候风速

climatic wind tunnel 全天候（汽车）风洞

climatological characteristics [ˌklaɪmətəˈlɒdʒɪkəl] 气候学特性

climbing support 助爬器

closed circuit 闭合电路

closed loop control 闭环控制

closed-loop controller 闭环控制器

cluster controller 群控制器（控制

风电机组群的）
cluster of wind farms 风电场群
clutches 离合器
CNC lathe 数控车床
CO_2 emission [ɪ'mɪʃən] CO_2 排放
coastal anti-typhoon type 沿海抗台风型风机
coating 喷涂
coaxial cable ['kəʊ'æksəl] 同轴电缆
coefficient of performance 特性系数
coefficient of torsional ['tɔːʃənəl] 扭转刚度系数
coherence function [kəʊ'hɪərəns] 相干函数
cold climate version 低温型机组
collar ['kɒlə] 凸缘；套环；卡圈；安装环
collar bolt [bəʊlt] 凸缘螺栓
collection line 集电线路
collector 收集器（油路用）
collector ring 集电环
color-match sealant 多色密封剂
combination pliers ['plaɪəz] 台钳
commissioning 调试
commissioning test [kə'mɪʃənɪŋ] 投运试验
common earthing system 共用接地系统
communication and control equipment 通信和控制设备
communication interface 通信接口
communication-port lightning protection 通信端口防雷
commutation 换向
commutator 换向器
commutator segment ['kɒmjʊteɪtə] 换向片
compensate misalignments of shaft 补偿轴对中偏差
complete drive systems 成套传动设备
complex terrain 复杂地形带
compliant tower (soft tower) [kəm'plaɪənt] 柔性塔架，软塔架
component inspection 部件检验
component localization [ˌləʊkəlaɪ'zeɪʃən] 本地化元件
component of force [kəm'pəʊnənt] 分力；力分量
component of turbulence ['tɜːbjʊl(ə)ns] 湍流分量
composite beam bridge 组合梁桥
composite blade 复合材料叶片
composite insulation ['kɒmpəzɪt] 组合绝缘；合成绝缘子
composite material 复合材料
composite plume [pluːm] 复合羽流
composite spacer（联轴器用）复合式轮毂
compressed air storage 压缩空气罐
compressible fluid 可压缩流体
compressive strength-to-weight ratio [kəm'presɪv] 抗压强度与重量比
compressor 压缩物；压缩机
computational aerodynamics [ˌkɒmpjʊ'teʃənl] 计算空气动力学
computational fluid dynamics (CFD)

计算流体力学
computational mesh 计算网格
concave len [kɒn'keɪv] 凹面镜
concave-convex len 凸凹镜
concentrated load 集中荷载
concentric 同轴的
concrete tower 混凝土塔筒
condition diagnostics system 状态诊断系统
condition monitoring system 状态监视系统
condition number 条件数
conductivity [ˌkɒndʌk'tɪvɪtɪ] 导电性
conductor clamp 卡线钳
conductor holder 夹线器
cone angle 锥顶角
conical ['kɒtɪkəl] 圆锥的；圆锥形的
coning hinge [hɪndʒ] 锥形铰链
coning the rotor ['rəʊtə] 锥形转子
connecting rod [kə'nektɪŋ] 连接杆
connecting wind farm to power network 风电场接入电网
connection flange 连接法兰
connection point 并网点
connection-wire drop 引线压降
connector（爬梯竖杆间用）连接件
console [kən'səʊl] 控制台
constant chord blade 等截面叶片
constant frequency AC [工] 恒频交流电
constant rotational speed [rəʊ'teɪʃənəl] 恒速
constant speed operation 恒速运行
construction administration expense 施工管理费
contact 触头
contact stress 接触应力
contactor 接触器
contamination monitoring system [kən'tæmɪˌneɪt] 污染物检测系统
continental climate ['klɑɪmɪt] 大陆性气候
continental wind 大陆风
continuous flow [kən'tɪnjʊəs] 连续流
continuous operation 持续运行
continuous power [能]（最坏风况下）持续功率
continuum [kən'tɪnjʊəm] 连续介质
continuum flow 连续介质流动
continuum model 连续介质模型
contour ['kɒntʊə] 轮廓，外形；等值线
contraction cone/section [kən'trækʃ(ə)n]（风洞）收缩段
contrary wind ['kɒntrərɪ] 逆风
control apparatus [ˌæpə'reɪtəs] 控制电器
control building 控制室
control by stalling 失速控制
control cabinet ['kæbɪtɪt] 控制柜
control circuit 控制电路
control desk 控制台
control device 控制装置
control gear 控制设备
control mechanism ['mekətɪzəm] 控制机构
control panel 控制板
control strategy ['strætɪdʒɪ] 控制

策略
control surfaces 控制翼面
control system (for wind turbines) 控制系统（风力机）
control-gear 控制齿轮
controller gain 控制器增益
conventional power plants capacity 常规电厂容量
converter 换流器；变换器；变流器；变频器
convertor feathering 变频顺桨
cooling system 冷却系统
coordinate systems 坐标系
copper loss [ˈkɒpə] 铜损
Coriolis force [ˌkɔːrɪˈəulɪs] 科里奥利斯力（科氏力）；地球自转偏向力
Coriolis parameter [pəˈræmɪtə] 科氏参数
corner vane（风洞）拐角导流片
corner-vane cascade [kæsˈkeɪd]（风洞）拐角导流片栅
correction 修正
correction factor 修正因子
correction for buoyancy [ˈbɒɪənsɪ] 浮力修正
correction for Reynolds number [ˈrenəldz] 雷诺数修正
correlation analysis 相关分析
correlation coefficient 相关系数
correlation function 相关函数
correlation length 相关长度
corrosion (offshore) [kəˈrəuʒən] 侵蚀（海上）
corrosion [kəˈrəuʒ(ə)n] 风蚀，动力侵蚀
corrosion resistance tests [rɪˈzɪstəns] 耐腐试验
corrosive climate 腐蚀性气候
cost of electricity 发电成本
cost per kilowatt hour of the electricity generated by WTGS [ˈkɪləuwɒt] 度电成本
cost-saving system 成本节约系统
cotter pin 开口销
counted reading 计量值
counter EMF 反电势
counter weight 配重重锤
countersunk [ˈkauntəsʌŋk] 埋头孔；暗钉眼
countersunk head screw/rivet [skruː] 埋头铆钉
counter-type cup anemometer [ˌænɪˈmɒmɪtə] 记数式风杯风速计
counterweight [ˈkauntəˌweɪt] 平衡锤
coupling 联轴器
coupling agent 耦合剂（检测用）
coupling bolt 联结；接合；耦合；耦合性
coupling capacitor [kəˈpæsɪtə] 耦合电容
coupling medium 耦合剂（检测用）
coverage [ˈkʌvərɪdʒ] 覆盖；敷层；有效区域
C_p performance curve [kɜːv] 风能利用系数曲线
C_p tracking 风能利用系数跟踪

C_p-λ curve 风机性能曲线
cramp 钳位（电路）；压（夹）板；卡子
crane 吊车；起重机
crank handle 手摇曲柄
crankcase 曲柄轴箱
crankshaft [ˈkræŋkˌʃɑːft] 曲轴；机轴
creep distance 爬电距离
creeping flow 蠕动流
criterion [kraɪˈtɪərɪən] 标准；判据；准则
critical breakdown voltage 临界击穿电压
critical buckling stress [ˈbʌklɪŋ] 临界扭曲应力
critical condition [ˈkrɪtɪk(ə)l] 临界条件；临界状态
critical damping 临界阻尼
critical design case 临界设计情况
critical divergence wind speed [daɪˈvɜːdʒ(ə)ns] 临界发散风速
critical frequency 临界频率
critical load case for tower base 塔基的临界载荷
critical overturning wind speed 临界倾覆风速
critical pressure gradient [ˈgreɪdɪənt] 临界压力梯度
critical regime [reɪˈʒiːm] 临界状态
critical Reynolds number [ˈrenəldz] 临界雷诺教
critical speed 临界转速
critical stresses 临界应力
cross correlation function 交叉相关函数
cross flow 横向流动，横流向的
cross headwind 侧逆风
cross mark 十字标记
cross plot 散点图
cross slotted screw 十字长孔
cross wind 横风；横风向的
cross-flow wind turbine 横流式风轮机
crosshead [机] 十字头；丁字头
cross-over frequency [ˈfrɪkwənsɪ] 交叉频率
cross-peen hammer 横头锤
cross-section 横断面；横切面；截面
cross-sectional area 截面积
cross-spectrum [ˈspektrəm] 交叉谱
cross-wind buffeting 横风抖振
cross-wind diffusion 横风扩散
cross-wind direction 横风向的
cross-wind-axis wind machine 横风轴风力机
crosswise 成十字状地；交叉地
crosswise ribbing 横纹（防滑）
crowbar 撬棒
C-S analyzer 碳硫分析仪
cube factor 风速立方因子
cumulative frequency distribution [ˈkjuːmjʊlətɪv] 累积频率分布
cup anemometer [ˌænɪˈmɒmɪtə] 风杯风速计
cup-counter anemometer 计数风杯风速计
cup generator anemometer 磁感风杯风速计
cupped wind machine 杯式风力机

current collecting line 集电线路
current distortion rate 电流畸变率
current ration [ˈkʌrənt] 电流定值
current transformer 电流互感器
curvature function of airfoil [ˈeəfɒɪl] 翼型弯度函数
customized software package 定制的软件包
cut-in wind speed 切入风速
cut-off frequency 截止频率
cut-off plate 节流板；挡板
cut-out wind speed 切出风速
cut-outs 挖空；开孔
cutter 刀具；切割机
cutting disk 切割盘
cyclic loads 循环载荷
cyclic pitch 周变桨距
cyclic stress limit 周期性应力极限
cyclically alternating vortex [ˈsɑɪklɪklɪ] [ˈɔːltəˌneɪtɪŋ] 周期性交替涡
cylinder block [ˈsɪlɪndə] 缸体
cylinder head 缸头
cylinder-head gasket [ˈgæskɪt] 缸头垫片；垫圈；接合垫
cylindrical [sɪˈlɪndrɪkəl] 圆柱形，圆柱体；柱面
cylindrical gear 圆柱齿轮

D

daily amplitude [ˈæmplɪtjuːd] 日变幅
daily extremes 日极端值
daily maximum temperature 日最高温度
daily mean value 日平均值
damage equivalent load [ɪˈkwɪvələnt] 损伤等效载荷
damage frequency [ˈfrɪkwənsɪ] 损坏频率
damper 风门；阻尼器；防震器；防振锤
damping 阻尼
damping coefficient [ˌkəʊɪˈfɪʃənt] 阻尼系数
damping ratio 阻尼比
Danish wind turbine concept 丹麦风力机概念
Darrieus rotor 达里厄风轮
data acquisition [ˌækwɪˈzɪʃən] 数据采集
data analysis [əˈnæləsɪs] 数据分析
data base 数据库
data circuit [ˈsɜːkɪt] 数据（传输）电路
data set 数据组
data set for power performance measurement [ˈmeʒəmənt] 功率特性测试数据组
data storage 数据存储
data terminal equipment （DTE） 数据终端设备
data viewing 数据查看
DC generators 直流发电机
DC-Link voltage 直流侧电压
deactivate [diːˈæktɪveɪt] 释放；使

无效

dead tank oil circuit breaker 多油断路器

de-aeration 排气孔（油箱）

debugging interface 调试界面

decentralized wind energy system [ˌdiːˈsentrəlaɪzd] 分散式风能系统

decision variable [ˈveərɪəbl] 决策变量

decouple 解列

decouple generator (from grid) 使机组退出电网；解列

deenergize 断电

defective [dɪˈfektɪv] 有缺陷的；欠缺的

deflection 偏向；偏斜；转向

deflection anemometer [ˌænɪˈmɒmɪtə] 偏转风速计

deflection angle 偏转角

deflection force of earth rotation 地球自转偏向力

deflection gauge [geɪdʒ] 偏转计；挠度计

deflector [dɪˈflektə] 导风板，导流板

deflector wind machine 导风器式风力机

deformation 变形；形变；畸变；失真

degrease [diːˈgriːs] 脱脂；除油污

degree celsius [ˈselsɪəs] 摄氏度

degree of curvature [ˈkɜːvətʃə] 弯度

degrees of freedom (DOF) 自由度

delay operator [ˈɒpəreɪtə] 延时操作数

delivery rate 输油速度；传输速度

delta connection 三角形接线

Delta winding 角形绕组

delta-wing vortex 三角翼涡

demagnetization [ˈdiːmægtɪtaɪˈzeɪʃən] 退磁；去磁

density of air 空气密度

depression 低气压；低压

depressurizes [diːˈpreʃəraɪz] 使减压；使降压

depth gauge [depθ] [geɪdʒ] 深度计

design condition 设计工况

design lifetime 设计寿命

design lift coefficient 设计升力系数

design limits 设计极限

design load 设计荷载

design optimation 设计优化

design parameter 设计参数

design philosophy [fɪˈlɒsəfɪ] 设计原则；设计原理

design pressure coefficient [ˌkəʊɪˈfɪʃ(ə)nt] 设计压力系数

design procedure [prəˈsiːdʒə] 设计程序

design situation 设计状况

design tip speed ratio 设计尖速比

design wind condition 设计风况

design wind load 设计风载

design wind speed 设计风速

detection impedance [ɪmˈpiːdəns] 检测阻抗

deterministic fluctuating wind speed [dɪˌtɜːmɪnˈɪstɪk] [ˈflʌktʃʊeɪtɪŋ] 主脉动风速

deterministic gust 主阵风
deterministic loads [dɪˌtɜːmɪˈtɪstɪk] 确定载荷
deterministic models 确定性模型
deterministic rotor loads 确定转子载荷
deviation [ˌdiːvɪˈeɪʃən] 偏差；偏移
dew point 露点
dial micrometer 千分尺
dielectric [ˌdaɪɪˈlektrɪk] 电介质；绝缘体
dielectric constant 介质常数
dielectric loss 介质损耗
dielectric test 介质试验
diesel generators 柴油发电机
diestock [ˈdaɪstɒk] 螺丝攻
differential gear [ɡɪə] 差速齿轮
differential protection 差动保护
diffuser [dɪˈfjuːzə] 扩散器
diffuser augmented wind turbine [ɔːɡˈmentɪd] 风力发动机
digital control 数字控制
digital control system 数字控制系统
digital controller 数字控制器
digital signal processing [prəʊˈsesɪŋ] 数字信号处理
digital terrain model [teˈreɪn] 数字地形模型
digitizing tablet [ˈtæblɪt] 数字面板
direct current 直流电
direct current machine 直流电机
direct driving mode 直驱式
direct grid connection [ɡrɪd] 直接栅极接线
direct solar radiation 直接太阳辐射
direct voltage 直流电压
direct-drive 直驱
direct-drive generators 直驱发电机
directivity (for WTGS) 方向性(风力发电机组)
directly coupled squirrel-cage 直接耦合鼠笼
disc brake 盘式制动器
discharge 卸下；放出；流注；放电
disconnector 隔离开关
discounted cash flow (DCF) 现金流量折现法
discrete controller 离散控制器
discrete Fourier transformation (DFT) 离散傅里叶变换
discrete gust models 离散阵风模型
discrete vortices [ˈvɔːtɪˌsiːz] 离散涡
discretized cantilever beams [ˈkæntɪliːvə] [biːm] 离散悬臂梁
discretized mechanical systems [mɪˈkætɪkəl] 离散机械系统
dismount [ˌdɪsˈmaʊnt] 拆卸；卸下
dispersed generation 分散发电
displacement amplitude [ˈæmplɪtjuːd] 位移幅值
display lamp 指示灯
distance constant 距离常数
distance ring 间隔环
distortion energy method [ˈmeθəd] 失真能量法
distortion rate 畸变率
distributed grid 分布栅极

distributing apparatus [ˌæpəˈreɪtəs] 配电电器
distribution automation system [ɔːtəˈmeɪʃən] 配电网自动化系统
distribution dispatch center 配电调度中心
distribution function 分布函数
distribution system 配电系统
distributor 分电盘；配电器
diurnal variation [daɪˈɜːnəl] [ˌveərɪˈeɪʃən] 日变化
divider ratio [ˈreɪʃɪəʊ] 分压器分压比
domestic load [dəʊˈmestɪk] 民用电
dormant failure [ˈdɔːmənt] [ˈfeɪljə] 潜在故障
double clamp 双卡头
double feed induction generator (DFIG) 双馈感应发电机
double phase [feɪz] 两相
double row deep groove ball bearing 双列深沟圆柱滚珠轴承
double-feds，doubly-fed 双馈
double-helical gear 人字齿轮
doubly fed generating 双馈发电系统
doubly fed variable speed constant frequency 双馈式变速恒频风电机组
doubly-fed generator 双馈发电机
doubly-fed induction generator (DFIG) 双馈感应发电机
doubly-fed wind turbine generator system 双馈式风电机组
down-conductor 引下线
downslope winds [ˈdaʊnsləʊp] 下坡风
downstream 下游；顺流
down-valley windflow 下坡风；出谷风
downwash [ˈdaʊnwɒʃ] 下降气流
downwind 下风向
downwind configuration [kənˌfɪgjʊˈreɪʃən] 下风式结构
downwind rotor 下风向风轮
downwind sector 下风向扇形区
downwind turbine 下风式风力机
downwind WGTS 下风向风电机组
drag (force) [dræg] 阻力
drag area [车] 风阻面积
drag brake 阻力刹车板
drag center 阻力中心
drag coefficient [ˌkəʊɪˈfɪʃənt] 阻力系数
drag crisis 阻力临界值
drag cup anemometer [ˌænɪˈmɒmɪtə] 阻力型风杯风速计
drag due to lift 升致阻力
drag friction [ˈfrɪkʃ(ə)n] 摩擦阻力
drag from pressure 压差阻力
drag hinge 阻力铰
drag lift ratio 阻升比
drag polar 阻力极曲线
drag reduction 减阻
drag spoiler 阻力板
drag translator 阻力型平移式风车
drag wind load 风阻荷载
drag-reducing device 减阻装置
drag-type rotor 阻力型风轮
drag-type wind machine 阻力型风

力机
drain 泄油
drip pan 油滴盘
drive control 驱动控制
drive motor 驱动电机
drive train 驱动力
driven gear [gɪə] 从动齿轮
driven yaw [jɔː] 偏航驱动
driver 驱动器
driving gear 主动齿轮
drizzle 油滴（雾滴）
drum brake 鼓状刹车
dry-type transformer 干式变压器
dual bus 双母线
ducted wind turbine 轴流式风轮机
dump load 甩负荷
dump load resistor 转储荷载电阻
duplex [ˈdjuːpleks] 双（向；重）；双工；二重
duplex transmission 双工传输
durability [ˌdjʊərəˈbɪlətɪ] 耐久性；耐用性
dust cap 防尘罩
dust-protected 防尘
duty ratio [ˈreɪʃɪəʊ] 负载比
dynamic analysis [daɪˈnæmɪk] [əˈnæləsɪs] 动态分析；动力特性分析
dynamic brakes 动态刹车
dynamic control 动态控制
dynamic coupling [ˈkʌplɪŋ] 齿啮式连接
dynamic factor 动载系数
dynamic load simulation 动载模拟
dynamic magnification [ˌmægtɪfɪˈkeɪʃən] 动力放大
dynamic models 动态模型
dynamic regulated 动态控制
dynamic response 动态响应；动力特性
dynamic stall 动态失速
dynamic structural models [ˈstrʌktʃərəl] 动态结构模型
dynamic wake models 动态尾涡模型
dynamo 发电机；直流发电机
dynamometer [ˌdaɪnəˈmɒmɪtə] 测力计；功率计

E

earth conductor; earth wire 接地线
earth electrode 接地体
earth fault 接地故障
earth resistance 接地电阻
earthed circuit 接地电路
earthing 接地
earthing reference points [ˈrefərəns] 接地基准点
earthing switch 接地开关
earth-termination system 接地装置
eccentric loading 偏心载荷
ecological assessment [ˌiːkəˈlɒdʒɪkəl] 生态评价
economies of scale [skeɪl] 规模经济
eddy correlation [ˌkɒrəˈleɪʃ(ə)n] 涡旋相关

eddy current [ˈkʌrənt] 涡流
eddy current damper 涡流阻尼器
eddy Prandtl number 涡流普朗特数
eddy resistance [rɪˈzɪst(ə)ns] 涡动阻力
eddy transfer theory 涡动传递理论
eddy viscosity [vɪˈskɒsətɪ] 涡黏性
eddy viscosity model 涡流黏度模型
edge effect 边缘效应
edge fairing 边缘整流
edgewise [ˈedʒweɪz] 沿边；把刀刃朝外
edgewise bending 扁弯；边缘弯曲
edgewise bending moments 沿层方向弯曲力矩
effect of altitude 高度影响；高度效应
effective angle of attack 有效迎角
effective buoyancy [ˈbɒɪənsɪ] 有效浮力
effective camber 有效弯度
effective chord length 有效弦长
effective ground level 有效地面高度
effective power 有效功率；水力功率
effective terrain height [teˈreɪn] 有效地形高度
effective wind speed 有效风速
effects of azimuth angle 方位角影响
effects of terrain [teˈreɪn] 地形影响
efficiency of WTGS [ɪˈfɪʃənsɪ] 风力发电机组效率
E-glass 玻璃纤维
Eiffel Polar 埃菲尔极地
eigenmode 本征模
eigenperiod 固有周期
eigenvalue 本征值
Ekman boundary layer 埃克曼边界层
elastic coupling [ˈkʌplɪŋ] 弹性连接
elastic modulus 弹性模量
elastic resistence 弹性阻力
elastic scaling 弹性缩尺
electric actuator [ˈæktjʊeɪtə] 电动执行机构
electric block 触电；电击
electric cable 电缆
electric charge 电荷
electric circuit 电路
electric coupler 电耦合器
electric current 电流
electric energy transducer [trænzˈdjuːsə] 电能转换器
electric field 电场
electric flicker 电闪变
electric grids 电网
electric independent drive pitch 电动独立变桨
electric machine 电机
electric outlet 电源插座
electric plier [ˈplɑɪə] 电气钳
electric screw-driver 电钻
electric shock 触电；电击
electric welding 电焊
electric wire [ˈwɑɪə] 电线；电缆
electrical arc [ɑːk] 电弧

electrical contact 电触头
electrical device 电气元件
electrical discharge 放电
electrical distribution networks 配电网络
electrical elements/parts 电气组件
electrical endurance [ɪn'djʊərəns] 电气寿命
electrical facilities 电器设备
electrical grid 电网
electrical protection 电气保护
electrical rotating machine 旋转电机
electrical systems 电气系统
electrical wiring 电气布线
electrochemical deterioration [dɪˌtɪərɪə'reɪʃən] 电化学腐蚀
electromagnetic braking system [ɪˌlektrəʊmæg'netɪk] 电磁制动系统
electromagnetic induction 电磁感应
electromagnetic interference (EMI) 电磁干扰
electromagnetism [ɪˌlektrəʊ'mægnɪtɪzəm] 电磁学
electromagnet [ɪˌlektrəʊ'mægnɪt] 电磁铁
electromotive force 电动势
electronegative gas 负电性气体
electronic bus-bar 电子汇流排；电子总线
electroplating [ɪ'lektrəʊˌpleɪtɪŋ] 电镀；电镀术
electrostatic voltmeter 静电电压表
electrostatics 静电学
ellipsoidal co-ordinates [ˌelɪp'sɒɪdəl] 椭圆坐标
embedded generation [ɪm'bedɪd] 嵌入式发电
emergency braking system [ɪ'mɜːdʒənsɪ] 紧急制动系统
emergency escape 紧急出口（机舱）
emergency feathering 紧急顺桨
emergency shutdown (for wind turbines) 紧急停机（风力机）
emergency stop plate 限位板
emergency stop push-button 紧急停车按钮
encode [en'kəʊd] 编码
encoder 编码器
end-cap 端头盖帽（爬梯用）
endurance limit [ɪn'djʊərəns] 疲劳极限
endurance test 耐久性试验
endwise 末端朝前或向上的；向前的
energy balance 能量平衡
energy consumption 能量损耗
energy demand 能量需求
energy extraction 能量提取；能量开采
energy gap [gæp] 能隙；能级距离
energy generation cost 发电成本
energy production after wake flow 尾流折减后发电量
energy rose 能量玫瑰图
energy storage 蓄能；储能器
energy yield [jiːld] 发电量
engagement mesh [ɪn'geɪdʒmənt]

[meʃ] 啮合
environment 环境
environment condition 环境条件
epoxy [ep'ɒksɪ] 环氧基树脂
epoxy resin 环氧树脂
equatorial region ['riːdʒən] 赤道地区
equilibrium wake [ˌiːkwɪ'lɪbrɪəm] 平衡尾涡
equipment failure information 设备故障信息
equipotential bonding 等电位连接
equivalent circuit 等效电路
equivalent continuous A-weighted sound pressure level 等效连续A计权声级
equivalent impedance 等效阻抗
equivalent loading 等效负荷
erection [ɪ'rekʃən] 安装
ethernet switch 以太网交换机
Euler's turbine equation 涡轮机欧拉方程
Eurocode 欧洲规范
event information 事件信息
excitation 励磁
excitation response 励磁响应
exciter 励磁机
exciting winding 激磁绕组
expenses for occupying construction land 建设用地费
expulsion gap [ɪk'spʌlʃən] 灭弧间隙
extendibility 扩展性
external conditions (for WTGS) 外部条件(风电机)
external field 外(部流)场
external flow 外部绕流
external gear 外齿轮
external lightning protection system [ɪk'stɜːnəl] 外部防雷系统
external load 外部荷载
external power supply 外部动力源
external pressure coefficient 外部压力系数
external pressure loss 外部压力损失
external suction 外部吸力
external wind load 外部风载
Extra-High Voltage (EHV) 超高压
extraneous loading [ɪk'streɪnɪəs] 附加荷载
extraordinary wind [ɪk'strɔːd(ə)n(ə)rɪ] 异常风
extrapolated power curve [kɜːv] 外推功率曲线
extreme 极端;极限
extreme annual wind speed [ɪk'striːm] 年极端风速
extreme atmospheric events [ætməs'ferɪk] 极端大气现象
extreme climate wind 极端梯度风
extreme environmental condition 极限环境条件
extreme gust [ɪk'striːm] 极端阵风
extreme lifetime wind speed 寿命极端风速
extreme loads 极端载荷
extreme maximum ['mæksɪməm] 极端最高

extreme return period（风速）极端重现期
extreme surface wind 地面极端风
extreme values 极值
extreme wind 极端风
extreme wind speed 极端风速
extruded aluminium blade [æl(j)ʊ'mɪnɪəm] 挤压铝叶片

F

face width 齿宽
face-on-attack 迎风面
factor of fatigue [fə'tiːg] 疲劳系数
factor of safety 安全系数
factored resistance [rɪ'zɪst(ə)ns] 设计风阻
factored wind load 设计风载
fag bolt 疲劳螺栓
fail safe principle 安全原则
failing load 破坏荷载
fail-safe 失效—安全；故障自动保护
fail-safe component 故障安全组件
fail-safe-system 失效系统
failure 失效
failure criterion [kraɪ'tɪərɪən] 破坏判据
failure load 破坏载荷
failure rate 事故频率
failure stress 破坏应力
fair wind 顺风
fairing 整流罩；（风洞支架）风挡；减阻装置
fall off 开始顺风
fall protection system 防坠落防护
fall wind 下坡风；下降风
fallout wind 沉降风
fan belt 风扇皮带
fan blade 风扇叶片
fan heater ['hiːtə] 风扇加热器；暖风机
fan hub 风扇轮毂
fan inlet 通风机进气口
fan power factor 风扇功率系数
far wake 远尾流
farm power curve 风电场功率曲线
farm server 风电场服务器
fast Fourier transform（FFT）快速傅立叶变换
fast reclosure 快速重合闸
fasteners ['fɑːstnəz] 紧固件
fastest mile wind speed 最大英里风速
fast-response instrument ['ɪnstrʊm(ə)nt] 快速响应仪表
fatigue criterion [fə'tiːg] [kraɪ'tɪərɪən] 疲劳判据
fatigue criticality [fə'tiːg] [ˌkrɪtɪ'kælətɪ] 疲劳临界
fatigue cycle counting ['saɪkl] 疲劳循环计数
fatigue damage 疲劳损伤
fatigue design 疲劳设计
fatigue evaluation [ɪˌvæljʊ'eɪʃən] 疲劳评定
fatigue failure ['feɪljə] 疲劳破坏

fatigue life 疲劳寿命
fatigue limit 疲劳极限
fatigue loading 疲劳载荷
fatigue properties 疲劳性能
fatigue spectra combination 疲劳光谱组合
fatigue strength [streŋθ] 疲劳强度
fatigue stress [stres] 疲劳应力
fatigue stress range 疲劳应力范围
fault 故障
fault conditions [fɔːlt] 故障条件
fault current [ˈkʌrənt] 故障电流
fault earthing 故障接地
fault finding 故障查找
fault ride through（FRT） 故障穿越
fault warning priority 优先级报警
favourable interference [ˈfeɪvərəbl] [ɪntəˈfɪər(ə)ns] 有利干扰
favourable pressure difference 顺压差
favourable pressure gradient [ˈgreɪdɪənt] 顺压梯度
feasibility [ˌfiːzəˈbɪlɪtɪ] 可行性
feather direction 顺桨方向
feather/feathering [ˈfeðə] 顺桨
feathering position 顺桨位置
feeder 馈电线
feeding 馈送
feeler gauge [ˈfiːlə] [geɪdʒ] 测隙规；塞尺
felling axe 外轮轴
FFT（fast Fourier transform） 快速傅立叶变换
fiber interface 光缆接口
fiberglass [ˈfaɪbəglɑːs] 玻璃纤维
fiberglass blades 玻璃钢叶片
Fiberglass Reinforced Plastic 玻璃纤维强化塑料（叶片）
field data 现场数据
field distortion [dɪsˈtɔːʃən] 场畸变
field gradient [ˈgreɪdɪənt] 场梯度
field reliability test 现场可靠性试验
field strength [streŋθ] 场强
field stress 电场力
field test with turbine 外联机试验
field testing 实地测试
filling plug 加油（注油）接头
filter 滤波器
fine pitch 精细节距
finite element method（FEM） [ˈfaɪnaɪt] [ˈmeθəd] 有限元法
finite-element analysis [əˈnæləsɪs] 有限元分析
fire extinguisher [ɪkˈstɪŋgwɪʃə] 灭火器
firebrick [ˈfaɪəbrɪk] 耐火砖
firmware 固件
first edgewise eigenmodes 第一边沿本征模
first flapwise eigenmodes 第一翼面本征模
fixation [fɪkˈseɪʃən] 固定杆；固定
fixed contact 静触头
fixed coordinate systems [kəʊˈɔːdɪnɪt] 固定坐标系统
fixed hub 固定套
fixed pitch 定桨距
fixed speed generator 恒速发电机

fixed speed operation 恒速运行
fixed-blade pitch 定桨距
fixed-speed fixed-pitch (FS-FP) 恒速定桨距
fixed-speed turbines 恒速风力机
fixed-speed variable-pitch (FS-VP) 恒速变桨距
fixing bolts 固定螺栓
flange connection/joint [flændʒ] 法兰连接
flange coupling 凸缘联轴器
flanged nut 凸缘螺母
flanged union 凸缘连接
flank [flæŋk] 侧面；侧翼；侧腹；肋
flap 襟翼；副翼
flap-lag 挥舞-摆振
flapping hinge 翼动铰
flapping model 拍动模型
flapwise bending 弯曲方向
flapwise bending moment 副翼方向弯曲力矩
flash counter 雷电计数器
flash disc 闪存
flash welding [ˈweldɪŋ] 闪光焊
flashover [ˈflæʃˌəʊvə] 闪络
flashpoint [ˈflæʃˈpɔɪnt] 闪点
flat-head rivet [ˈrɪvɪt] 平头铆钉
flat nut 平螺母
flat plate aerodynamics [ˌeərəʊdaɪˈnæmɪks] 平板空气动力学
flat plate flutter [ˈflʌtə] 平板颤振
flat terrain 平坦地形
flat-plate drag 平板阻力
flat-plate flow 平板绕流
flatwise [ˈflætwɑɪz] 平放地
flatwise bending 板状弯曲；(旋转平面) 面外弯曲
flatwise direction 平向
Flettner rotor [ˈrəʊtə] 弗莱特纳转子
flex [ˈfleks] 弯曲 (四肢)；伸缩；折曲
Flexible AC Transmission System (FACTS) [ˈfleksɪbl] 柔性交流输电系统
flexible blade [ˈfleksɪb(ə)l] 柔性叶片
flexible communication interface 弹性通信界面
flexible gear [gɪə] 柔性齿轮
flexible pliers [ˈplɑɪəz] 万向套筒扳手
flexible rolling bearing 柔性滚动轴承
flexible rotor [ˈrəʊtə] 柔性转子
flexible shaft 柔性轴
flexible spanner band [ˈspænə] 柔性套筒扳手
flexural and torsional loads [ˈflekʃərəl] [ˈtɔːʃənəl] 弯曲和扭转载荷
flexural oscillation [ˈflekʃərəl] [ˌɒsɪˈleɪʃən] 弯曲振动
flexural rigidity [rɪˈdʒɪdətɪ] 抗弯刚度
flexural stress 挠曲应力
flexure [ˈflekʃə] 弯曲；挠曲
flexure torsion flutter [ˈtɔːʃ(ə)n] 弯

扭颤振
flicker [ˈflɪkə] 闪烁；闪变
flicker coefficient 闪变系数
flicker step factor 闪变阶跃系数
float chamber 浮子
flow angularity [ˌæŋgjʊˈlærətɪ] 气流偏角
flow blockage [ˈblɒkɪdʒ] 流动阻塞
flow calibration [ˌkælɪˈbreɪʃən] 流场校测
flow characteristics [ˌkærəktəˈrɪstɪk] 流动特性
flow distortion [dɪsˈtɔːʃən] 气流畸变
flow field 流场
flow separation [ˌsepəˈreɪʃən] 气流分离
flow states 流态
fluctuating aerodynamic force 脉动气动力
fluctuating load 脉动荷载
fluctuating moment 脉动力矩
fluctuating plume model 脉动羽流模式
fluctuating pressure 脉动压力
fluctuations [ˌflʌktjʊˈeɪʃəns] 波动
fluid damping 流体阻尼
fluid dynamic damping [daɪˈnæmɪk] 流体动力阻尼
fluid dynamics 流体动力学
fluid viscosity [vɪˈskɒsətɪ] 流体黏性
flutter [ˈflʌtə] 颤振
flux 磁通；通量；焊剂；流动
fly-ball force 离心力
fly-ball governor [ˈgʌvənə] 离心调节器
fly-ball weight 离心重量
follower piston 随动活塞
footpump 脚泵
forbidden zones [fəˈbɪdən] 禁区
force balance 力平衡；测力天平
force coefficient [ˌkəʊɪˈfɪʃ(ə)nt] 力系数
force derivative [dɪˈrɪvətɪv] 力导数
forced circulation [sɜːkjʊˈleɪʃ(ə)n] 强制环流
forced convection [kənˈvekʃ(ə)n] 强制对流
forced flow 强迫流动
forced lubrication 强制润滑
forge [fɔːdʒ] 锻造；炼炉；熔炉
fork-lift truck [trʌk] 叉架式运货车；铲车
forming lubricant 成型润滑剂
forward-scattering [ˈfɔːwəd] [ˈskætərɪŋ] 前身散射
foundation [faʊnˈdeɪʃən] 基础；地基
foundation earth electrode [ɪˈlektrəʊd] 基础接地体
foundation ring 基础环
foundation works of power generation equipment 发电设备基础
foundry [ˈfaʊndrɪ] 铸造；翻砂；铸工厂；玻璃厂；铸造厂
four quadrant operation 四象限运行
Fourier analysis [ˈfʊrɪər] 傅立叶分析
Fourier coefficient 傅立叶系数

Fourier component [kəm'pəunənt] 傅立叶分量
Fourier phase spectrum ['spektrəm] 傅立叶相位谱
Fourier series ['sɪəriːz] 傅立叶级数；傅立叶序列
Fourier transform 傅立叶变换
four-jaw chuck [tʃʌk] 四爪卡盘
four-stroke 四冲程
fracture ['fræktʃə] 断裂
fragment ['frægm(ə)nt] 碎片；片段
free atmospheric wind [ætməs'ferɪk] 自由大气风
free boundary ['baʊnd(ə)rɪ] 自由边界
free convection [kən'vekʃ(ə)n] 自由对流
free stand tower 独立式塔架
free stream boundary 自由流边界
free stream surface 自由流面
free stream turbulence ['tɜːbjʊl(ə)ns] 自由来流湍流（度）
free stream velocity 自由流速度
free stream wind speed 自由流风速
free streamline 自由流线
free wheeling ['hwiːlɪŋ] 单向离合器
free wind 自由风；顺风
free yaw [jɔː] 自由偏航
freewheeling 惯性滑行
free-yaw rotor 定向风轮
freezing rain 冻雨
frequency ['frɪkwənsɪ] 频率
frequency converter [kən'vɜːtə] 变频器
frequency inverter 变频器
frequency distribution [dɪstrɪ'bjuːʃ(ə)n] 频率分布
frequency distribution histogram (diagram) 风能频率分布直方图
frequency domain [dəʊ'meɪn] 频域
frequency drift 频移
frequency effect 频率效应
frequency meter anemometer [ˌænɪ'mɒmɪtə] 频率表式风速计
frequency modulation (FM) [ˌmɒdjʊ'leɪʃən] 调频
frequency of gust 阵风频数
frequency of occurrence [ə'kʌr(ə)ns] 出现频率
frequency of sampling 取样频率
frequency of turbulence ['tɜːbjʊl(ə)ns] 湍流频率
frequency of vortex shedding 旋涡脱落频率
frequency of wind direction 风向频率
frequency of wind speed 风速频率
frequency response 频率响应
frequency spectrum ['spektrəm] 频谱
frequency-spectral-density 频谱密度
frequent wind speed 常现风速
friction coefficient [ˌkəʊɪ'fɪʃənt] 摩擦系数
friction drag 摩擦阻力

frictional resistance [rɪ'zɪst(ə)ns] 摩擦阻力
friction layer 摩擦层
friction point 摩擦点
friction velocity 摩擦速度
frictional boundary layer ['baʊnd(ə)rɪ] 摩擦边界层
frictional moment 摩擦力矩
frictional stress 摩擦应力
frictional torque [tɔːk] 摩擦扭矩
frictionless wind 无摩擦风，理想风
front end processor ['prəʊsesə] 前置机；前端处理器
front panel 面盘（仪表盘）
front spoiler ['spɒɪlə] 前扰流板
frontal area ['frʌntəl] 迎风面积
frontal drag 迎面阻力
frontal resistance [rɪ'zɪst(ə)ns] 迎面阻力
frontal projected area 迎面投影面积
FRP 纤维增强塑料
fuel cells 燃料电池
fuel saving 节约燃料
fuel tank 燃料箱
full converter 全功率变流器
full feather ['feðə] 完全顺桨
full gale 强风
full grid compatibility 电网兼容性
full load 满负荷；满载
full power variable flow mode 全功率交流技术
full-load equivalent hours [ɪ'kwɪvələnt] 等效满负荷小时数
full-power converter 全功率变流器
full-scale measurement ['meʒəm(ə)nt] 全尺寸测量
full-scale model 全尺寸模型
full-scale prototype ['prəʊtətaɪp] 全尺寸原型物
full-scale Reynolds number ['renəldz] 全尺寸雷诺数
full-scale time 全尺寸时间
full-span pitch control 全范围桨距控制
function indicator 功能指示灯
fuse 熔断器
fuzzy controllers ['fʌzɪ] 模糊控制器

G

gain margin [geɪn] ['mɑːdʒɪn] 增益裕度
gain schedule ['ʃədjuːəl] 增益调度
gale [geɪl] 强风；大风
galvanized steel 镀锌钢材
gamma function ['gæmə] 伽马函数
gas insulated substation (GIS) ['ɪnsəˌleɪtəd] 气体绝缘变电站
gas mask [mɑːsk] 防毒面具
gaseous insulation ['gæsiəs] 气体绝缘
gasket ['gæskɪt] 垫片；垫圈；接合垫
gasometer [gæ'sɒmɪtə] 气量计

gate valve [vælv] 门阀
gauge [geɪdʒ] 标准尺；规格；量规
Gauss distribution [gaʊs] 高斯分布
gear [gɪə] 齿轮
gear cutting 齿加工
gear lever 变速杆
gear meshing 齿轮啮合
gear motor [ˈməʊtə] 齿轮马达
gear pair 齿轮副
gear pair with parallel axes [ˈpærəlel] [ˈæksiːz] 平行轴齿轮副
gear pump 齿轮泵
gear ratio [ˈreɪʃɪəʊ] 齿轮齿数比
gear stage 齿轮级
gear train 轮系；齿轮传动链
gear volume [ˈvɒljuːm] 齿轮体积
gear wheel 齿轮
gearbox [ˈgɪəbɒks] 变速箱
geared part 啮合部件
gearless [ˈgɪəlɪs] 无齿轮的；无传动装置的
gears with addendum modification [əˈdendəm] 变位齿轮
Gedser wind turbine 盖兹风电机
Geiger counter [ˈgaɪgə] 盖格计数器
gel coat 凝胶漆；涂层
gel coated 胶衣
generalized coordinates [kəʊˈɔːdɪneɪts] 广义坐标
generalized force vector [ˈvektə] 广义力向量
generalized load 广义载荷
generalized mass 广义质量
generator 发电机
generator behavior 发电机性能
generator mounting [ˈmaʊntɪŋ] 发电机安装
generator name plate rating 发电机铭牌评级
generator shortage [ˈʃɔːtɪdʒ] 发电机短路
generator topology 发电机拓扑结构
geographical variation [dʒɪəˈgræfɪkəl] [ˌveərɪˈeɪʃən] 地理变异
geometric chord of airfoil [dʒɪəʊˈmetrɪk] [ˈeəfɒɪl] 翼型几何弦长
geo-strophic winds [ˈstrɒfɪk] 地转风
GFRP blade [bleɪd] 玻璃纤维增强塑料叶片
gimbals [ˈdʒɪmbəls] 平衡环；平衡架
gimbal-type 万向型
gimlet [ˈgɪmlɪt] 手钻；螺丝锥
girder [ˈgɜːdə] 梁；钢桁的支架
GIS (gas insulated substation; geographic information system) 气体绝缘变电站；地理信息系统
glass fiber [ˈfaɪbə] 玻璃纤维
glass fibre reinforced plastic 玻璃钢
glass insulator [ˈɪnsjʊleɪtə] 玻璃绝缘子
glaze 雨凇
global winds 全球风
globe valve [vælv] 球形阀

gloss paint 光滑涂料

glow discharge 辉光放电

goggles [ˈgɒglz] 风镜；护目镜

Goodman diagram [ˈgʊdmæn] [ˈdaɪəgræm] 古德曼曲线

Goodman relation 古德曼关系

governor [ˈgʌvənə] 调节器；控制器

grader [ˈgreɪdə] 分类机；分级机

gradient wind [ˈgreɪdɪənt] 梯度风

gradient wind height 梯度风高度

gradient wind speed 梯度风速

grading ring 均压环

graphite flakes [ˈgræfaɪt] [fleɪks] 石墨片

gravitational acceleration [ˌgrævɪˈteɪʃənəl] 重力加速度

gravitational field 重力场

gravitational loads [ˌgrævɪˈteɪʃənəl] 重力载荷

gravitational potential energy 重力势能

gravity foundation (offshore) 重力桩（海上风电）

gravity wind 重力风

grazing angle [ˈgreɪzɪŋ] 掠射角

grease [griːs] 油膏；润滑油

grease gun 注油枪；滑脂枪

green energy 绿色能源

greenhouse effect [ɪˈfekt] 温室效应

grid (electrical) [grɪd] 电网

grid codes [ˈkəʊdz] 电网码

grid connected time 并网时间

grid connected/connection [kəˈnekʃən] 并网

grid connection system 并网系统

grid failure [ˈfeɪljə] 电网毁坏；电网故障

grid frequency [ˈfrɪkwənsɪ] 电网频率

grid loss 电网损失

grid penetration [ˌpenɪˈtreɪʃən] 电网渗透率

gridding method [ˈmeθəd] 网格化方法

grid-dip protection 电网电压跌落保护

grid-side lightning protection 网侧防雷

Griggs-Putnam index 格里戈—普特南级数（植物风力指标）

grinding disk 摩擦盘

grinding machine 磨床

grinding wheel 砂轮

groove [gruːv] 凹槽

ground boundary [ˈbaʊndərɪ] 地面边界层

ground roughness [ˈrʌftɪs] 地面粗糙度

ground wire 接地线

grounding 接地

grounding capacitance [kəˈpæsɪtəns] 对地电容

growth of boundary layer 边界层增长

GRP (Glass Reinforced Plastic) 玻璃纤维强化塑料（叶片）

grub screw 自攻螺丝

guard rail 护栏

guide bars 导向棍

guide blade 导流（叶）片
guide block 导向块
guide ring 导向绳
guiding shaft 导向轴
guidling vane 导流片，导向舵
gull wing sail 海鸥翼式帆
gumming 树胶分泌
gun excitation [ˌeksɪˈteɪʃ(ə)n] 阵风激励
gun recorder 阵风记录仪
Gurney flap 格尼襟翼
gust 阵风
gust amplitude [ˈæmplɪtjuːd] 阵风变幅
gust and lull 阵风阵息
gust anemometer [ˌænɪˈmɒmɪtə] 阵风风速计
gust averaging time [ˈævərɪdʒɪŋ] 阵风平均时间
gust component [kəmˈpəʊnənt] 阵风分量
gust decay time [dɪˈkeɪ] 阵风衰减时间
gust downwash [ˈdaʊnwɒʃ] 阵风下洗
gust duration [djʊˈreɪʃ(ə)n] 阵风持续时间
gust effect 阵风效应
gust energy favor 阵风能量因子
gust environment [ɪnˈvaɪrənm(ə)nt] 阵风环境
gust factor [ˈgʌst] 阵风系数
gust factor approach [əˈprəʊtʃ] 阵风因子法
gust frequency [ˈfriːkw(ə)nsɪ] 阵风频数
gust generator 阵风发生器
gust influence [ˈɪnflʊəns] 阵风影响
gust lapse rate [læps] 阵风递减率
gust lapse time 阵风递减时间
gust loading 骤风载荷
gust measuring anemometer [ˌænɪˈmɒmɪtə] 阵风风速计
gust of rain 阵雨
gust peak speed 阵风最大风速
gust response factor 阵风响应因子
gust scale 阵风尺度
gust size 阵风尺寸
gust spectrum [ˈspektrəm] 阵风谱
gust speed 阵风速度
gust structure 阵风结构
gust volume 阵风容积
gust-effect factor 阵风效应因子
gustiness [ˈgʌstɪnɪs] 阵风性
gustiness effect 阵风效应
gustiness factor 阵风因子
gustiness wind 阵风
gusty 阵风的；疾风的
guy 拉索
guy cable anchor 拉索地锚
guy clip 线卡子
guy rope/wire 拉线
guyed cantilever [ˈkæntɪliːvə] 拉索支撑悬臂
guyed structure 拉索（支撑）结构
guyed tower 拉索式塔架
guy-ropes 拉绳
gyromill 板翼风车（直叶片竖轴风轮机）
gyroscope [ˈdʒaɪərəskəʊp] 陀螺仪；

回旋装置；回转仪
gyroscopic effect [ˌdʒaɪərəsˈkɒpɪk] 陀螺效应
gyroscopic force 陀螺力
gyroscopic loads 陀螺载荷
gyroscopic motion 陀螺运动

H

hacksaw [ˈhæksɔː] 可锯金属的弓形锯；钢锯
hail 冰雹
hairspring [ˈheəsprɪŋ] 细弹簧；游丝
half-round file 半圆锉
half shaft 半轴
half-duplex transmission [ˈdjuːpleks] 半双工传输
half-moulds 半模
Halogen free cable 无卤电缆
hanger [ˈhæŋə] 吊架
hardened and tempered steel [ˈhɑːdənd] [ˈtempəd] 调质钢
hardener 固化剂；硬化剂
hardware platform [ˈhɑːdweə] [ˈplætfɔːm] 硬件平台
harmonic [hɑːˈmɒtɪk] 谐波
harmonic current 谐波电流
harmonic distortion [dɪsˈtɔːʃən] 谐波失真
harmonic filter 谐滤器
harness 全身式安全带
hatch 舱口；舱口盖；开口；进人孔
headscrew 主轴螺杆
headstock [ˈhedstɒk] 主轴承
heat exchanger 热交换器
heating blanket 加热毯
helical gear [ˈhelɪkəl] 斜齿轮
helical gear single-helical gear 斜齿单螺旋齿轮
helical gearing 螺旋齿轮；斜齿轮
helical planetary gear (2-stage)（二级）行星传动
helical vortex [ˈvɔːteks] 螺旋涡
helical wake 螺旋形尾涡
helicoidal vortex sheet 螺旋涡片
helix 螺旋；螺旋状物
Helmholtz theorem [ˈhelmhəults] 亥姆霍兹定理
herring-bone gear [ˈherɪŋ] [bəun] 双曲面齿轮
hexadecimal [ˌheksəˈdesɪməl] 十六进制
hexagon spanner [ˈheksəgən] [ˈspænə] 六方扳手
hexagonal nut [hekˈsægənəl] 六角螺母
high exponent [ɪkˈspəunənt] 高次幂指数
high harmonic 高次谐波
high profile [ˈprəufaɪl] 高轮廓线
high speed brake 高速闸
high speed shaft 高速轴
high temperature superconductor cable 高温超导电缆
high viscosity fluid 高黏度液体

high voltage DC transmission [ˈvəʊltɪdʒ] 高压直流输电
high wind speed 大风速
high-speed rotor 高速转子
high-tension cable [ˈtenʃən] 高压电缆
highvoltage engineering [ˌendʒɪˈnɪərɪŋ] 高电压工程
highvoltage testing technology [ˈtestɪŋ] 高电压试验技术
hinged blades [hɪndʒd] 铰接叶片
histogram [ˈhɪstəgræm] 柱状图
hoist [hɒɪst] 起重机（台、架等）；提升机；卷扬机
Holand windmill [ˈwɪndmɪl] 荷兰风车
hollow shaft [ˈhɒləʊ] 空心轴
hook [hʊk] 挂钩
hook bolt 吊耳
hook spanner 钩；弯脚扳手
Hooke's law 胡克定律
hook-up wire 架空电线
horizon axis [həˈrɑɪzən] [ˈæksɪs] 水平轴
horizon axis rotor 水平轴转子
horizontal axis wind turbine(HAWT) 水平轴风力发电机
horizontal axis windmill 横轴风车；水平轴风力机
horizontal buoyancy correction [hɒrɪˈzɒnt(ə)l] [ˈbɒɪənsɪ] 水平浮力修正
horizontal coherence [kə(ʊ)ˈhɪər(ə)ns] 水平相干性
horizontal turbulent diffusion [ˈtɜːbjʊl(ə)nt] [dɪˈfjuːʒ(ə)n] 水平湍流扩散
horizontal wind 水平风
horizontal wind field 水平风场
horizontal wind shear 水平风切
horseshoe magnet [ˈhɔːsʃuː] [ˈmægnɪt] 马蹄形磁铁
horseshoe vortex 马蹄涡
horseshoe vortex system 马蹄涡系
hose 软管；胶皮管；蛇管
hose assembly 胶管总成
hose clip 管夹
hot wire anemometer [ˌænɪˈmɒmɪtə] 热丝风速仪
hot-dip 浸镀；热浸；热浸镀
hot-film anemometer 热膜风速计
hot-wire anemometer 热线风速计
hot-wire direction meter 热线风向计
hourly mean wind speed 每小时平均风速
hourly wind speed 每小时风速
hours of wind 刮风小时数
household 家庭用户
hovercraft [ˈhɒvəkrɑːft] 水翼船
hub [hʌb] 轮毂
hub height [hɑɪt] 轮毂高度
hub precone 桨毂预锥角
hub rigidity [rɪˈdʒɪdətɪ] 轮毂刚度
hubcap 轮毂罩
hull [hʌl] 外壳；船体
humanized computer operation interface 人性化的计算机操作界面
humidity [hjuːˈmɪdətɪ] 湿气；潮

湿；湿度

humidity sensitive element 湿敏元件

hurricane [ˈhʌrɪk(ə)n] 飓风（十二级以上）

hurricane boundary layer [ˈbaʊnd(ə)rɪ] 飓风边界层

hurricane core 飓风核心

hurricane eye 飓风眼

Hutter wind turbine 赫特风轮机

Hybrid systems 混合系统

hydraulic actuator [haɪˈdrɔːlɪk] 液压执行器

hydraulic brake 液压制动器

hydraulic braking system 液压制动系统

hydraulic crane system 液压吊车系统

hydraulic cylinder [ˈsɪlɪndə] 液压缸

hydraulic damper 液压阻尼器

hydraulic filter 液压过滤器

hydraulic fluid [ˈflu(ː)ɪd] 液压油

hydraulic hatch system 液压舱口系统

hydraulic hoses 液压软管

hydraulic motor 液压马达

hydraulic pitch axis 液压变桨轴

hydraulic power pack 液压联动机构

hydraulic pump 液压泵

hydraulic ram 液压活塞

hydraulic system 液压系统

hydraulic transmission 液压传动

hydraulic turbine 水轮机

hydro plant [ˈhaɪdrəʊ] 水电站

hydro power station 水力发电站

hydro-elastic modeling 液动弹性模型

hydrogen storage 储氢

hydrogenerator 水轮发电机

hydrological study 水文研究

hygrographs [ˈhaɪgrəgrɑːfs] 自动湿度记录计

hygrometer [haɪˈgrɒmɪtə] 湿度计

hysteresis 滞后

Hz（Hertz） 赫兹

I

IAWE（International Association for Wind Engineering） 国际风工程学会

IBL（Internal boundary layer） 内边界层

IC（integrated circuit） 集成电路

ice detector 冰凌探测仪

iced blade 叶片结冰

ideal flow [ɑɪˈdɪəl] 理想流动

ideal fluid [ˈfluːɪd] 理想流体

ideal gas 理想气体

ideal power curve [kɜːv] 理想功率曲线

ideal rotors 理想转子

ideal wind turbines 理想风力机

identity matrix [ɑɪˈdentɪtɪ][ˈmeɪtrɪks] 单位矩阵

idler gear 空转齿轮；换向齿轮

idler pulley [ˈpʊlɪ] 惰轮；空转轮；导轮
idling [ɑɪdlɪŋ] 空转；怠速；空载
IEC（international Electrotechnical Commission）国际电工（技术）委员会
IEC wind class IEC 风况类别
IEE（Institution of Electrical Engineers）电气工程师学会（英）
IEEE（Institute of Electrical and Electronic Engineers）电气与电子工程师学会（美）
immersion heater 浸入式加热器
impact factor 撞击系数；影响系数
impedance 阻抗
impedance angles [ɪmˈpiːdəns] 阻抗角
impedance voltage 阻抗电压
impeller 叶轮
impinge [ɪmˈpɪndʒ] 撞击；冲击
impulse current [ˈkʌrənt] 冲击电流
impulse flashover [ˈflæʃˌəʊvə] 冲击闪络
impulse load test 冲击动载荷试验
impulsive loads [ɪmˈpʌlsɪv] 冲击载荷
in line displacement [dɪsˈpleɪsm(ə)nt] 顺风向位移
incidence angle 入射角
incompressible flow [ˌɪnkəmˈpresəbl] 不可压缩流
independent pitch 独立变桨
independent pole operation 分相操作
indirect grid connection [ˌɪndɪˈrekt] 间接并网
individual pitch control [ˌɪndɪˈvɪdjʊəl] 变桨距控制
indoor climate 室内气候
induced velocity [vɪˈlɒsətɪ] 诱导速度
inductance [ɪnˈdʌktəns] 感应系数；自感应；电感
induction coil [kɒɪl] 电感线圈
induction factor 感应系数
induction generator 感应发电机
induction machine 感应电机
inductive components [kəmˈpəʊnənt] 电感元件
industry standard 行业标准
inertia forces [ɪˈnɜːʃɪə] 惯性力
inertia loads 惯性载荷
inertia switch 惯性开关
inertial sub-range 湍流惯性负区
infinite plane/slab source [ˈɪnfɪnɪt] 无限平面源
infinite span [spæn] 无限翼展；无限展长
infinite vortex street 无限涡街
infinite wake 无限尾流
inflammable [ɪnˈflæməbl] 易燃的
inflation [ɪnˈfleɪʃ(ə)n] 充气
inflow 入流
inflow angle（flow angle） 入流角
influence area [ˈɪnflʊəns] 影响区；影响面积
influence by the wind shear 风切变影响
influence coefficient [ˌkəʊɪˈfɪʃ(ə)nt] 影响系数

influence of the tower shadow 塔影效应
influence zone 影响区
influx [ˈɪnflʌks] 流入量
infrequent wind [ɪnˈfriːkw(ə)nt] 异常风
ingress moisture [ˈɪngres] [ˈmɒɪstʃə] 水分浸入
inhomogeneous flow [ˌɪnhɒmə(ʊ)ˈdʒiːnɪəs] 非均匀流动
inhomogeneous turbulence [ˈtɜːbjʊl(ə)ns] 非均匀湍流
inhomogenous field 不均匀场
initial concentration [ɪˈnɪʃəl] [kɒns(ə)nˈtreɪʃ(ə)n] 起始浓度
initial condition 初始条件
initial disturbance [dɪˈstɜːb(ə)ns] 初始扰动
initial plume dimension [pluːm] [dɪˈmenʃ(ə)n] 初始羽流尺度
initial plume rise 初始羽流抬升
initial release rate 初始释放速率
initial tunnel turbulence [ˈtɜːbjʊl(ə)ns] 风洞初始湍流度
initial velocity [vɪˈlɒsɪtɪ] 初速度
initial wind speed 初始风速
initialize 初始化
inlet opening 入口开启度
inlet port 入口
inlet valve 入口阀
in-line array 行阵
in-line response 顺风向响应
inmost layer [ˈɪnməʊst] 最内层
inner boundary layer 内边界层
inner rotor 内转子
inoperative [ɪnˈɒpərətɪv] 不起作用的；无效的
in-plane 面内
in-plane bending 面内弯曲
in-plane fatigue loads [fəˈtiːg] 面内疲劳载荷
input power 输入功率
input shaft 输入角
inshore 沿海；沿岸
inshore wind 向岸风
in-site 现场
inspection 审查；检查
inspection earthing 检修接地
instability [ɪnstəˈbɪlɪtɪ] 不稳定性；不稳定度
installation 安装
installation and operation 安装与运行
installation costs [ˌɪnstəˈleɪʃən] 安装费
instantaneous [ˌɪnstənˈteɪnjəs] 瞬间的；即刻的；即时的
instantaneous area source [ˌɪnst(ə)nˈteɪnɪəs] 瞬时面源
instantaneous incidence [ˈɪnsɪd(ə)ns] 瞬时迎角
instantaneous measured [ˈmeʒəd] 瞬时测值
instantaneous overturning moment 瞬时倾覆力矩
instantaneous point source 瞬时点源
instantaneous power 瞬时功率
instantaneous pressure 瞬时压力
instantaneous turbulent energy 瞬时湍流能量
instantaneous value 瞬时值

instantaneous volume source 瞬时体源
instantaneous wind speed 瞬时风速
instrument transducer 测量互感器
insulant [ˈɪnsjʊlənt] 绝缘材料
insulating boots 绝缘靴
insulating bushing 绝缘套管
insulating gloves [glʌvz] 绝缘手套
insulation 绝缘
insulation coordination [kəʊˌɔːdɪˈneɪʃən] 绝缘配合
insulation parameter [ɪnsjʊˈleɪʃ(ə)n] [pəˈræmɪtə] 日照参数
insulation ratio 绝缘比
insulation resistance 绝缘电阻
insulation tester 绝缘测试器
insulator 绝缘子；瓷瓶；瓷珠
integrated control technology 一体化控制技术
integrated coupling [ˈɪntɪgreɪtɪd] 固定连接
integrated gearbox 集成变速箱
integrated value 累计值
integration design 集成设计
intelligent control system 智能控制系统
intensive 加强器
interaction (with grid) 交互作用（与电网）
interannual variation 年际变化
interchange 互换；交换；交换机
interconnection (for WTGS) [ˌɪntəkəˈnekʃən] 互连（风力发电机组）
interface [ˈɪntəfeɪs] 接口
interference regions [ˌɪntəˈfɪərəns] 干扰区域
interlocker [ˌɪntə(ː)ˈlɒkə] 联锁装置
internal calipers [ˈkælɪpəz] 内卡钳
internal discharge 内部放电
internal gear pair 内齿轮副
internal lightning protection system 内部防雷系统
internal-combustion engine 内燃机
International Electro-technical Commission 国际电工技术委员会
International Energy Agency (IEA) 国际能源署
interval 间隔；时间间隔；区间
intervene 干涉；干预；插入；介入
intrared temperature measuring instrument 远红外线测温仪
intrinsic rotor damping [ɪnˈtrɪnsɪk] [ˈdæmpɪŋ] 内在转子阻尼
inverse discrete Fourier transformation [ˈfʊrɪər] 离散傅里叶逆变换
inverter [ɪnˈvɜːtə] 逆变器；反用换流器；反相器
inverter station 换流站
investigation and design expense 勘察设计费
investment cost 投资成本
inviscid flow [ɪnˈvɪsɪd] 非黏性流
iron alloy [ˈɑɪən] 铁合金
iron core 铁芯
iron loss 铁损
irregular fluctuation [ɪˈregjʊlə] [ˌflʌktjʊˈeɪʃən] 不规则脉动
irregular grading 不规则级配

irregular terrain [te'reɪn] 不平地形
irregularing in wind 风多变
irreversibility ['ɪrɪˌvɜːsə'bɪlətɪ] 不可逆性
irreversible motion [ɪrɪ'vɜːsɪb(ə)l] 不可逆运动
irrotational flow [ˌɪrəʊ'teɪʃ(ə)n(ə)l] 无旋流
isallobaric wind [ɑɪˌsælə'bærɪk] 等变压风
isanemone 等风速线
isentropic [ˌɑɪsen'trɒpɪk] 等熵线
isentropic flow 等熵流
isentropic process 等熵过程
island operation 岛上运行
ISO (international standardization organization) 国际标准化组织
ISO certified 通过 ISO 体系认证
isobar ['ɑɪsə(ʊ)bɑː] 等压线
isochoric process [ˌɑɪsəʊ'kɔːrɪk] 等容过程
isoclines ['ɑɪsə(ʊ)klɑɪn] 等倾线
isogradient [ˌɑɪsəʊ'greɪdɪənt] 等梯度线
isohel ['ɑɪsəʊhel] 等日照线
isohyets [ˌɑɪsə(ʊ)'hɑɪɪt] 等雨量线
isolated aerofoil ['eərəfɔɪl] 孤立翼型
isolated area 孤立地区
isolated operation 孤立运行
isoline ['ɑɪsə(ʊ)lɑɪn] 等值线
isometric projection [ˌɑɪsəʊ'metrɪk] 等角投影；正等侧投影
isopiestics [ˌɑɪsəʊpɑɪ'estɪk] 等压线
isopleths ['ɑɪsə(ʊ)pleθ] 等值线
isopycnic [ˌɑɪsəʊ'pɪknɪk] 等密度线
isosceles triangle [ɑɪ'sɒsɪliːz] 等边三角形
iso-surface 等值面
isotach ['ɑɪsəʊtæk] 等风速线
isotherm ['ɑɪsə(ʊ)θɜːm] 等温线
isothermal line 等温线
isothermal atmosphere [ˌɑɪsəʊ'θɜːməl] 等温大气
isothermal energy storage 等温贮能
isothermal layer 等温层
isothermal process 等温过程
isotropic dispersion [ˌɑɪsə(ʊ)'trɒpɪk] [dɪ'spɜːʃ(ə)n] 各向同性弥散
isotropic point source 各向同性点源
isotropic turbulence ['tɜːbjʊləns] 各相同性湍流
isovel (isovelocity) ['ɑɪsəvel] 等速线
iteration [ɪtə'reɪʃ(ə)n] 迭代
iterative analysis ['ɪt(ə)rətɪv] 迭代分析
iterative earth 重复接地

J

jack 千斤顶
jet 喷嘴；射流
jointing compound 复合填料
jubilee clip ['dʒuːbɪliː] 连接螺旋夹
jumper 跨接线；跨接；跨(短) 接片
jumper clamp 跳线线夹

K

Kalman power spectrum 卡尔曼功率谱
Kalman filter 卡尔曼滤波器
key-operate 键盘操作
keyway 键槽
killed steel 脱氧钢
kinetic (potential) energy 动（势）能
knuckle joint [ˈnʌkl] 万向接头
knurled nut 凸螺母
Kutta condition 库塔条件
Kutta-Joukowski equation 库塔—儒科学夫斯基方程
Kutta-Joukowski theorem 库塔—儒科学夫斯基定理

L

labyrinth [ˈlæbərɪnθ] 迷宫
labyrinth packing 汽封
labyrinth seal 迷宫式密封
ladder bracket 爬梯固定件
ladder equipment 攀爬设备（用于风电塔）
ladder with preassembled fall arrest rail 防坠落轨道（爬梯）
lag time 滞后时间
lagged incidence [ˈɪnsɪd(ə)ns] 滞后迎角
lagging 绝缘层材料
lagging motion 摆振运动
Lagrangian autocorrelation tensor [ləˈgrændʒɪən] [ˈɔːtəʊˌkɒrəˈleɪʃən] [ˈtensə] 拉格朗日相关张量
Lagrangian autocovariance 拉格朗日自协方差
Lagrangian correlation 拉格朗日相关
Lagrangian covariance [kəʊˈveərɪəns] 拉格朗日协方差
Lagrangian equation [ɪˈkweɪʒ(ə)n] 拉格朗日方程
Lagrangian integral time scale [ˈɪntɪgr(ə)l] [ɪnˈtegr(ə)l] 拉格朗日积分间尺度
Lagrangian similarity theory [sɪməˈlærətɪ] [ˈθɪərɪ] 拉格朗日相似理论
Lagrangian spectral function [ˈspektr(ə)l] 拉格朗日谱函数
Lagrangian strain 拉格朗日应变
lake breeze 湖风
laminar aerofoils [ˈlæmɪnə] 层流翼型
laminar boundary layer [ˈlæmɪnə] [ˈbaʊnd(ə)rɪ] 层流边界层
laminar cellular convection [ˈseljʊlə] [kənˈvekʃ(ə)n] 层流环型对流
laminar convection 层流对流
laminar drag 层流阻力

laminar Ekman boundary layer [ˈbaʊnd(ə)rɪ] 埃克曼层流边界层
laminar flow 层流
laminar flow airfoil [ˈeəfɒɪl] 层流翼型
laminar plume [pluːm] 层流羽流
laminar separation [sepəˈreɪʃ(ə)n] 层流分离
laminar sub layer 层流底层
laminar viscosity [vɪˈskɒsɪtɪ] 层流黏性
laminar vortex shedding [ˈvɔːteks] [ˈʃedɪŋ] 层流旋涡脱落
laminar vortex street 层流涡街
laminar wake 层流尾流
laminar-turbulent transition [ˈtɜːbjʊl(ə)nt] 层流向湍流转换
laminate 薄板（片）
LAN (local area network) 局域网
land and sea breezes 海陆风
land Beaufort scale [ˈbəʊfət] 陆地蒲福风级
land breeze 陆风
land spill 地面溢浸
land wind 陆地风
landform [ˈlæn(d)fɔːm] 地貌，地形
landscape assessment [ˈlændskeɪp] 景观评估
land-sea breeze 海陆风
land-use impacts 土地利用的影响
landward wind 海风
lap joint 搭接
Laplace 拉普拉斯
Laplace equation [lɑːˈplɑːs] 拉普拉斯方程
Laplace transform 拉普拉斯变换
lapse rate 温度垂直梯度
large power drill [drɪl] 大型钻
large-scale circulation [sɜːkjʊˈleɪʃ(ə)n] 大尺度环流
large-scale eddy 大尺度涡旋
large-scale model 大缩尺比模型
large-sized wind machine 大型风力机
laser shaft alignment instrument 激光对中仪
laser tracker 激光跟踪仪
latent fault 潜在故障
lateral buckling 侧向屈曲
lateral correlation function [ˌkɒrəˈleɪʃ(ə)n] 横向相关函数
lateral deflection [dɪˈflekʃ(ə)n] 侧向挠度
lateral diffusion [dɪˈfjuːʒ(ə)n] 横向扩散
lateral displacement [dɪsˈpleɪsm(ə)nt] 侧向位移
lateral flapping 侧向挥舞
lateral force 侧（向）力
lateral gust 横向阵风
lateral length scale 横向（积分）长度尺度
lateral load 横向荷载
lateral oscillation [ˌɒsɪˈleɪʃən] 横向振动
lateral resistance [rɪˈzɪst(ə)ns] 横向阻力
lateral scale 横向尺度
lateral shear 横向剪切（力）

lathe [leɪð] 车床
lathe tool 车床工具
lattice steel towers ['lætɪs] 格子形钢塔架
lattice tower 网格塔架
leading edge 前缘
lead-lag moment 超前—滞后时刻
leaf spring 叠簧
leakage current ['liːkɪdʒ] 泄漏电流
leakage current insulating version 漏电流绝缘型
leakage flux [flʌks] 漏磁通
learning curves 学习曲线
LED (light emitting diode) 发光二极管
lee side of the tower 塔的背风侧
left-hand thread 左旋螺纹
legal representative 法人代表
length of blade 叶片长度
length scales 长度尺度
lengthwise ['leŋθwaɪz] 纵长的
lengthwise ribbing 纵向防滑纹
level class 层
level switch 油位开关
leverage ['liːvərɪdʒ] 杠杆作用
licensing of wind turbine technology 风电机组技术许可证
life cycle costing 寿命周期成本
life factor 寿命系数；使用年限因数
life test 寿命试验
life-cycle 寿命周期
lifespan 寿命
lifetime 使用寿命；寿命周期
lifetime estimations 寿命估算
lift and drag (L/D) 升力和阻力
lift coefficient [ˌkəʊɪ'fɪʃ(ə)nt] 升力系数
lift curve slope [kɜːv] 升力曲线斜率
lift effect 升力效应
lift force 升力
lift type wind machine 升力型风力机
lift (to) drag ratio ['reɪʃɪəʊ] 升阻比
lift-dependent drag 升致阻力
lifting 提升
lifting line 升力线
lifting line theory ['θɪərɪ] 升力线理论
lifting plane theory 升力面理论
lifting surface 升力面
lift-to-drag ratio ['reɪʃɪəʊ] 升阻比
lift-type device [dɪ'vaɪs] [能] 升力型装置
lift-type rotor 升力型风轮，升力型转子
light air 高空大气；软风（一级风）
light breeze/wind 轻风（二级风）
light metal spirit level ['metəl] 光金属水平仪
lighting conductor 光导体
lightning ['laɪtnɪŋ] 闪电
lightning arrester 避雷器
lightning current 雷电流
lightning overvoltage [ˌəʊvə'vəʊltɪdʒ] 雷电过电压
lightning protection system 防雷

系统
lightning protection zone 防雷区
lightning rod 避雷针
lightning strike 雷击
lightning stroke 雷电波
lightning surges 雷击电涌
limit gauge 极限量规
limit speed switch 限速开关
limit state 极限状态
limit switch 限位开关
limited amplitude response[ˈæmplɪtjuːd] [rɪˈspɒns] 限幅相应
limited current circuit 限流电路
limiting value [ˈvæljuː] 极限值
limit-state design 极限状态设计
line blow 强风
line graph 折线图
line trap 线路限波器
linear acceleration method [ˈlɪnɪə] [ˈmeθəd] 线性加速度法
linear correlation [ˌkɒrəˈleɪʃ(ə)n] 线性相关
linear dynamic response [daɪˈnæmɪk] 线性动态响应
linear elasticity [elæˈstɪsɪtɪ] 线性弹性
linear energy transfer 线性能量传递
linear equation 线性方程
Linear fractional transformation (LFT) 分式线性变换
linear function 线性函数
linear interpolation 线性插值
Linear matrix inequality (LMI) [ˌɪnɪˈkwɒlətɪ] 线性矩阵不等式
Linear parameter varying (LPV) 线性变参数
linear relation 线性关系
linear scale 线性比例
Linearised LPV models 线性变参数模型
linearity [ˌlɪnɪˈærətɪ] 线性；线性度
linearization 线性化
linearized aerodynamics [ˌeərə(ʊ)daɪˈnæmɪks] 线化空气动力学
linearized aerodynamics model 线性空气动力学模型
linearized hot wire anemometer [ˌænɪˈmɒmɪtə] 线化热线风速计
linearized model 线性化模型
linearized potential field [pə(ʊ)ˈtenʃ(ə)l] 线化势流场
linearized theory 线（性）化理论
line-commutated 线换向
line-drop compensation (LDC) 线路压降补偿
lining 内衬；衬里
live tank oil circuit breaker 少油断路器
load 负载
load bed 荷载试验台
load case 载荷状况；负载状况
load characteristic [ˌkærəktəˈrɪstɪk] 负载特性
load flow reversal （电网）潮流反转
load leveling 负荷调整
load line 载重线
load paths 载荷路径

load presetting 预设负荷
load relief [rɪˈliːf] 减载
load resistor 负荷电阻
load shedding 甩负荷
load spectrum [ˈspektrəm] 荷载谱
load-bearing 承重
load-bearing structure 承载结构
load-carrying capability [ˌkeɪpəˈbɪlətɪ] 承载能力
load-duration curves 负荷曲线
local angle of latitude 当地方位角
local angle-of-attack 当地迎角
local blade chord 当地叶片弦（长）
local breeze 局部风
local circulation [sɜːkjʊˈleɪʃ(ə)n] 局部环流
local climate 局部气候
local climate condition 局部气候条件
local concentration 当地浓度
local control 现地控制；就地控制
local diffusivity [ˌdɪfjuːsɪvətɪ] 当地扩散率
local drag 当地阻力
local flow 局部气流；当地流动
local gust 局部阵风
local lift 当地升力
local Math number 当地马赫数
local power network 地方电网
local pressure 当地压力
local separation [sepəˈreɪʃ(ə)n] 局部分离
local speed ratio 当地速度比
local topography 局部地形
local turbulence [ˈtɜːbjʊl(ə)ns] 当地湍流（度）
local wind 局部风
local wind characteristics [ˌkærəktəˈrɪstɪk] 局部风特性
local wind environment 局部风环境
local wind profile 局部风廓线
local wind regime [reɪˈʒiːm] 局部风况
local wind resource 当地风力资源
localized length scale 当地积分长度尺度
lock 加锁；锁（紧、定）；自动跟踪
Lock number 洛克数
locking device 锁定装置
locking nut 自锁螺母
locknut [ˈlɒknʌt] 防松螺母；对开螺母
log dec 对数衰减
log linear distribution 对数线性分布
log paper（半）对数坐标纸
logarithmic decrement [ˌlɒgəˈrɪðmɪk] 对数衰减
logarithmic distribution 对数分布
logarithmic distribution law 对数分布律
logarithmic frequency spectrum [ˈfriːkw(ə)nsɪ] [ˈspektrəm] 对数频谱
logarithmic law 对数（变化）律
logarithmic profile [ˈprəʊfaɪl] 对数剖面
logarithmic scale 对数尺度
logarithmic velocity profile [vɪˈlɒsɪtɪ]

对数速度廓线
logarithmic wind profile 对数风（速）廓线
logarithmic wind shear law 对数风切变律
log-linear wind profile 对数线性风廓线
log-log paper 双对数坐标纸
log-log plot 双对数坐标图
log-normal distribution 对数正态分布
long term flicker severity 长时间闪变值
longitudinal pressure gradient [ˌlɒn(d)ʒɪˈtjuːdɪn(ə)l] 纵向压力梯度
longitudinal response 纵向响应
longitudinal stability 纵向稳定性
longitudinal stiffness modulus 纵向刚模量
longitudinal turbulence component [ˈtɜːbjʊl(ə)ns] 纵向湍流分量
longitudinal turbulence spectrum [ˈspektrəm] 纵向湍流谱
longitudinal wind 纵向风；经向风
long-term effect 长期效应
long-term mean wind speed 长期平均风速
long-term observation system [ɒbzəˈveɪʃ(ə)n] 长期观测系统
long-term prediction 长期预报
long-term wind data 长期风资料
long-term wind speed average 长期平均风速
looking downwind 顺风观察
loop system 环网系统
loss 损耗
loss angle（介质）损耗角
loss of head 压力头损失
lost-head nail [neɪl] 断头钉
low carbon chrome-nickel steel [ˈnɪkəl] 低碳铬镍钢
low density wind tunnel [ˈtʌnl] 低密度风洞
low drag configuration [kənˌfɪgəˈreɪʃ(ə)n] 低阻构型
low emission [ɪˈmɪʃ(ə)n] 低（污染）排放
low energy-content wind 低（含）能风
low frequency end 低频端
low level jet 低空急流
low pass filter 低通滤波器
low pollution energy source 低污染能源
low speed brake 低速闸
low spend wind tunnel [ˈtʌnl] 低速风洞
low temperature climate protection 低温天气防护
low voltage apparatus [ˌæpəˈreɪtəs] 低压电器
low wind speed 低风速
low-drag airfoil 低阻翼型
low-drag profile 低阻外形
lower atmosphere 低层大气
lowest atmosphere layer 最低层大气，贴地大气
low-level wind 低层风
low-speed aerodynamics

[ˌeərə(ʊ) daɪˈnæmɪks] 低速空气动力学
low-speed aerofoil [ˈeərəfɔɪl] 低速翼型
low-speed characteristics [ˌkærəktəˈrɪstɪks] 低速特性
low-speed shaft 低速轴
low-speed stability 低速稳定性
low-turbulence wind tunnel [ˈtɜːbjʊl(ə)ns] 低湍流度风洞
LPV affine model [əˈfaɪn] 线性变参数仿射模型
LPV controller 线性变参数控制器
LPV gain scheduling techniques 线性变参数变增益调节技术
LPV gain-scheduled controller [ˈsʃedjuːəld] 线性变参数变增益控制器
LPV model of fixed-pitch WECS 定桨式风力机线性变参数模型
LPV model of variable-pitch WECS [ˈveərɪəbl] 变桨式风力机线性变参数模型
lub point 润滑点
lubricant [ˈluːbrɪkənt] 润滑剂
lubricant with high solid content 高固成分润滑（风电机组润滑系统用语）
lubrication [ˌluːbrɪˈkeɪʃən] 润滑
LVRT (low voltage ride through) 低电压穿越
LWECS (Large scale wind energy conversion system) 大型风能转换系统
Lyapunov function 李雅普诺夫函数

M

Mach number 马赫数
Mach number effect 马赫数效应
machine control 整机控制
machine elements 机械零件；机械元件
machine productivity curve 机器生产率曲线
machinery adhesive 机械粘合剂
macro scopic convection 大范围对流
macro turbulence [ˈtɜːbjʊl(ə)ns] 大尺度湍流
macro viscosity [vɪˈskɒsɪtɪ] 宏观黏性
macroclimate [ˌmækrəʊˌˈklaɪmɪt] 大气候
macro-meteorological [ˈmækrəʊ] [miːtɪərəˈlɒdʒɪkəl] 宏观气象的
macrometeorology [ˌmækrəʊˌmiːtjəˈrɒlədʒɪ] 大尺度气象学
macrorelief 大起伏；大地形
macroscale [ˈmækrəskeɪl] 宏观尺度
magnet bonding 磁性粘合
magnetic fields [mægˈnetɪk] 磁场
magnetic flux density 磁感应强度；磁通密度
magnetic particle test 磁粉探伤

magnetic valves 电磁阀
magnifier [ˈmægnɪfɑɪə] 放大器
magnifying glass [ˈmægnɪfɑɪŋ] 放大镜
Magnus effect [ˈmægnəs] 马格纳斯效应
Magnus effect rotor 马格努斯效应转子
Magnus force 马格努斯力
main and transfer busbar 单母线带旁路
main bearing 主轴承
main bearing housing 轴承座
main box 主控柜
main circuit [ˈsɜːkɪt] 主电路
main components 主要元部件
main contact 主触头
main control unit 主控系统
main flow air 主气流
main frame 主机架
main loads 主载荷
main powerline 电力干线
main spar 主梁
main stream 主流
main stream wind 主流风
main vortex [ˈvɔːteks] 主旋涡
main wind direction 主风向
main wind energy 主风能
mainframe 主机架
maintenance [ˈmeɪntənəns] 维修
maintenance Building 维修间
maintenance test 维护试验
major oscillation [ˌɒsɪˈleɪʃən] 主振
malfunction [mælˈfʌŋkʃən] 失灵
mandrel [ˈmændrəl] 心轴
manhole [ˈmænhəʊl] 检修孔
man-made source 人工源
manometer [məˈnɒmɪtə] 气压表，压力计
manometer 流体压力计
manometer pressure 表压力
manual friction brake [ˈfrɪkʃ(ə)n] 手动摩擦制动器
marine aerosol [məˈriːn] [ˈeərəsɒl] 海洋气溶胶
marine air mass 海洋气团
marine atmosphere 海洋大气
marine engine [ˈendʒɪn] 海用引擎
marine environment 海洋性环境
marine environment [ɪnˈvɑɪrənm(ə)nt] 海洋环境
marine meteorology [ˌmiːtɪəˈrɒlədʒɪ] 海洋气象学
marine wind regime [reɪˈʒiːm] 海洋风况
maritime air mass 海洋气团
Markov matrix [ˈmeɪtrɪks] 马尔可夫矩阵
Maskell method [ˈmeθəd] 马斯克尔（风洞阻塞）修正法
masonry drill [ˈmeɪsənrɪ] 石钻
mass conservation [kɒnsəˈveɪʃ(ə)n] 质量守恒
mass flow 质量流量
mass flow continuity [ˌkɒntɪˈnjuːɪtɪ] 质量流量连续性
mass flow rate/ratios 质量流率
mass flux 质量通量
mass inertia [ɪˈnɜːʃə] 质量惯性
mass matrix 质量矩阵

mass moment of inertia 质量惯性矩
mass sink 质量汇
mass-damper-spring system 质量阻尼器弹簧系统
mass-distribution [dɪstrɪˈbjuːʃ(ə)n] 质量分布；群分布
master cylinder [ˈsɪlɪndə] 主液压缸
master frequency 主（振）频（率）
mat 垫块
material damping [məˈtɪərɪəl] 材料阻尼
material properties 材料属性
material safety factors 材料的安全因素
mathematical model [mæθ(ə)ˈmætɪk(ə)l] 数学模型
mathematical simulation [ˌsɪmjʊˈleɪʃən] 数学模拟
matrix element [ˈmeɪtrɪks] 矩阵元
matrix method 矩阵法
matrix reduction 矩阵化简
maximum lift coefficient [ˌkəʊɪˈfɪʃ(ə)nt] 最大升力系数
maximum allowable level [ˈmæksɪməm] 最大容许速度
maximum bare table acceleration 空载最大加速度
maximum blockage [ˈblɒkɪdʒ] 最大阻塞
maximum designed wind speed 最大设计风速
maximum gust lapse interval [læps] 阵风最大递减时段
maximum instantaneous wind speed [ˌɪnst(ə)nˈteɪnɪəs] 最大瞬时风速
maximum measured power 最大测量功率
maximum mixing depth（MMD）最大混合厚度
maximum permissible release [pəˈmɪsɪb(ə)l] 最大容许排放
maximum permitted power 最大允许功率
maximum power 最大功率
maximum power of wind turbine 风力机最大功率
maximum rotational speed 最大转速
maximum rotor blade envelop [ɪnˈveləp] 风轮叶片最大包线
maximum shear method 最大剪法
maximum suction point [ˈsʌkʃ(ə)n] 最大吸力点
maximum theoretical power coefficient [ˌkəʊɪˈfɪʃənt] 理论最大功率系数
maximum torque coefficient [tɔːk] 最大力矩系数
maximum turning speed of rotor 风轮最高转速
maximum wind speed 最大风速
Maxwell distribution 马克斯韦尔分布
mean aerodynamic center [eərə(ʊ)daɪˈnæmɪks] 平均气动中心
mean aerodynamic chord [kɔːd] 平均气动弦（长）
mean annual precipitation [ˈænjʊəl]

[prɪˌsɪpɪˈteɪʃ(ə)n] 平均年降水量

mean annual range of temperature 年平均温度差

mean annual wind power density 年平均风能密度

mean blade chord 平均叶片弦长

mean camber line（翼型）中弧线

mean daily temperature 日平均温度

mean environmental wind [ɪnˌvaɪrənˈmentl] 平均环境风

mean external wind loading [ɪkˈstɜːn(ə)l] 平均外风载荷

mean free path 平均自由行程

mean geometric chord [ˌdʒɪəˈmetrɪk] 平均几何弦（长）

mean geometric chord length of airfoil 平均翼型几何弦长

mean grain size 平均粒径

mean hourly wind speed 平均小时风速

mean internal wind loading 平均内部风载荷

mean life 平均寿命

mean line 中弧线

mean molecular velocity [məʊˈlekjʊlə] 平均分子速度

mean monthly temperature 月平均温度

mean monthly wet-bulb temperature 月平均湿球温度

mean response 平均响应

mean sea level 平均海平面

mean solar time 平均太阳时

mean square departure/deviation [dɪˈpɑːtʃə] [diːvɪˈeɪʃ(ə)n] 均方偏差

mean square response 均方响应

mean square wind speed 均方风速

mean strain 平均应变

mean stress 平均应力

mean velocity defect [vɪˈlɒsɪtɪ] 平均速度缺（损）

mean velocity profile 平均速度廓线

mean wetted length 平均浸湿长度

mean wind 平均风

mean wind profile 平均风速廓线

mean wind speed 平均风速

mean wind speed model 平均风速模型

measure-correlate-predict (MCP) [ˈkɔːrəˌleɪt] 相关预测

measured hole [ˈmeʒəd] 测量孔；测压孔

measured power curve [kɜːv] 测量功率曲线

measured reading 测量读数；测量值

measureing scale [skeɪl] 测量刻度

measurement mast [ˈmeʒəm(ə)nt] 测风杆

measurement parameters 测量参数

measurement period [ˈpɪərɪəd] 测量周期

measurement seat 测量位置

measurement sector 测量扇区

measuring element 定量块；测量单元

measuring equipment 测试设备

measuring instrument 测试仪器
measuring tape 卷尺
mechanical admittance [mɪ'kænɪk(ə)l] [əd'mɪt(ə)ns] 力导纳
mechanical agitation [ædʒɪ'teɪʃ(ə)n] 机械搅拌
mechanical balance 机械天平
mechanical brake [mɪ'kætɪkəl] 机械制动
mechanical braking system 机械制动系
mechanical cooling tower 机械通风冷却塔
mechanical coupling 机械耦合
mechanical damper 机械阻尼器
mechanical damping 机械阻尼
mechanical draft [drɑːft] 机械通风
mechanical drawing 机械图
mechanical endurance/life [ɪn'djʊərəns] 机械寿命
mechanical flywheel energy storage 飞轮储能
mechanical loads 机械载荷
mechanical magnification factor [ˌmægnɪfɪ'keɪʃ(ə)n] 机械放大因子
mechanical mixing 机械混合
mechanical noise 机械噪声
mechanical sand control 机械固沙
mechanical spacer 机械隔振子
mechanical stirring ['stɜːrɪŋ] 机械搅拌
mechanical transfer function ['fʌŋ(k)ʃ(ə)n] 力传递函数
mechanical turbulence ['tɜːbjʊl(ə)ns] 机械湍流
mechanical ventilation [ˌventɪ'leɪʃ(ə)n] 机械通风
median wind speed 中值风速
medium sized wind machine 中型风力机
medium voltage 中等电压
medium-scale wind energy conversion system 中型风能转换系统
membrane ['membreɪn] 薄膜；隔板；表层
memory stick 记忆棒
mercury barometer ['mɜːkjʊrɪ] [bə'rɒmɪtə] 汞（水银）气压计
meshing 接合；相合；啮合
meshing interference [ˌɪntə'fɪərəns] 啮合干涉
metal alloy ['ælɒɪ] 金属合金
metal fatigue [fə'tiːg] 金属疲劳
metal grating shield 金属网屏蔽
metal oxide arrester(MOA) [ə'restə] 氧化锌避雷器
metallographic analyzer 金相分析仪
meteorological element [ˌmiːtɪərə'lɒdʒɪkəl] ['elɪm(ə)nt] 气象要素
meteorological forecast [ˌmiːtɪərə'lɒdʒɪkəl] 气象预报
meteorological model 气象模型
meteorological parameter [pə'ræmɪtə] 气象参数
meteorological reference wind speed 气象参考风速
meteorological standard condition

标准气象条件
meteorological station 气象站
meteorological symbol 气象符号
meteorological tower 气象测量塔；气象塔
meteorological wind runnel [ˈrʌn(ə)l] 气象风洞
meteorology [ˌmiːtɪəˈrɒlədʒɪ] 气象学
metered reading 测量读数；测量值
method of Bins [ˈmeθəd] [bɪnz] 比恩法
method of image 镜像学
method of least square 最小平方法
method of partial safety factors 部分安全系数法
Michel one-step method [ˈmaiːkəl] 米歇尔一步法
micrometer screw gauge [maɪˈkrɒmɪtə] [geɪdʒ] 螺旋测位器
microprocessing unit 微处理器
microscope [ˈmaɪkrəskəʊp] 显微镜
micrositing 微观选址
microstructure microscope 金相显微仪
microswitch 微动开关
milling machine 铣床
minicomputer program 微机程控
minimum wind to yaw [ˈmɪnɪməm] 最小调向风
missing 故障；损失；遗漏
mitre block [ˈmaɪtə] 斜接；斜面接合块
mitre joint 斜接
mixed divider [dɪˈvaɪdə] 混合分压器
mixing distributor 混合分配器
mixture 混合；混合物；混合剂
mobile wind speed unit 流动式风速测量车
modal amplitude [ˈæmplɪtjuːd] 模态幅值
modal analysis [əˈnæləsɪs] 模态分析
modal damping 模态阻尼
modal damping coefficient [ˌkəʊɪˈfɪʃənt] 模态阻力系数
modal displacement 模态位移
modal force 模态力
modal frequency 模态频率
modal mass 模态质量
modal response 模态响应
modal stiffness [ˈstɪftɪs] 模态刚度
modal wind load 模态风载
mode shape 模态形状
mode velocity [vɪˈlɒsɪtɪ] 模态形状
model wake blocking 模型尾流阻塞
moderate breeze 和风（四级风）
moderate flagging [ˈflægɪŋ] 中度旗状（Ⅲ级植物风力指示）
moderate gale [geɪl] 疾风（七级风）
moderate gust 中等阵风
modification 改造；改变；变型；变体
modification of topography [ˌmɒdɪfɪˈkeɪʃ(ə)n] [təˈpɒgrəfɪ] 地形改造

modification of wind 风场改造；风控制

modified shape 改型

modular concept 模块式设计概念

modulator-demodulator [diːˈmɒdjʊleɪtə] 调制解调器

module 模数

module design 模块设计

module-based design 模块化设计

modulus [ˈmɒdjʊləs] 模量；系数；模数

modulus of elasticity [ˌelæsˈtɪsətɪ] 弹性模量；弹性模数

moisture [ˈmɒɪstʃə] 湿气；湿度；潮湿

moisture content 水分含量

moisture ingress 水分渗入

mold release agent 脱模剂

Mole wrench [məʊl] [rentʃ] 莫尔扳手

moment coefficient 力矩系数

moment derivative [dɪˈrɪvətɪv] 力矩导数

moment of couple 力偶矩

moment of deflection [dɪˈflekʃ(ə)n] 弯矩；挠矩

moment of friction [ˈfrɪkʃ(ə)n] 摩擦力矩

moment of inertia [ɪˈnɜːʃɪə] 惯性矩，转动惯量

moment of momentum 动量矩

moment of probability distribution 概率分布矩

moment of spectral density function 谱密度函数矩

moment of stability 稳定力矩，稳性力矩

moment of torsion [ˈtɔːʃ(ə)n] 扭（转力）矩

momentum boundary layer 动量边界层

momentum budget 动量收支

momentum conservation [kɒnsəˈveɪʃ(ə)n] 动量守恒

momentum equation [ɪˈkweɪʒ(ə)n] 动量方程

momentum flux 动量通量

momentum method 动量方法

momentum theory 动量理论

momentum thickness [ˈθɪknɪs] 动量厚度

monitored information 监视信息

monitoring 监控

mono-pile foundation 单桩基

monthly average wind speed 月平均风速

monthly variation of windspeed [veərɪˈeɪʃ(ə)n] 月风速变化

morphological characteristics [ˌmɔːfəˈlɒdʒɪkəl] [ˌkærəktəˈrɪstɪk] 地貌特征

most frequent wind direction [dɪˈrekʃ(ə)n] 盛行风向

motor 电动机

motor brake status 电动机刹车状态

motor start 马达启动

moulded case circuit breaker (MCCB) [ˈsɜːkɪt] 塑壳线路断路器

moulding sand 沙模

mountain air 山地空气
mountain and valley breeze 山谷风
mountain barrier [ˈbærɪə] 山地障碍物
mountain breeze 山风
mountain chain 山脉
mountain effect 山地效应
mountain pass 山隘口
mountain valley wind 山谷风
mountain wind 山风
movable crane [ˈmuːvəbl] [kreɪn] 桥式吊车
movable load 移动荷载
moving blade 可动叶片
moving contact [ˈkɒntækt] 动触头
moving coordinate system [kəʊˈɔːdɪneɪt] 移动坐标系
moving reference system 移动参考系
multi-bladed rotor 多叶片风轮
multibladed windmill [ˈwɪn(d)mɪl] 多叶片风车
multi-degree of freedom 多自由度
multi-directional wind 多向风；不定向风
multi-element model 多段（气动弹性）模型
multilayered Blow [ˈmʌltɪˌleɪəd] 多层流动
multilevel user access control 多级用户管理权限
multimanometer 多管压力计
multi-megawatt [ˈmegəwɒt] 兆瓦级
multimeter [mʌlˈtɪmɪtə] 万用表；多用途计量器
multimode [ˈmʌltɪˌməʊd] 多模态
multi-MW class 数兆瓦级（风电机组）
multinomial distribution 多项式分布
multiple stage planetary gear train [ˈmʌltɪpl] 多级行星齿轮系
multiplication [ˌmʌltɪplɪˈkeɪʃən] 乘法；增加
multipressure measuring system [ˈmeʒərɪŋ] 多点测压系统
multistage axial-flow fan [ˈmʌltɪsteɪdʒ] [ˈæksɪəl] 多级轴流通风机
MV transmission 中压输出
MW class WTGS 兆瓦级（风电机组）
Myklestad method [ˈmeθəd] 莫克来斯塔德法

N

NACA airfoil NACA 翼型
nacelle [nəˈsel] 机舱
nacelle acceleration 机舱加速度
nacelle bedplate 机舱底座
nacelle cover 机舱罩
nacelle loading 机舱装载
narrow range variable speed [ˈveərɪəbl] 窄幅变速

national grid [grɪd] 国家电网
national wind resource 国家风力资源
natural activity 天然放射性
natural aerodynamics [ˌeərə(ʊ)daɪˈnæmɪks] 自然空气动力学
natural air cooling 自然风冷
natural atmosphere 天然大气
natural attenuation [əˌtenjʊˈeɪʃən] 自然衰减
natural background 天然本底
natural boundary-layer wind 自然边界层风
natural convection 自然对流
natural draft [drɑːft] 自然通风
natural draft cooling tower 自然通风冷却塔
natural environment [ɪnˈvaɪrənm(ə)nt] 自然环境
natural frequency [ˈfriːkwənsɪ] 固有频率
natural gusty wind 天然阵风
natural hazard [ˈhæzəd] 自然危害
natural oscillation [ˌɒsɪˈleɪʃən] 固有振荡
natural turbulence [ˈtɜːbjʊl(ə)ns] 自然湍流
natural turbulent wind [ˈtɜːbjʊl(ə)nt] 自然湍流风
natural undamped frequency [ˌʌnˈdæmpt] 无阻尼固有频率
natural ventilation [ˌventɪˈleɪʃ(ə)n] 自然通风
natural vibration [vaɪˈbreɪʃ(ə)n] 固有振动
natural wind 自然风
natural wind boundary layer 自然风边界层
natural wind environment 自然风环境
natural wind field 自然风场
natural wind gust 自然阵风
natural wind hazard [ˈhæzəd] 自然风害
Navier-Stokes equations 纳维尔—斯托克斯方程
near wake 近尾流
near wake region [ˈriːdʒən] 近尾流区
near-surface layer 近地面层
near-surface wind 近地风
needle file 针锉
needle valve 针形阀；针状活门
negative aerodynamic damping [ˌeərə(ʊ)daɪˈnæmɪk] 负气动阻尼
negative aerodynamic stiffness [ˈstɪfnɪs] 负气动刚度
negative angle of attack [ˈæŋg(ə)l] 负迎角
negative buoyancy [ˈbɒɪənsɪ] 负浮力
negative correlation coefficient [ˌkɒrəˈleɪʃ(ə)n] [ˌkəʊɪˈfɪʃ(ə)nt] 负相关系数
negative damping 负阻尼
negative frequency 负频率
negative ions 负离子
negative mass 负质量

negative phase sequence [feɪz] [ˈsiːkwəns] 负相序
negative phase sequence equivalent circuit [ɪˈkwɪvələnt] 负相序等效电路
negative pitch control 负变桨控制
negative pressure 负压
negatively buoyant plume [ˈbɒɪənt] 负浮力羽流
net electric power output 净电功率输出
net load 净荷载
net present value 净现值
net radiation 净辐射
net radiation flux 净辐射通量
net radiometer [ˌreɪdɪˈɒmɪtə] 净辐射计
net sectional area [ˈsekʃ(ə)n(ə)l] 净截面积
network 电网；网络
network connection point (for WTGS) [kəˈnekʃən] 电网连接点（风力发电机组）
network impedance phase angle [ɪmˈpiːdəns] 电网阻抗相角
network loss 电网损失
neural networks 神经网络
neutral atmosphere 中性天气
neutral point 中性点
neutral stratification [ˌstrætɪfɪˈkeɪʃən] 中性分层
Newton's second law 牛顿第二定律
no follow current 无续流
nodular cast iron structure 球墨铸铁材料
noise 噪声
noise sensitive building 噪声敏感建筑物
no-load 空载
no-load operation 空载运行
nominal power 标称功率；额定功率
non Newtonian flow [njuːˈtəʊnɪən] 非牛顿流
non Newtonian viscosity [vɪˈskɒsɪtɪ] 非牛顿黏性
non-destructive testing [dɪˈstrʌktɪv] 非破坏性试验
non-dimensional [dɪˈmenʃənəl] 无量纲的
non-dimensional time 无因次时间
non-eddying flow 无旋流
nonhomogeneous [ˈnɒnˌhɒməˈdʒiːnjəs] 非均匀性的；非均质的非均质的
noninertial coordinate system [kəʊˈɔːdɪneɪt] 非惯性坐标系
non-isotropic turbulence 非均匀性湍流
non-lift balloon 无升力气球
non-lifting body 非升力体
nonlinear aerodynamic force [nɒnˈlɪnɪə] [ˌeərə(ʊ)daɪˈnæmɪk] 非线性气动力
nonlinear aerodynamics [ˌeərə(ʊ)daɪˈnæmɪks] 非线性空气动力学
non-linear gains 非线性增益
non-linearity 非线性
non-load bearing element 非承重

部件
non-load current 空载电流
non-load voltage 空载电压
non-renewable fuels; non-renewable energy source [rɪˈnjuːəbl] [fjʊəlz] 非再生能源
non-return-flow wind tunnel [ˈtʌnl] 开路式风洞
non-shrinking [ʃrɪŋkɪŋ] 非收缩性
nonstationary aerodynamic derivative [ˈnɒnˈsteɪʃənərɪ] [ˌeərə(ʊ)daɪˈnæmɪk] [dɪˈrɪvətɪv] 非定常气动导数
non-stationary flow 非平稳流动
non-steady aerodynamics [ˌeərə(ʊ)daɪˈnæmɪks] 非定常空气动力学
non-steady lifting surface theory 非定常升力面理论
non-steady motion 非定常运动
non-structural module [ˈmɒdjuːl] 非结构模量
non-uniform field 不均匀场
nonuniform medium [nɒnˈjuːtɪfɔːm] 不均匀介质
nonuniform topography [təˈpɒgrəfɪ] 不均匀地形
non-working flank 非工作齿面
normal [ˈnɔːməl] 正常;普通;标准
normal air 标准空气
normal atmosphere [ˈætməsfɪə] 标准大气
normal braking system 正常制动系
normal condition 正常状态
normal displacement [dɪsˈpleɪsm(ə)nt] 法向位移
normal distribution 正态分布
normal force 法向力
normal induced velocity [vɪˈlɒsətɪ] 法向诱导速度
normal mode 正规模态
normal operation 正常运行
normal pitch 法向齿距
normal pressure gradient [ˈgreɪdɪənt] 法向压力梯度
normal probability curve [kɜːv] 正态概率曲线
normal shutdown 正常停机
normal state 标准状态
normal temperature and pressure (NTP) 常温常压
normal turbulence model [ˈtɜːbjʊləns] 法向湍流模型
normal wind 正常风
normality [nɔːˈmælətɪ] 当量浓度
normalization [ˌnɔːməlaɪˈzeɪʃən] 归一化,正态化
normalization condition 归化条件
normalization constant 归一化常数
normalized co-spectrum 归共同谱
normalized cross spectrum 归一化互谱
normalized power spectrum function [ˈspektrəm] 归一化功率谱函数
normalized spectrum 归一化谱
normalized coefficient [ˌkəʊɪˈfɪʃ(ə)nt] 归一化系数
northeast trades [ˈtreɪdz] 东北信

风带
nose cone 整流罩；头锥体
nose fairing [ˈfeərɪŋ] 头部整流罩
nose-up moment 抬头力矩
no-slip condition 无滑流条件
no-wind period [ˈpɪərɪəd] 无风期
NPL type wind tunnel [ˈtʌnl] NPL 式（开路闭口）风洞
nuclear accident 核事故
nuclear environment 核环境
nuclear pollution 核污染
nuclear power 核能
nuclear power station 核电站
nuclear reactor 核反应堆
nuclide [ˈnjuːklaɪd] 核素
numerical forecast [njuːˈmerɪk(ə)l] 数值预报
numerical method 数值法
numerical model 数值模型
numerical simulation [ˌsɪmjʊˈleɪʃən] 数值模拟
Nusselt number 奈斯勒数
nut 螺母；螺帽
nut tap 螺母丝锥
N-year wind speed N 年一遇风速
nylon 尼龙

O

O ring O 形环
O&M（operation and maintenance）运行与维修（费用）
objective function [əbˈdʒektɪv] 目标函数
oblique angle [əˈbliːk] 斜角
oblique wind 斜风
observation balloon 观测气球
observation error 观测误差
observed wind sounding 测风探空
obstacle [ˈɒbstək(ə)l] 障碍物
obtuse-angled triangle [əbˈtjuːs] [ˈæŋgld] [ˈtraɪæŋgl] 钝角三角形
ocean climate [ˈəʊʃən] 海洋性气候
ocean wind 海洋风
octagon wind tunnel [ˈɒktəg(ə)n] 八角形试验段风洞
octagonal tower [ɒkˈtægənəl] 八角形的塔
octal [ˈɒktəl] 八进制
Odc & Owc Cargo service（over dimensional cargo）大件与超重货物运输
off shore area 近海区
off-shore discharge 近海［离岸］排放
off shore engineering [endʒɪˈnɪərɪŋ] 近海工程
off shore environment 近海环境
off-axis [ˈæksɪs] 偏离轴线
off-design condition 非设计工况
offset [ˈɒfset] 偏离偏心；主距
off-shore oil rig 海上平台
off-shore structure 近海结构物
offshore wind [ˈɒfˈʃɔː] 离岸风；陆风
offshore wind farm 近海（海上）风电场

offshore wind turbines 海上风力发电机

off-site 风场（外）

oil circulation system 稀油循环系统

oil cooler 油冷却器

oil line is choked 堵塞油路

oil property 油品质

oil purity 润滑油纯度

oil ring 给油环

oil seal 油封

oil switch 油开关

oil-filled power cable 充油电力电缆

oil-immersed type transformer [ɪ'mɜːst] 油浸式变压器

oil-impregnated paper 油浸纸绝缘

Oldham coupling 欧氏联轴节；滑块连接

OLTC (on load tap changer) 有载调压变压器

omni-directional wind 全向风

omni-directionality [dɪrekʃə'nælətɪ] 不定向性；全向性

on-board data storage 板（盘；屏）上数据存储

once-in-50-year (return) wind 50年一遇风

oncoming flow 来流，迎面流

oncoming turbulence ['tɜːbjʊl(ə)ns] 来流湍流（度）

oncoming wind 迎面风，来流风

one dimensional flow [dɪ'menʃənəl] 一维流动

one-to-one control mode 一对一控制方式

one-velocity diffusion theory [dɪ'fjuːʒ(ə)n] 单速扩散理论

online monitoring system 在线实时监控系统

online-access to the software 在线访问软件

on-load indicator ['ɪndɪkeɪtə] 有载指示器

on-load operation 有载运行

on-load tap changers 有载分接

onset wind speed ['ɒnset] 起始风速

on-shore area 海岸区

on shore wind 向岸风

onshore WTGS 陆上风机

on-site 厂区内；现场

on-site energy storage 就地贮能

on-site measurment 现场测量，实测

on site observation 现场观测；实测

open air 露天；户外

open air climate 露天气候

open area 开口面积；开阔地带

open circuit 断开电路

open circuit wind tunnel ['sɜːkɪt] 开路式风洞

open condition 开敞状态

open floor and roof wind tunnel 无上下壁风漏

open jet wind tunnel 开口试验段风洞

open ring wrench 开口扳手

open terrain [te'reɪn] 开阔地形

open test section [ˈsekʃ(ə)n] 开口试验段
open toothing 外露轮齿
open web girder [ˈgɜːdə] 空腹梁
open working section 开口试验段
open-circuit characteristic [ˌkærəktəˈrɪstɪk] 开路特性
open-circuit operation 开路运行
open-ended wind tunnel 开路式风洞
open-loop 开环的
open-phase protection 断相保护
open-throat wind tunnel 开口试验段风洞
open-truss tower 开放式桁架塔；桁架式塔架
operating condition [ˈɒpəreɪtɪŋ] 工况
operating lifetime 运行寿命
operating point 工作点
operating speed 运行速度
operating speed range 运行转速范围
operating states 工作状态
operating time 工作时间
operation amplifier [ˈæmplɪfaɪə] 运算放大器
operation and maintenance [ˈmeɪntənəns] 运行与维护
operation and maintenance costs 运行维护费用
operation in severe climates 恶劣气候运行
operation management [ˈmænɪdʒmənt] 运行管理
operation mechanism [ˈmekənɪzəm] 操动机构
operation start-up expense 生产准备费
operational envelope 运行包络线
operational environment 工作环境
operational wind speed 运行风速
opportunity cost [ˌɒpəˈtjuːnɪtɪ] 机会成本
opposing wind [əˈpɒzɪŋ] 逆风，迎面风
optimal blade design [ˈɒptɪməl] 最优叶片设计
optimal feedback 最优反馈
optimized chord distribution [ˈɒptɪmaɪzd] 叶弦优化分布
optimum blade geometry [ˈɒptɪməm] [dʒɪˈɒmɪtrɪ] 最佳叶片几何形状
optimum gear ratio 最佳齿轮比
optimum power coefficient [ˌkəʊɪˈfɪʃənt] 最佳功率系数
optimum rated wind speed 最佳额定风速
optimum rotational speed [rəʊˈteɪʃənəl] 最佳转速
optimum rotor performance 最佳转子性能
optimum site [ˈɒptɪməm] 最佳场址
optimum speed to windward 最佳迎风速度
optimum system design 系统优化设计
optimum tip speed ratio 最佳叶尖速比

orientation [ˌɔːriən'teɪʃ(ə)n] 方位；朝向

orientation control 调向；朝向控制

orientation mechanism ['mekənɪzəm] 迎风机构

orifice ['ɒrɪfɪs] 孔

orifice static tap ['stætɪk] 静压测量孔

orographic condition [ˌɔːrəʊ'græfɪk] 地形条件

orographic effect 地形影响

orographic lifting 地形抬升

orographic upward wind 地形性上升风

orthogonal mode shapes [ɔː'θɒgənəl] 正交模态

Orthogonality Condition 正交法则

oscillation [ˌɒsɪ'leɪʃən] 振荡；震荡

oscillator 振荡器

oscilloscope [ɒ'sɪləuskəup] 示波器

out-of-door hydrodynamics [ˌhaɪdrə(ʊ)daɪ'næmɪks] 室外流体动力学

out of-plane bending 面外弯曲

out yawing 外偏航

outgoing(incoming)line 出（进）线

outlet velocity [vɪ'lɒsɪtɪ] 出口速度；排出速度

out-of-phase [feɪz] 异相

out-of-plane bending moment 平面弯矩

output characteristic of WTGS [ˌkærəktə'rɪstɪk] 风力发电机组输出特性

output coupling 输出连接

output power 输出功率

output shaft 输出轴

oval nail ['əʊvəl] [neɪl] 椭圆钉

over current ['kʌrənt] 过电流

over current protection 过电流保护

over current protection trip 过流保护跳闸

over current protective device 过电流保护装置

over frequency 过频

over humidity protection 过湿保护

over lubrication 过度润滑

over power 过载功率

over speed 过速

over speed protection 过速保护

over temperature protection 过温保护

over voltage 过电压

over voltage protection 过电压保护

overall design 总体设计

overall dimension [dɪ'menʃ(ə)n] 外形尺寸，轮廓尺寸

overhead line 架空电缆

overhead power line; overhead wire 高架输电线

overheat [ˌəʊvə'hiːt] 加热；过度；使过热

overlap [ˌəʊvə'læp] 重叠；覆盖

overload 超载；过载；过负荷

overpressure 过压

overproduction 生产过剩

overspeed 过速

overspeed control 超速控制

overspeed device 超速保护装置
overspeed protection 超速保护
overspeed spoiler [ˈspɒɪlə] 超速保护扰流板
overspeed trial 超速试验
overturning 倾覆，倾翻
overturning moment [əʊvəˈtɜːnɪŋ] 倾翻力矩
overturning wind speed 倾覆风速
overvoltage [ˌəʊvəˈvəʊltɪdʒ] 过电压
over-voltage protection 过电压保护
oxidation [ˌɒksɪˈdeɪʃən] 氧化 [作用]；氧化层
oxide conversion coating 氧化膜
oxyacetylene welding [ˌɒksɪəˈsetɪliːn] 氧乙炔焊

P

parallel connection [ˈpærəlel] 并联
parallel key 平面键
parallel shaft arrangement 平行轴布置
parallelogram [ˌpærəˈleləgræm] 平行四边形
parked wind turbine 风力发电机停机
parking 停机
parking brake 停机制动
partial discharge 局部放电
partial safety factors 部分安全系数
partial-span pitch control 部分跨度间距控制
passive filter [fɪltə] 无源滤波器
passive pitch control 被动变桨控制
passive stall 被动失速
passive stall control 被动失速控制
passive yawing 被动偏航
pawl [pɔːl] 掣子；棘爪；制转杆
payback period [ˈpeɪbæk] [ˈpɪərɪəd] 偿还期
peak aerodynamic torque [ˌeərəʊdaɪˈnæmɪk] 气动力矩峰值
peak factor 峰值因数
peak reverse voltage 反向峰值电压
peak value 峰值
peak voltmeter [ˈvəʊltˌmiːtə] 峰值电压表
peak-load 峰荷
pedestal [ˈpedɪstəl] 基架；底座；基础
performance analysis 性能分析
performance measurement 性能测量
period 周期
periodic coefficients [ˌpɪərɪˈɒdɪk] 周期系数
periodic loads 周期载荷
periodic vibration 周期振动
permanent magnet [ˈpɜːmənənt] [ˈmægtɪt] 永久磁铁
permanent magnet generator 永磁发电机
permitted noise levels 允许噪音

水平
persistence forecast 持续性预报
personal protection equipment 个人保护装备
perspective drawing [pə'spektɪv] 透视图
per-unit system 每单位系统
petrochemical industry [ˌpetrəʊ'kemɪkəl] 石油化学工业
petrol engine ['petrəl] 汽油机
phase [feɪz] 阶段；状态；相；相位
phase advance 相位超前
phase compensation [ˌkɒmpen'seɪʃən] 相位补偿
phase displacement (shift) 相移
phase lead (lag) 相位超前（滞后）
phase margin ['mɑːdʒɪn] 相位裕度
phase sequence 相序
phase shifter ['ʃɪftə] 移相器
phase-to-phase voltage 线电压
photoelectric device 光电器件
photoelectric emission [ˌfəʊtəʊɪ'lektrɪk] 光电发射
photomontage [ˌfəʊtəmɒn'tɑːʒ] 蒙太奇照片
photovoltaic power [ˌfəʊtəʊvɒl'teɪɪk] 光伏发电
PI controller PI 控制器
PID controller PID 控制器
pile foundation 桩基
piled foundation [pɑɪld] 打桩基础
pile-driver 打桩机
pin 柱销
pinion ['pɪnjən] 小齿轮
pinion gear 小齿轮；行星齿轮
piston pump 柱塞泵
piston rod ['pɪstən] 活塞杆
pitch 齿距
pitch actuation systems 变桨驱动系统
pitch actuator 变桨驱动器
pitch angle 桨距角
pitch angle to aerodynamic torque gain [ˌeərəʊdɑɪ'næmɪk] 桨距角气动力矩增益
pitch bearing 变桨轴承
pitch circle 节距圆；中心圆
pitch control 桨距控制
pitch drive 变桨驱动
pitch point 节点
pitch rate limits 变桨速率限制
pitch regulated wind turbines 变桨风力机
pitch regulation 桨距调节
pitch setting angle 安装角
pitch speed 变桨速度
pitch system 变桨距系统
pitch the blade 调整桨距
pitching moment 变桨力矩；仰俯力矩
pitching moment coefficient 变桨力矩系数
pitching to feather; pitch-to-feather（变桨）顺桨
pitching to stall; pitch-to-stall（变桨）失速
pitch-regulated machines 变桨式机组
pitting 点蚀

pitting resistance 抗点蚀
plane of rotation 旋转平面
planet carrier 行星架
planet gear 行星齿轮
planet wheel bearing 行星轮轴承
planetary ['plænɪtərɪ] 行星的
planetary gear drive mechanism ['mekənɪzəm] 行星齿轮传动机构
planetary gear train 行星齿轮系
planetary helical gearbox 斜齿面行星齿轮箱
planning application 规划申请
platform 平台
plug connecting structure（变流单元）插拔式结构
plug-and-play system 即插即用界面
plug-in 快插式（开关）
PM（permanent magnet）synchronous generator 永磁同步发电机
pneumatic [njuː'mætɪk] 气动的
pneumatic pump 气动泵
point of common coupling 公共供电点
point plane gap 针板间隙
polarity effect [pəʊ'lærətɪ] 极性效应
porcelain insulator ['pɔːsəlɪn] 陶瓷绝缘子
position indicator 位置指示器
positive displacement pump 容积式泵
potential stress [pəʊ'tenʃəl] 电位应力（电场强度）
potential transformer（PT）电压互感器
power balance 功率平衡
power cable 电力电缆
power capacitor 电力电容
power coefficient [ˌkəʊɪ'fɪʃənt] 功率系数
power collection system（for WTGS）电力汇集系统（风力发电机组）
power control 功率控制
power curve [kɜːv] 功率曲线
power duration curve 功率持续曲线（负荷曲线）
power electronic converter [ˌɪlek'trɒnɪk] 电力电子变流器
power electronics 动力电子设备
power excursions 功率失常激增
power factor 功率因数
power factor correction（PFC）功率因数校正
power factor correction capacitor [kə'pæsɪtə] 功率因数校正电容器
power flows 功率流；电流量
power fluctuations [ˌflʌktjʊ'eɪʃəns] 电网波动
power law for wind shear 风切变对数幂次法则
power law profile ['prəʊfaɪl] 幂律廓线
power line carrier（PLC）电力线载波（器）
power measurement ['meʒəmənt] 功率测量；电力参数测量
power network 电力网络

power output 功率输出
power panel 控制盘
power performance 功率特性
power quality 电能质量
power ramp rate 功率变化率
power smoothing 电源平滑
power spectra containing periodic components [ˌpɪərɪˈɒdɪk] 含有周期分量的功率谱
power spectral density (PSD) 功率谱密度
power spectrum [ˈspektrəm] 功率谱
power swings 功率摇摆
power system 电力系统
power transducer [trænzˈdjuːsə] 电量变送器
power transformation and distribution works 变配电工程
power transmission 电力传输
power versus wind speed curve [ˈvɜːsəs] 功率与风速曲线
power-flow current 工频续流
Prandtl logarithmic law [ˌlɒgəˈrɪðmɪk] 普郎特对数法则
Prandtl's tip loss factor 普郎特叶尖损耗因子
prebend blade 预弯叶片
precipitation [prɪˌsɪpɪˈteɪʃən] 降水
pre-cut fibre sheet 预切纤维板
preflex blade 预弯叶片
preloaded flanges [priːˈləʊdɪd] 预装法兰
preloading 预加载荷
pressure angle 压力角
pressure coefficient [ˌkəʊɪˈfɪʃənt] 压力系数
pressure connection 压力接头
pressure control valve 压力控制器
pressure drag 压差阻力；压力阻力
pressure drop 压力下降；压强下降
pressure fed lubrication [ˌluːbrɪˈkeɪʃən] 压流润滑法
pressure gauge 压力表
pressure relay 压力继电器
pressure sensitive switch 压敏开关
pressure switch 压力开关
pressure test 耐压试验
pressurized oil supply 强制润滑
prevailing wind direction 主风向；盛行风向
preventive maintenance [ˈmeɪntənəns] 维护
primary current 一次电流
primary voltage 一次电压
primary (backup) relaying 主（后备）继电保护
prime grid substation 主网变电站
prime mover [prɑɪm] 原动机
principle of virtual work [ˈprɪnsəpl] 虚功原理
principles of electric circuits 电路原理
probability density functions [ˌprɒbəˈbɪlətɪ] 概率密度分布函数
probability distribution 概率分布
probability distribution of wind speeds 风速概率分布

profile correction [kə'rekʃən] 齿廓修形
profile modification 齿廓修正
programmable control [ˌprəʊ'græməbl] 可编程序控制
programming languages 编程语言
progressive system 递进式系统
project appraisal [ə'preɪzəl] 项目评估
projected area of blade [bleɪd] 叶片投影面积
proof voltage [pruːf] 耐电压
properties of composites [kəm'pəʊzɪt] 复合材料性质
properties of wood laminates 木质层压板属性
proportional valve 比例阀
protected against dropping water 防滴
protected against splashing 防溅
protected against the effects of immersion [ɪ'məːʃən] 防浸水
protection degree/lever 防护等级
protection spectacles ['spektəklz] 护目镜
protection system (for WTGS) 保护系统(风力发电机组)
protective circuit ['səːkɪt] 保护电路
protective earthing 保护接地
protective relay 保护继电器
protective relaying 继电保护
protocol ['prəʊtəkɒl] 协议
prototype turbines ['prəʊtətaɪp] 原型风轮机
proximity sensor 距离(接近)传感器
proximity switch 接近开关
pulsating current [pʌl'seɪtɪŋ] ['kʌrənt] 脉动电流
pulsating voltage 脉动电压
pulse width modulation (PWM) [ˌmɒdjʊ'leɪʃən] 脉冲宽度调制
pumped storage power station 抽水蓄能电站
punch machine 冲床
PWM (pulse width modulation) 脉宽调制

Q

quadrant 四象限
quality check 质量检验
quality factor 品质因数
quasistatic bending moments 准静态弯曲力矩
quasistatic response [rɪ'spɒns] 准静态响应
quasi-steady aerodynamics ['kwɑːzɪ] [ˌeərə(ʊ)daɪ'næmɪks] 准定常空气动力学
quasi-uniform field 准均匀场
quenching ['kwentʃɪŋ] 淬火;熄

R

rack and pinion [ˈpɪnjən] 齿轮齿条
radial pin coupling [ˈreɪdɪəl] [ˈkʌplɪŋ] 径向销连接
radial velocity [vɪˈlɒsətɪ] 径向速度
radiant flux [flʌks] 辐射通量
radio interference [ˌɪntəˈfɪərəns] 无线干扰
rain deflector 防雨罩
rainflow counting 雨流计数
rainflow cycle counting 雨流周期分析
rainflow method [ˈmeθəd] 雨流法
rammer 撞锤
ramp rate control 变化率控制
random load 随机荷载
random vibration 随机振动
ratchet spanner [ˈrætʃɪt] [ˈspænə] 棘轮扳手
rated (rotating) speed 额定转速
rated apparent power 额定视在功率
rated condition 额定工况
rated current 额定电流
rated frequency 额定频率
rated load 额定载荷
rated load torque [tɔːk] 额定转矩
rated operational current [ˈkʌrənt] 额定工作电流
rated operational voltage 额定工作电压
rated power 额定功率
rated power output 额定输出功率
rated reactive power [rɪˈæktɪv] 额定无功功率
rated speed of rotor 风轮额定转速
rated tip-speed ratio 额定叶尖速度比
rated torque 额定转矩
rated torque coefficient 额定力矩系数
rated turning speed of rotor 风轮额定转速
rated value 额定值
rated voltage 额定电压
rated wind speed 额定风速
rating of equipment [ɪˈkwɪpmənt] 设备额定值
ratio of over load 过载度
ratio of tip-section chord to root-section chord [kɔːd] 叶片根梢比
ratio of transformation 变换比
Rayleigh distribution [ˈreɪlɪ] 瑞利分布
reactance [rɪˈæktəns] 电抗
reactance voltage 电抗电压
reactive current 无功电流
reactive power 无功功率
reactive power charges 无功功率收费
reactive power compensation 无功补偿
reactor 电抗器
read out 读出
readjust 重新调整，再调整

real time 实时
real time monitoring 实时监控
reamer 钻孔器；刀；铰床
rear bearing 后轴承
rear lights 尾灯
reciprocal [rɪˈsɪprəkəl] 互逆（反）；可逆；倒数
reciprocating engine [ˈendʒɪn] 往复式发动机
reclosing 重合闸
recover [rɪˈkʌvə] 恢复；再现
recovery voltage 恢复电压
rectified 整流
rectifier 整流器
reduced frequency [ˈfrɪkwənsɪ] [力] 简约频率；换算频率；约化频率；斯德鲁哈尔数
reducing valve 减压阀
reduction gear 减速齿轮；减速装置
redundancy [rɪˈdʌndənsɪ] 冗余技术
reference distance 基准距离
reference height 基准高度
reference roughness length [ˈrʌfnɪs] 基准粗糙长度
reference wind speed 参考风速
regulating characteristics [ˌkærəktəˈrɪstɪks] 调节特性
regulating mechanism [ˈmekənɪzəm] 调速机构
regulating mechanism by adjusting the pitch of blade 变桨距调速机构
regulating mechanism of turning wind rotor out of the wind sideward [ˈsaɪdwəd] 风轮偏侧式调速机构
reinforced concrete 钢筋混凝土
reinforcement 钢筋
relative humidity [hjuːˈmɪdətɪ] 相对湿度
relative thickness of airfoil [ˈeəfɒɪl] 翼型相对厚度
relative velocity 相对速度
relative voltage change 相对电压变化
relay 继电器
relay circuit board 继电器板
relay panel 继电器屏
relay protection 继电保护
releasing coating 脱模涂层
reliability [rɪˌlɑɪəˈbɪlətɪ] 可靠性
reliability determination test 可靠性测定试验
relief valve 溢流阀；安全阀
remote access 远程访问
remote communities [rɪˈməʊt] 偏远社区
remote control 遥控；遥控装置；遥控操作
remote display service 远程显示服务
remote monitoring system 远程监控系统
remote terminal unit（RTU） 远动终端
renewable energy [rɪˈnjuːəbl] 再生能源
repair sleeve [sliːv] 补修管
repair time 修复时间

report generation 报告生成
reserve capacity 备用容量
reservoir [ˈrezəvwɑː] 容器；蓄水池；贮存器
reset 复位；重置；清除
residual capacitance [rɪˈzɪdjʊəl] [kəˈpæsɪtəns] 残余电容
residual current 残余电流
residual current circuit-breaker 漏电断路器
residual pressure 余压
residual voltage 残压
resilient mountings 柔性支撑
resin 树脂
resist 抗蚀剂；保护层
resistance 阻抗
resistance of an earthed conductor 接地电阻
resistance voltage 电阻电压
resistivity [ˌriːzɪsˈtɪvətɪ] 电阻率
resistor 电阻器
resolver 分解器
resonance [ˈrezənəns] 谐振；共振
resonance-induced loads 共振引起的载荷
resonant bending moment 共振弯矩
resonant frequency 谐振频率
resonant response 共振响应
resonant root bending moment 谐振根弯矩
resonant size reduction factor 谐振规模缩减因子
resonant tip response 共振尖响应
response time 响应时间
rest 停机
restart 重新启动
restoring force 回复力
restriking 电弧重燃
retaining ring 护环
return flow 回油（口）
return information 返回信息
revolution counter 旋转计数器
Reynolds number（Re）[ˈrenəldz] 雷诺数
Reynolds stress 雷诺应力
RF（radio frequency） 无线电射频
rheostat [ˈriːəustæt] 变阻器
ribbed [rɪbd] 有罗纹的；有棱条纹的
ribbed tubular heating element 螺纹管状加热元件
Richardson number [ˈrɪtʃədsən] 理查森数
rigging 索具；绳索；传动装置；装备
right angle 直角
right-angled screw driver 平口起子
rigid body planar rotation 刚体平面旋转
rigid coupling [ˈkʌplɪŋ] 刚性连接；刚性联轴器
rigidity [rɪˈdʒɪdətɪ] 刚度
rigidity gear 刚性齿轮
rime 雾凇
ring bus 环形母线
ring circuit 回路
ring earth external [ɪkˈstɜːnəl] 环形接地体
ring gear 内齿圈；环形齿轮

ring spanner 环形扳子
ring vortices [ˈvɔːtɪˌsiːz] 环涡
ripsaw [ˈrɪpsɔː] 粗齿锯
rivet 铆钉；固定
rms (root mean square) 均方根值
robust control [rəʊˈbʌst] 鲁棒控制
robust performance 鲁棒性能
robust stability 鲁棒稳定性
rocker arm 摇臂；摇杆；往复杆
Rogowski coil 罗可夫斯基线圈
roller 柱销套；滚轮（轴、筒）
roller bearing 滚子轴承
root circle 齿根圆
root end 叶片根部
root locus plot [ˈləʊkəs] 根轨迹图
root of blade 叶根
root vortex [ˈvɔːteks] 根涡
rope suspensions 锚固点
Rose diagram 风向玫瑰图
rotary transformer 旋转变压器
rotating center 转动中心
rotating union 旋转体；旋转接头
rotational sampling [rəʊˈteɪʃənəl] 轮换抽样
rotational speed 旋转速度
rotationally sampled wind velocity 旋转采样风矢量
rotor 转子；风轮
rotor axis tilt angle 风轮轴仰角
rotor blade theory 旋翼理论
rotor diameter 风轮直径
rotor disc 转盘
rotor hub 转子轮毂
rotor locking 转子锁锭
rotor mass imbalance 风轮质量不平衡
rotor over-current protection 转子过流保护
rotor power coefficient 风能利用系数
rotor side rated capacity 机侧额定容量
rotor solidity [səˈlɪdɪtɪ] 风轮实度
rotor speed 风轮转速
rotor swept area 风轮扫掠面积
rotor thrust [θrʌst] 旋翼拉力
rotor torque [tɔːk] 旋翼扭矩
rotor wake 风轮尾流
rotor whirl modes [hwɜːl] 转子旋转模式
rotor-schub 风轮推力
rotor-swept area 叶轮扫掠面积
roughness length [ˈrʌfnɪs] 粗糙长度
roughness measuring machine 粗糙度测量仪
round-head rivet 圆头铆钉
round-head screw 圆头螺丝（钉）；螺旋体
routine test 常规试验
rpm (revolution per minute) 转/分
rubber elastic component 橡胶弹性元件
rubber profile 橡胶构件
run-away 飞逸
rung 横档（爬梯用）
Runge-Kutta method [ˈmeθəd] 容库法
running loop 活结

running-in wear 磨合期磨损
rural electrification [ɪˌlektrɪfɪˈkeɪʃən] 农村电气化
rustling 沙沙声

S

safe life 安全寿命
safety belt 安全带
safety class (IEC WTGS Class) 安全等级(IEC 风电机组类别)
safety color 安全色
safety concept [ˈkɒnsept] 安全方案
safety distance 安全距离
safety factors 安全系数
safety helmet [ˈhelmɪt] 安全帽
safety impedance [ɪmˈpiːdəns] 安全阻抗
safety isolating transformer [ˈaɪsəleɪtɪŋ] 安全隔离变压器
safety loop 安全链路
safety marking 安全标志
safety stop position 安全停止位置
safety switch 安全开关
safety system 安全系统
safety valve 安全阀
salient-pole [ˈseɪljənt] [pəʊl] 凸极
sall regulation 失速调节
salt fog/spray 盐雾
salt spray test 盐雾测试
salty climate 盐雾气候环境
sampling test 抽样试验
sand blasting 沙暴
sand storm climate 风沙气候
sandwich construction 夹层构造
sandwich material 夹心材料
saturate [ˈsætʃəreɪt] 使饱和;浸透;使充满
saturation characteristic [ˌkærəktəˈrɪstɪk] 饱和特性
SCADA (supervisory control and data acquisition) [ˈsjuːpəˌvaɪzərɪ] [ˌækwɪˈzɪʃən] 监测控制和数据采集
scaffolding [ˈskæfəldɪŋ] 脚手架
scalene trangle [ˈskeɪliːn] 不等边三角形
scanning electron microscope 扫描电镜
scraper [ˈskreɪpə] 刮器;擦具;刮漆刀
screen display 屏幕显示
screw [skruː] 螺丝钉;螺杆;螺孔;螺旋桨
screw cylinder 螺杆缸
screw jack 螺旋千斤顶
screw tap 螺丝攻
screw thread 螺纹
screwdriver 螺丝刀;起子
scriber [ˈskraɪbə] 划线器;描绘标记的用具
sealant 密封剂
sealing 密封
seam welding 接合焊
seam welding machine 缝焊机
secondary current [ˈsekəndərɪ] 二次电流

secondary voltage 二次电压
section modulus 断面系数；剖面模数
security 安全（性）；保密（性）；安全措施
security coupling 安全联轴器
segregation [ˌsegrɪˈgeɪʃən] 分离；分凝法
self-commutated；self-excitation [ˈkɒmjuːteɪtɪd] 自励；自激
self-diagnosing 自诊断
self-induction 自感
self-locking nut 自锁螺母
self-starter 自动点火装置；自动启动器
self-tapping screw 自攻螺丝
self-tuning controllers 自调节控制器
semiconductor [ˌsemɪkənˈdɒktə] 半导体；半导体器件
semi-permanent 半永久
sensitivity [ˈsensɪˈtɪvɪtɪ] 灵敏（度）；灵敏性
sensitivity of following wind 顺风调向灵敏度
sensor 传感器
sequence [ˈsiːkwəns] 顺序；时序；排序；序列
sequential order of the phase 相序
series （shunt） compensation 串（并）联补偿
series circuit 串联电路
series connection 串联
series product 系列产品
service condition 使用条件
service life 使用寿命
service lift 升降梯
serviceability limit states 使用极限状态
servomechanism [ˌsɜːvəʊˈmekənɪzəm] 伺服系统
servomotor [ˈsɜːvəʊˌməʊtə] 伺服电动机
servomotor drive 伺服电机驱动
set pressure 设定压力
set value 设定值
setting angle （of blade） 叶片安装角
setting pressure 设定压力
settle 稳定；固定；调整；调度
severe gust 强阵风
severe storm [sɪˈvɪə] 强风暴
severity of flicker 闪烁严重程度
shackle [ˈʃækl] 束缚物；U 形挂环
shadow 阴影
shadow flicker [ˈflɪkə] 影闪烁
shaft brake 轴制动
shaft gravity moment [ˈgrævɪtɪ] 轴重力矩
shaft speed 轴速
shaft test 测试主轴
shaft tilt 轴倾斜
sharp draft [drɑːft] 强力通风
sharp-edged gust [edʒd] 突发阵风
shear loading 剪切荷载
shear center [ʃɪə] 剪切中心
shear core structure 剪切核心结构
shear deformation 剪切变形，切变
shear flow 剪切流
shear generator [ˈdʒenəreɪtə]（风）

切变发生器
shear layer 剪切层；切变层
shear loads 剪切负荷
shear modulus [ˈmɒdjʊləs] 剪切模量
shear of wind 风切变
shear stress 剪应力
shear turbulence 剪切湍流
shear velocity [vɪˈlɒsɪtɪ] 剪切速度
shear wall structure 剪力墙结构
shear webs 抗剪腹板
shear wind 切边风
shed roof 单坡屋顶
shed vortex [ˈvɔːteks] 脱落旋涡
sheet metal workshop 钣金车间
shell structure 薄壳结构
shield [ʃiːld] 屏蔽；罩；防护
shield wire 避雷线
shielded metal arc welding 自动保护金属极电弧焊；埋弧焊
shielding ring 屏蔽环
shim 垫片；填隙片
shock 冲击；震动
shock absorber [əbˈsɔːbə] 减震器；震动吸收器
shock current 触电电流
shock load 冲击荷载
shore wind 海岸风
short circuit 短路
short circuit current 短路电流
short circuit impedance [ɪmˈpiːdəns] 短路阻抗
short circuit testing 短路试验
short duration loading [djʊˈreɪʃ(ə)n] 短期荷载
short-circuit characteristic [ˌkærəktəˈrɪstɪk] 短路特性
short-circuit operation 短路运行
short-circuit ratio 短路比
short-duration gust [djʊˈreɪʃ(ə)n] 短期阵风
short-term cut-out wind speed 短时切出风速
short-term severity of flicker 短时间闪变值
short-term wind data [ˈdeɪtə] 短期风资料
shrink [ʃrɪŋk] 缩小；收缩
shrink-fitted 收缩装配
shrinking ratio [ʃrɪŋkɪŋ] [ˈreɪʃɪəʊ] 收缩比
shroud [ʃraʊd] 罩；(风轮) 集风罩
shrouded rotor 有罩风轮
shrouded rotors 鼠笼转子
shunt reactor 并联电抗器
shut down wind speed; shut-off wind speed 刹车风速
shutdown 停止系统运行；停工(机)；关机
shutdown for wind turbine 风轮机停机
shutting down 停机
SICV 螺纹插装阀
side force 侧力
side gust 侧阵风
side headwind 侧逆风
side leading wind 侧顺风
side rake 侧向垧斜 (度)
sidewind 侧风；横风
signal circuit 信号电路

silicium [sɪˈlɪsɪəm] 硅
silicon carbide [ˈkɑːbaɪd] 碳化硅
silicon oil sealed 硅油密封
silicon rubber 硅橡胶
silicone heater 硅胶加热器
silicone heating pad 硅胶加热垫
similarity coordinate [kəʊˈɔːdɪneɪt] 相似坐标
similarity criteria [kraɪˈtɪərɪə] 相似准则
similarity law 相似律
similarity parameter [pəˈræmɪtə] 相似参数
similarity theory 相似理论
similitude [sɪˈmɪlɪtjuːd] 相似物
simplex transmission 单向传输
simulated wind 模拟风
simulations model 仿真模型
sine wave 正弦波
single bladed windmill [ˈbleɪdɪd] [ˈwɪn(d)mɪl] 单叶片风车
single bus 单母线
single clamp 单卡头
single core cable 单芯电缆
single degree of freedom systems [ˈfriːdəm] 单自由度系统
single helical gear [ˈhelɪkəl] 单螺旋齿轮
single line system 单线系统（风机润滑系统用语）
single phase [feɪz] 单相
single planetary gear train [ˈplætɪtərɪ] 单级行星齿轮系
single point lub 单点润滑
single-row roller bearings 单排滚珠轴承
single-weather rotor 单朝向风轮；定向风轮
sinusoidal loading 正弦式载荷
siphon [ˈsaɪfən] 虹吸管
site analysis 厂址分析
site assessment 厂址评估
site electrical facilities 风场电器设备
site evaluation 场址评价
site investigations 现场调查；实地勘查
site location 安装地点
site measurement 现场实测
site selection 选址
site survey 场址勘测
siting 选址
siting criteria [kraɪˈtɪərɪə] 定址准则
skew angle 倾斜角；偏转角
skidding 轴承打滑
skin effect 集肤效应
skin friction 表面摩擦
skin thickness 表面厚度
sky radiation 天空辐射
slab foundation 板式基础
sledge hammer 大锤
sleeve [sliːv] 套筒；套管
slewing ring bearings 回转支承轴承
slewing rings 回转环
slide 滑动触头；使滑动
slide caliper [ˈkælɪpə] 游标卡尺
slide valve [vælv] 滑阀
sliding shoes 滑动制动器

slight breeze 轻风（二级风）
slight flagging 轻微旗状（Ⅱ级植物风力指示）
slim size 纤细型；纤细设计
slip 转差率
slip ring 滑环；集流环
slip speed 转差率
slipping clutch [klʌtʃ] 滑动离合器
slipstream 滑流
slope 斜率；坡度；山坡
slope angle [ˈæŋg(ə)l] 坡度角
slope downwind 下坡风
slope flow 山坡气流
slope ground 斜坡
slope of lift curve 升力曲线斜率
slope up-wind 上坡风
slope wind 山坡风
slow separation 分流；汽流分离
slow-speed rotor 低速风轮
small disturbances method [dɪˈstɜːb(ə)nsɪz] 小扰动方法
small perturbance theory 小扰动理论
small right-of-way 道路占地少
small-scale WECS（SWECS） 小型风能转换装置
small scale wind tunnel 小型风洞
small size eddy [ˈedɪ] 小尺寸涡旋
small-scale gust 小尺度阵风
small-scale model 小缩尺比模型
small-scale turbulence [ˈtɜːbjʊl(ə)ns] 小尺度湍流
small-sized wind machine [məˈʃiːn] 风力机
smart grid 智能电网
Smith-Putnam wind turbine 史密斯—普特南大型风轮机
smoke risk 烟雾险情
S-N curve [kɜːv] S—N 曲线
snips [snɪps] 剪断
snow load 雪载
socket [ˈsɒkɪt] 插座（槽、孔、口）
socket spanner 套筒扳手
soft brake system 软制动
soft cut in 软切入
soft towers 柔性塔
soft-start unit 软启动
software package 软件包
software platform [plætfɔːm] 软件平台
solar constant 太阳常数
solar radiation [ˌreɪdɪˈeɪʃən] 太阳辐射
solar spectrum [ˈspektrəm] 太阳光谱
soldering [ˈsɒldərɪŋ] 焊料，焊锡；焊接，结合
soldering iron [ˈɑɪən] 烙铁
solenoid [ˈsəulənɔɪd] 螺线管
solenoid valve 电磁阀
solidity losses 实度损失
solidly earthed 直接接地
solitary [ˈsɒlɪtərɪ] 单独；唯一
solvent [ˈsɒlvənt] 溶媒；溶剂；解决方法
sound level 声级
sound level meter 声级计
sound power level（LW） 声功率级
sound pressure level（LP） 声压级

source code 源代码
space charge 空间电荷
spanner 扳手；活络扳手；扳子
spar 翼梁
sparking plug [plʌg] 火花塞
sparkover 放电
special purpose spanner 专用扳手
specific power [spɪˈsɪfɪk] 功率系数
specified breakaway torque 规定的最初起动转矩
spectral analysis [ˈspektrəl] 频谱分析
spectral gap 谱隙
spectrum [ˈspektrəm] 光谱
speed fluctuations [ˌflʌktjʊˈeɪʃəns] 速度波动
speed increasing gear 增速齿轮副
speed increasing gear train 增速齿轮系
speed increasing/up ratio 增速比
sphere gap 球隙
spherical roller bearing [ˈsferɪkəl] 球面滚子轴承；调心滚珠轴承
spheroidal graphite iron [sfɪəˈrɒɪdəl] 球状石墨铸铁
spigot [ˈspɪgət] 插口；插头；塞子；（水）龙头
spindle [ˈspɪndl] 主轴；心轴
spiral gear [ˈspɑɪərəl] 螺旋齿
spirit level 水平仪
splash lubrication 飞溅润滑
splice plates 拼接板
splicing sleeve [ˈsplɑɪsɪŋ] 接续管
spline 方栓；齿条；止转楔；花键
splined coupling 花键式连接
splined shaft 花键轴
splint [splɪnt] 夹板；托板
split bearing 分离轴承
spoilers 扰流器；扰流板
spoiling flap 阻尼板
spot corrosion [kəˈrəʊʒən] 点腐蚀
spot power price 实时电价
spot welding 点焊
spray lubrication 喷射润滑
spray nozzle 喷嘴
sprayer [ˈspreɪə] 喷雾；喷雾器
spring [sprɪŋ] 弹簧；弹出
spring balance 弹簧秤
spring-bow compass 弹簧小圆规
spring constant 弹簧常数
spring washer 弹垫
sprocket [ˈsprɒkɪt] 链轮齿
spur gear [spɜː] 正齿轮；直齿圆柱齿轮
spur wheel gear 正齿轮
square nut [skweə] 四方螺母；螺帽
square thread 矩形螺纹
square-head bolt 方头螺栓
squirrel cage motor [ˈskwɜːrəl] 鼠笼式电机
stability of following wind 调向稳定性
stabilizer [ˈsteɪbɪlɑɪzə] 稳定器；平衡器
stable stratification [ˌstrætɪfɪˈkeɪʃən] 稳定分层
stagnation boundary layer [stægˈneɪʃən] 滞止边界层
stagnation flow 滞止流动

stagnation line 滞止线
stagnation point 驻点
stagnation pressure 驻压
stagnation streamline [ˈstriːmlɑɪn] 滞止流线
stagnation temperature 驻温
stainless 无瑕疵的；不锈的
stall 失速；使失速
stall angle 失速迎角；临界迎角
stall control 失速控制
stall flutter [ˈflʌtə] 失速颤振
stall hysteresis 失速延迟（滞后）
stall hysteresis excitation [ˌhɪstəˈriːsɪs] [ˌeksɪˈteɪʃ(ə)n] 失速迟滞激励
stall incidence [ˈɪnsɪd(ə)ns] 失速迎角
stall regulated machines 失速调节机组
stall regulated wind turbine 失速调控的风机
stall regulation 失速调节
stall regulation [regjʊˈleɪʃ(ə)n] 失速式限速
stalled area 失速区
stalled flow 失速气流
stalling angle of attack 失速迎角；临界迎角
stand-along system 独立运行系统
standard air pressure [ˈstændəd] 标准大气压
standard atmosphere 标准大气
standard atmosphere pressure 标准大气压
standard atmospheric state [ˌætməsˈferɪk] 标准大气状态
standard density altitude 标准密度高度
standard deviation 标准差
standard error 标准误差
standard uncertainty 标准误差
standard visual range [ˈvɪʒjʊəl] 标准视程
standardized wind speed [ˈstændəˌdaɪzd] 标准风速
standby source of power 备用动力源
standby time 待命时间
standstill 静止状态；停顿；静止；停机
star connection 星形连接
start up 发动；开动；启动
starter motor 马达启动
starting signal 启动信号
starting torque [tɔːk] 启动力矩；启动转矩
starting torque coefficient [ˈkəʊɪˈfɪʃənt] 启动力矩系数
state estimator [ˈestɪmeɪtə] 状态估计器
state information 状态信息
static loads 静载荷
static switch 静态开关
static synchronous compensator 静态同步补偿器
Static var compensation（SVC） 静止无功补偿
stationary blade 固定叶片
Statistical energy analysis [əˈnæləsɪs] 统计能量分析

statistical methods [stəˈtɪstɪkəl] 统计方法

statistical prediction 统计预测

stator 定子

stator windings 定子绕组

steady loads 稳定负荷

steady state analysis 稳态分析

steady-state voltage 稳态电压

steady-state voltage regulation 稳态电压调节

steam engine [ˈendʒɪn] 蒸汽机

steam turbine 蒸汽轮机

steel sheet scissor [ˈsɪzə] 钢板剪刀

steel tubular tower [ˈtjuːbjʊlə] 钢管塔筒

steel wool 钢丝线

steel-reinforced aluminum conductor [əˈljuːmɪnəm] 钢芯铝绞线

steering box [ˈstɪərɪŋ] 舵盘

steering gear 方向齿

steering wheel 舵轮；方向盘

step response 阶跃响应

step voltage 跨步电压

step-by-step 步进

step-by-step dynamic analysis [daɪˈnæmɪk] 逐步动态分析

step-by-step solution 逐步求解法

step-down transformer 降压变压器

step-up substation 升压变电站

step-up transformer 升压变压器

stiff shaft 刚性轴

stiff towers 刚性塔

stiffness-to-weight ratio 刚度重量比

stipulate [ˈstɪpjʊleɪt] 规定；保证

stochastic loads [stɒˈkæstɪk] 随机载荷

stochastic rotor thrust fluctuations [ˌflʌktjʊˈeɪʃəns] 随机转子推力波动

stochastic tower bending moments 随机塔弯矩

Stokes' flow 斯托克斯流

stopcock [ˈstɒpkɒk] 管闩；活塞；活栓；旋塞阀

storage condition 贮存条件

straight connection 直接头

strain clamp 耐张线夹

strain gauges 应变仪

strain-life regression lines 应变—寿命回归线

stray capacitance [kəˈpæsɪtəns] 杂散电容

stray inductance [ɪnˈdʌktəns] 杂散电感

stream angle 气流（偏）角

stream pattern [ˈpæt(ə)n] 流谱

stream surface 流面

stream temperature 气流温度

stream tube 流管

stream turbulence [ˈtɜːbjʊl(ə)ns] 气流湍流（度）

streamer [ˈstriːmə]（流动显示）飘带

streamer breakdown 流注击穿

streamline angle [ˈstriːmlaɪn] 流线偏角

streamline contour [ˈkɒntʊə] 流线型外廓

streamline curvature effect [ˈkɜːvətʃə]

流线弯曲效应
streamline fairing 流线型罩
streamline shape 流线型
streamline squeezing effect 流线汇集效应
stream-tube 流管
streamwise component [kəm'pəunənt] 流向分量
streamwise pressure gradient ['greɪdɪənt] 流向压力梯度
streamwise vorticity [vɔː'tɪsətɪ] 流向涡量
strength 强度
strength-to-stiffness ratio 强度刚度比
strength-to-weight ratio 强度重量比
stress 应力
stress concentration 应力集中
stress tensor ['tensə] 应力张量
strict compliance with ISO 严格执行 ISO 标准/符合 ISO 标准要求
stringer（叶片）纵梁
strip heater 条状加热器
stroboscopic effect [ˌstrəubəu'skɒpɪk] 频闪效应
strong breeze 强风（六级风）
strong current control 强电控制
strong gale [geɪl] 烈风（九级风）
strong-wind boundary layer 强风边界层
structural adhesive 结构黏合剂
structural damping ['strʌktʃərəl] 结构阻尼
structural loads 结构载荷
structural properties 结构性质
stud bolt 双头螺栓
subaudible vibrations [ˌsʌb'ɔːdəbl] 音频振动
subcritical flow regime [sʌb'krɪtɪk(ə)l] [reɪ'ʒiːm] 亚临界流动状态
subcritical Reynolds number ['renəldz] 亚临界雷诺数
subcritical tower 软塔架；柔性塔架
subcritical vortex shedding ['vɔːteks] 亚临界旋涡脱落
subcritical wind speed 亚临界风速
submarine cable 水下电缆
submarine fibre optic cable system 海底光纤电缆
substation ['sʌbsteɪʃən] 变电站
suck-down wind tunnel 下吸式风洞
suction 吸力；负压；吸气
suction anemometer [ˌænɪ'mɒmɪtə] 吸管式风速计
suction coefficient [ˌkəuɪ'fɪʃ(ə)nt] 负压系数；吸力系数
suction load 吸力荷载
suction performance 抽吸油脂功能
sun gear 恒星齿轮；中心齿轮；太阳轮
superconductor 超导
super-finishing 超光；超精磨加工
superheater ['sjuːpəˌhiːtə] 过热器
supervisory control 监控
supervisory control and data acquisition（SCADA） 监控与数据采集

support structure for wind turbine 风轮机支撑结构
surface breakdown 表面击穿
surface roughness [ˈrʌftɪs] 表面粗糙度
surface roughness ratio 表面粗糙度比率
surface temperature 表面温度
surge [sɜːdʒ] 冲击；过电压
surge attester [əˈtestə] 避雷器
surge diverters 浪涌分流器；避雷器
surge impedance [ɪmˈpiːdəns] 波阻抗
surge protection 电涌保护
surge protective devices 电涌保护装置
surge suppressor 电涌保护器
surveillance system 观测系统；监视系统
survival ambient temperature 安全环境温度
survival wind speed [səˈvɑɪvəl] 安全风速；极限（生存）风速
suspension clamp [səˈspenʃən] 悬垂线夹
suspension insulator [ˈɪnsjʊleɪtə] 悬式绝缘子
sustained discharge [səˈsteɪnd] 自持放电
swept area 扫掠面积
switch board 配电盘；开关屏
switchbox 变电箱
switch-fuse 开关熔丝
switchgear [ˈswɪtʃɡɪə] 开关设备；开关柜
switching 切换；通断
switching current 闭合电流
switching operation 切换运行；开关操作
switching overvoltage [ˌəʊvəˈvəʊltɪdʒ] 操作过电压
switching pressure 转换压力
symmetric laminate [sɪˈmetrɪk] [ˈlæmɪneɪt] 对称层板
synchronous (rotational) speed [ˈsɪŋkrənəs] 同步转速
synchronous coefficient [ˌkəʊɪˈfɪʃənt] 同步系数
synchronous condenser 同步调相机
synchronous generator 同步发电机
synchronous machine 同步电机
synchronous speed 同步转速
system design 系统设计
system integration 系统集成
system software 系统软件
system with effectively earthed neutral [ɪˈfektɪvlɪ] 中性点有效接地系统

T

tachometer [tæˈkɒmɪtə] 转速计；转速器
tail wagon [ˈwæɡ(ə)n] 尾车
tail wind 顺风

tailing edge 后缘
tail-vane [veɪn] 尾舵
tangential [tæn'dʒenʃəl] 相切的；正接的
tangential flow 切向流
tangential flow induction factor 切向诱导因子
tangential force 切向力
tangential induced velocity 切向诱导速度
tangential wind 切向风
tap position information 分接头位置信息
taped transformer 多级变压器
taper ratio ['reɪʃɪəʊ] 梢根比，尖削比
tapered blade ['teɪpəd] 尖削叶片
tapered structure 尖削结构
tapping hole 侧压孔
tarpaulin covering (for wind turbines) [tɑː'pɔːlɪn] 防水篷布罩(风机用)
Taylor's hypothesis [haɪ'pɒθɪsɪs] 泰勒假说
T-connector T形线夹
technical issues ['ɪʃjuːz] 技术问题
teeter 跷板式结构
teeter angle 跷跷板角
teeter brake [breɪk] 跷板式制动器
teeter hinge 跷跷板铰链
teeter hinge [hɪn(d)ʒ] 跷板铰链
teeter loads 跷跷板载荷
teeter natural frequency ['frɪkwənsɪ] 跷跷板固有频率
teeter stability 跷跷板稳定
teetered hub 跷跷板式轮毂
teetering angle 跷板角
teetering hub 跷板式桨毂
teetering rotor 跷板式风轮
telecommunication cable [ˌtelɪkəˌmjuːtɪ'keɪʃən] 通信电缆
telemeter [tɪ'lemɪtə] 遥测
telemonitoring 远程监视
telescopic shaft [ˌtelɪ'skɒpɪk] 伸缩轴
temperature coefficient 温度系数
temperature gradients 温度梯度
temperature range 温度范围
temperature rise 温升
temperature sensor 温度传感器
tempering 回火
template 样板；金属模片
temporary works 施工辅助工程
tender design 投标设计
tensile stresses 张应力
tension 压力；张力；牵力；电压
tension loading 拉伸负荷
terminal ['tɜːmɪnəl] 接线端子
terminal block 端子台
terminal connector [kə'nektə] 设备线夹
terminal point 端子
terminal voltage 端电压
tertiary winding ['tɜːʃərɪ] 第三绕组
test data 试验数据
test nipple ['nɪpl] 测试嘴
test object 被试品
test on bed 台架试验
test site 试验场地
test stand for gear 传动试验台
test-bed 试验台

the Betz limit 贝兹极限
the family of airfoil 翼型族
T-head bolt T形头螺栓
theodolite [θɪ'ɒdəlɑɪt] 经纬仪
Theodorsen's function 泰奥多森函数
thermal breakdown ['θɜːməl] 热击穿
thermal power station 火力发电站
thermistor [θɜː'mɪstə] 热敏电阻；热控管
thermocouple ['θɜːməʊˌkʌpl] 温差电偶；热电偶
thermograph ['θɜːməʊgrɑːf] 温度记录器；热录像仪
thermostat ['θɜːməustæt] 自动调温器；恒温器
thickness function of airfoil 厚度函数
thickness of airfoil 翼型厚度
thin-walled steel tubular towers ['tjuːbjʊlə] 薄壁钢管塔筒
threaded bushes 螺纹衬套
threaded rods 牙条
three core cable 三心电缆
three way switch 三路开关
three-coordinates measuring machine 三坐标测量仪
three-jaw chuck 三爪卡盘
threshold 阈值
throttle valve 节流阀
thrust bearing 推力轴承
thrust coefficient 推力系数
thrust load 推力负载；轴向载荷
thunderstorm ['θʌndəstɔːm] 雷暴
thyristor [θɑɪ'rɪstə] 硅控整流器；半导体闸流管
thyristor soft-start unit 晶闸管软启动装置
tidal current 潮流
tilt angle 倾斜角
tilt angle of rotor shaft 风轮仰角
tilting 倾斜
tilting moment 倾覆力矩
time domain 时域
time response 时间响应
time scale [skeɪl] 时间尺度
time server 时间服务器
time step 时间步长
time switch 定时开关
time-dependent flow 非定常流动；随时间变化流动
time-dependent load 非定常荷载；随时间变化荷载
time-domain [dəʊ'meɪn] 时域
time-invariant windspeed [ɪn'veərɪənt] 稳定风速
time-varying windspeed ['veərɪŋ] 随时间变化风速
timing chain 调速链
timing gear 调速齿
tip brake 叶尖制动
tip circle 齿顶圆
tip loss 叶尖损失
tip loss factor 叶尖损失因子
tip of blade 叶尖
tip section pitch [pɪtʃ] 叶尖段桨距
tip speed 叶尖速度
tip speed coefficients 叶尖速度系数
tip speed ratio 尖速比

tip vane augmented wind turbine [ɔːgˈmentɪd] [ˈtəːbaɪn] 叶尖舵片增力型风轮机
tip vortex [ˈvɔːteks] 叶尖涡
tipping 仰倾，倾翻
toggle joint [ˈtɒgl] 肘接；弯头接合
tolerance [ˈtɒlərəns] 公差；容差（限、许）；允许误差
tommy bar 撬棒；螺丝钻
tonality 音值（风机）
tool box 工具箱
tool post 工具站
tooth bending stress 轮齿弯曲应力
tooth depth [depθ] 齿高
tooth flank 齿面
tooth space 齿槽
tooth thickness 齿厚
toothed gear wheels [tuːðd] 齿轮
top and bottom hatches 上下盖（升降梯轿厢入/出口）
top ring 顶环（风机顶环）
top windspeed 最大风速
topoclimate 地形气候
topoclimatology 地形气候学
topographic effect [ˌtɒpəˈgræfɪk] 地形效应
topographic element 地形因素
topographic exposure factor [ɪkˈspəʊʒə] 地形开敞度
topographic feature 地形特征
topographic interference [ɪntəˈfɪər(ə)ns] 地形干扰
topographic modification [ˌmɒdɪfɪˈkeɪʃ(ə)n] 地形改造
topographic relief 地形起伏
topography [təˈpɒgrəfɪ] 地形；地势
topography channel wind [ˈtʃæn(ə)l] 地形狭道风
topologies 拓扑结构
tornado cyclone 龙卷风气旋
tornado [tɔːˈneɪdəʊ] 龙卷风；陆龙卷
tornado alley [ˈælɪ] 龙卷风通道
tornado axis [ˈæksɪs] 龙卷风轴
tornado belt 龙卷风带
tornado center 龙卷风中心
tornado core 龙卷风核
tornado effect 龙卷风效应
tornado F-scale 龙卷风 F 等级
tornado parameter [pəˈræmɪtə] 龙卷风参数
tornado path 龙卷风路径
tornado simulation model [ˌsɪmjʊˈleɪʃən] 龙卷风模拟模型
tornado-affected area [tɔːˈneɪdəʊ] 龙卷风影响区
tornado-like vortex [ˈvɔːteks] 龙卷风式旋涡
tornado-resistant design [rɪˈzɪstənt] 抗龙卷风设计
torque arm 力臂
torque characteristic [ˌkærəktəˈrɪstɪk] 转矩特性
torque coefficient 转矩系数；力矩系数
torque control loop [luːp] 转矩闭环控制
torque spanner/wrench [rentʃ] 力矩扳手

torque-endurance curves 扭矩耐力曲线
torque-limiting 限制扭矩
torsion angle [ˈtɔːʃən] 扭转角
torsion bar 力矩杆
torsion coefficient 扭转刚度系数；扭转系数
torsion frequency 扭转频率
torsional flexibility [ˌfleksɪˈbɪlɪtɪ] 扭转柔度；扭转挠度
torsional rigidity 扭转刚度
torsional vibration [vaɪˈbreɪʃən] 扭转振动
total harmonic distortion [hɑːˈmɒtɪk] 总谐波失真
touch screen 触摸屏
touch voltage [ˈvəʊltɪdʒ] 接触电压
tower 塔架
tower base critical load case 塔基临界载荷
tower base fore-aft bending moment 塔基前后弯矩
tower crane 塔式吊车
tower displacement 塔位移
tower height 塔高
tower shadow 塔影
tower shadow effect 塔影效应
towing plate 牵引板
track rod 磁棒
trailing edge 后缘
trailing edge stall 后缘失速
trailing vortices 翼梢尾涡
train of gears 齿轮系
transfer function 传递函数
transfer of wind turbine technology 风轮机技术转让
transfer ratio 传递比
transfer switching 倒闸操作
transformation matrices [ˈmeɪtrɪsiːz] 变换矩阵
transformer 变压器
transformer fitted with OLTC 有载调压变压器
transient loads [ˈtrænzɪənt] 瞬态载荷
transient rotor 瞬态电流
transistor 晶体管
transitional processes 转换过程
transmission accuracy [trænzˈmɪʃən] 传动精度
transmission chain 传动链
transmission error 传动误差
transmission line 传输线；送出线路
transmission ratio 传动比
transmission system 输电系统
transmission technology 传输技术；传动技术
transmission test bed 传动试验台
transportation condition 运输条件
transversal cut [trænzˈvɜːsəl] 横剖
transverse 横切的；横轴
transverse engine 水平机
trapezium [trəˈpiːzɪəm] 梯形
travel limit switch 限位开关
travelling crane [kreɪn] 移动式吊车
treeing 树枝放电
triac [ˈtrɑɪæk] 端双向可控硅开关元件

trigger (relay) 跳闸
trigger electrode [ɪ'lektrəʊd] 触发电极
trimming ['trɪmɪŋ] 清理焊缝
trip circuit 跳闸电路
trip coil [kɔɪl] 跳闸线圈
trip wire 激紊线
trip-free 自由脱扣
triple phase [feɪz] 三相
tripping 跳闸
try square 曲尺；检验角尺
T-square 丁字尺
tubular tower ['tjuːbjʊlə] 塔筒；管状塔
tuned circuit 调谐电路
turbine of megawatt class 兆瓦级风电机组
turbine state 风电机组工况
turbogenerator [ˌtɜːbəʊ'dʒenəreɪtə] 汽轮发电机
turbojet ['tɜːbəʊdʒet] 涡轮喷气飞机
turbulator strip 扰流带，噪音条
turbulence intensity ['tɜːbjʊləns] 湍流强度
turbulence scale parameter [pə'ræmɪtə] 湍流尺度参数
turbulent boundary layers ['baʊndərɪ] 湍流边界层
turbulent flows 紊流
turn buckle 紧螺器；花篮螺栓
turn ratio 匝比；变比
turntable ['tɜːnˌteɪbl] 转车；转盘
turret lathe [leɪð] 六角车床
twist of blade 叶片扭角
twisting 缠绕
twisting of the power cable 扭缆
two level braking 双级制动
two-stroke engine 二冲程发动机
two-speed operation 双速运行

U

U-bolt U 形挂钩；U 形螺栓
UHF (Ultra High Frequency) 超高频
ultimate limit state ['ʌltɪmət] 最大极限状态；极限限制状态
ultimate loads 极限载荷
Ultra-high voltage (UHV) 特高压
ultrasonic [ˌʌltrə'sɒnɪk] 超声波
ultrasotıc flaw detector 超声波探伤仪
uncertainty in measurement [ˌʌn'sɜːtəntɪ] ['meʒəmənt] 测量误差
undamped vibration [ˌʌn'dæmpt] 无阻尼振荡
under frequency 欠频
under load switch 低载荷开关
under speed 欠速
under speed protection 欠速保护
under voltage 低电压；欠电压
underground cable 地下电缆
underwater cable 水下电缆
undulate ['ʌndjʊleɪt] 起伏

utiform field [ˈjuːtɪfɔːm] 均匀场
uniform steady wind 均匀定常风
uniform wind 均匀风
uninterruptible power supply 不间断电源
unit control 单元控制
unit transformer 机组变压器
universal coupling [ˌjuːtɪˈvɜːsəl] 万向联轴节；万向连接器；万向联轴器
universal joint 万向接头；万向节
universal pliers [ˈplɑɪəz] 万能手钳
unlockable barrier 可锁可解装置（爬梯）
unmanned operation 无人操作
unscrew [ˌʌnˈskruː] 旋开；旋松
unshrouded rotor [ˌʌnˈʃrɑʊd] 无罩风轮
unstable atmosphere 不稳定大气
unstable stratification 不稳定分层
unsteady aerodynamics [ˌeərəʊdɑɪˈnæmɪks] 非定常气动力
unsteady airfoil theory 非定常翼型理论
unsteady BEM model 非定常边界元法
unsteady Bernoulli's equation [bɜːˈnuːlɪ] 非定常伯努利方程
unsteady boundary layer 非定常边界层
unsteady flow 非定常流
unstreamlined 非流线型的
unsymmetry [ˈʌnˈsɪmɪtrɪ] 不匀称
untwist [ˌʌnˈtwɪst] 解缆
up wash 上洗（流）
updraft（updraught）[ˈʌpdrɑːft] 上升气流；向上通风
upflow 上升气流
uplift coefficient [ˌkəʊɪˈfɪʃ(ə)nt] 上升（扬压）系数
uplift wind load 上吸风载
upslope wind 上坡风
upstream 上游
upstream influence [ˈɪnflʊəns] 上游影响
upstream windspeed 上游风速
up-valley wind 进谷风
upwind 迎风；顶风；上升气流；上风向；逆风
upwind angle 迎风角；对风角
upwind configuration [kənˌfɪgjʊˈreɪʃən] 上风式结构
upwind effect 上吹效应
upwind fetch 上风吹程
upwind machines 上风式机组
upwind orientation 上风风向
upwind rotor 上风式风轮
upwind turbine 上风式风力机；迎风机组
USB USB 接口；U 盘
user application software 用户应用程序
user-friendliness 用户友好性（界面）
user-port lightning protection 用户端口防雷
using data 运用数据
using wind atlases [ˈætləs] 使用风地图集
utility grid/network [juːˈtɪlətɪ] 电

网，输电系统

V

V-thread V形螺纹
vaccum fan 抽风机；真空机
vacuum bag sealing 真空袋密封
vacuum brake [ˈvækjʊəm] 真空制动闸
vacuum circuit breaker 真空断路器
vacuum gauge [geɪdʒ] 真空计
vacuum pump 真空泵；真空抽气机
vacuum tube [tjuːb] 真空管；电子管
validity [vəˈlɪdɪtɪ] 有效性
valley breeze [briːz] 谷风
valley floor 谷底
valley flow 山谷气流
valley slope wind 谷坡风
valley wind 谷风
valley wind circulation 谷风环流
vane 风向标；（风洞）导流片；尾舵
vane-type anemometer [ˌænɪˈmɒmɪtə] 翼式风速计
variable chord blade [ˈveərɪəbl] 变弦长叶片
variable frequency AC 变频交流电
variable geometry Musgrove machine [dʒɪˈɒmɪtrɪ] [məˈʃiːn] 马斯格卢夫可变几何型风力机
variable geometry vertical-axis wind turbine 可变几何型竖轴风轮机
variable resistor [rɪˈzɪstə] 可变电阻
variable speed drive system 变速驱动系统
variable speed operation 变速运行
variable speed wind turbine 变速风力机；变速风电机组
variable transformer 调压变压器
variable-frequency generator [ˈdʒenəreɪtə] 变频发电机
variable-pitch control scheme [skiːm] 变桨距控制方案
variable-pitch rotor 可变桨距风轮
variable-speed constant frequency generator 变速恒频发电机
variable-speed control scheme 变速控制方案
variable-speed fixed-pitch（VS-FP）变速定桨距
variable-speed rotor 变速风轮
variable-speed turbines 变速风力机
variable-speed variable-pitch（VS-VP）变速变桨距
variable-wind 多变风
vector wind field 向量风场
veering 调向，对风
veering wind 顺转风
velocity amplitude [ˈæmplɪtjuːd] 速度幅值
velocity duration curve 速度曲线
velocity head [vɪˈlɒsɪtɪ] 速度头
velocity potential 速度势
velocity pressure 速压

velocity profile 速度廓线
velocity triangle [ˈtraɪæŋg(ə)l] 速度三角形
Venn diagram 维恩图
ventilation [ˌventɪˈleɪʃən] 通风装置；通风
Venturi tube [venˈtjʊərɪ] 文氏管；文丘里管
vernier calliper gauge [ˈkælɪpə] 游标尺
vertical axis Darrieus turbine [ˈtɜːbaɪn] 达里厄竖轴风轮机
vertical-axis rotor 竖轴风轮
vertical-axis wind machine 竖轴风力机
vertical axis wind turbine 立式风力机；垂直轴风力发电机
vertical axis WTGS 垂直轴风电机组
vertical wind shear 垂直风切变
vertical wind tunnel 立式风洞
VHF（Very High Frequency）特高频
vibration damper 振动阻尼器
vibration detector 振动检测仪
vibration frequency [vaɪˈbreɪʃən] [ˈfrɪkwənsɪ] 振动频率
vibration isolator [vaɪˈbreɪʃən] [ˈaɪsəleɪtə] 减震器
vibration level 振动等级
vibration sensor 振动传感器
vibration switch 振动开关
vibration test 振动试验
vibratory finishing machine 振动抛光机
violent storm 暴风（十一级风）
violent tornado [tɔːˈneɪdəʊ] 强龙卷
virtual airfoil 虚翼型
virtual angle of attack 虚迎角
virtual camber 虚弯度
virtual camber effect 虚弯度效应
virtual incidence 虚迎角
virtual work 虚功
viscosity coefficient [vɪˈskɒsɪtɪ] 黏性系数
viscosity effect 黏性效应
viscosity fluid 黏性流体
viscosity friction [ˈfrɪkʃ(ə)n] 黏性摩擦
viscosity index [ˈɪndeks] 黏度指数
viscosity resistance [rɪˈzɪst(ə)ns] 黏滞阻力
viscous drag [ˈviskəs] 黏性阻力
viscous flow 黏性流动
visual impact 视觉冲击
visual inspection 外观检查
visualizing signal processing 可视信号处理
voltage change factor [ˈvəʊltɪdʒ] 电压变化系数
voltage dip 电压跌落；电压突降
voltage divider [dɪˈvaɪdə] 分压器
voltage drop 电压降
voltage range 电压范围
voltage to earth 对地电压
voltage variations 电压变化
voltmeter [ˈvəʊltˌmiːtə] 电压表；伏特计
von Karman constant 冯卡曼常数
vortex [ˈvɔːteks] 涡流

vortex cylinder [ˈsɪlɪndə] 涡流柱
vortex shedding 脱流
vortex strength 涡强
vortex systems 涡系
VSCF (variable speed constant frequency) 变速恒频风机

W

Wagner function [ˈwɑːgnə] 瓦格纳函数
wake 尾涡；尾流
wake effects [ɪˈfekts] 尾流效应
wake flow 尾流
wake loss 尾流损失
wake models 尾流模型
wake turbulence [ˈtɜːbjʊləns] 尾流扰动
washer [ˈwɒʃə] 垫片；垫圈
waste oil container [kənˈteɪnə] 废油容器
water wheel 水轮
water-antioxidant [ˌæntɪˈɒksɪdənt] 抗氧化剂
water-based cleaner 水性清洗剂
water-cooled engine 水冷内燃机
water-resistant [rɪˈzɪstənt] 水电阻
wave 波
wave band switch 波段开关
wave front 波头
wave tail 波尾
waveform acquisition 波形采集
waveguide [ˈweɪvgaɪd] 波导；波导管
wavelength [ˈweɪvleŋθ] 波长
weak current control 弱电控制
weather station 气象站
weather window 天气窗口
weaving 摆动
WECS 风力发电系统
weelbarrow [ˈhwiːlˌbærəʊ] 手推车
weelbase 轴距
weft [weft] 纬线
Weibull distribution 威布尔分布
Weibull function 威布尔函数
Weibull probability function [ˌprɒbəˈbɪlətɪ] 威布尔概率函数
Weibull scale factor 威布尔比例因子
weighing machine 称量器
weighted sound pressure level 声级
weighting functions 权重函数
weights 重量；砝码
welding 焊接；焊缝
welding imperfectionz [ˌɪmpəˈfekʃənz] 焊接缺陷
wheel brace 齿轮架
white body 白体
wighbridge [ˈweɪbrɪdʒ] 称量台；台秤；桥秤
winch 绞车；绞盘
wind abeam 侧风
wind abrasion [əˈbreɪʒ(ə)n] 风磨蚀
wind aloft [əˈlɒft] 高空风
wind angle 风迎角，风角
wind area 迎风面积
wind arrow 风矢

wind assessment 风场评价
wind axes system 风轴系；坐标系
wind azimuth [ˈæzɪməθ] 风方位角
wind break 风障
wind channel [ˈtʃæn(ə)l] 风洞；风道
wind characteristic [ˌkærəktəˈrɪstɪk] 风特性
wind chart 风图
wind circulation [sɜːkjʊˈleɪʃ(ə)n] 风环流；气流循环
wind class 风（力）级
wind climate 风气候
wind cock 风向标
wind col 风坳
wind comfort level 风舒适度
wind concentrator 集风器
wind cone 风向袋
wind conversion efficiency 风能转换效率
wind corrosion [kəˈrəʊʒ(ə)n] 风蚀
wind current [ˈkʌr(ə)nt] 风流；气流
wind data 风数据；风资料
wind deflection [dɪˈflekʃ(ə)n] 风向偏转
wind description [dɪˈskrɪpʃ(ə)n] 风描述
wind desiccation [ˌdesɪˈkeʃən] 风干作用
wind diffuser [dɪˈfjuːzə] 扩风器
wind direction fluctuation [ˌflʌktjʊˈeɪʃən] 风向波动
wind direction frequency 风向频率
wind direction measurement 风向测量
wind direction sensor [ˈsensə] 风向感受器
wind discontinuity 风不连续性
wind dispersion [dɪˈspɜːʃ(ə)n] 风弥散（作用）
wind divide 风向界线
wind drag 风阻
wind drag load 风阻荷载
wind drift [drɪft] 吹流；风积
wind dynamic load [dɑɪˈnæmɪk] 动态风载
wind eddy [ˈedɪ] 风涡
wind effect 风效应
Wind Energy Collection Systems (WECS) 风能收集系统；风力发电系统
wind energy collector 风能收集装置
wind energy content 风能蕴藏量
wind energy conversion 风能转换
Wind Energy Conversion System (WECS) 风能转换系统
wind energy converter 风能转换装置
wind energy density 风能密度
wind energy engineering [endʒɪˈnɪərɪŋ] 风能工程
wind energy extraction limit [ɪkˈstrækʃ(ə)n] 风能获取极限值
wind energy farm 风能田；风车田
wind energy flux 风能通量
wind energy potential [pə(ʊ)ˈtenʃ(ə)l] 风能潜力
wind energy region [ˈriːdʒ(ə)n] 风能区别

wind energy resource [rɪ'sɔːs; rɪ'zɔːs] 风能资源
wind energy rose 风能玫瑰（图）
wind energy spectrum ['spektrəm] 风能谱
wind energy system 风能转换系统
wind energy/power 风能
wind engineer 风工程师
wind engineering 风工程
wind exposure area [ɪk'spəʊʒə] 受风面积
wind farm 风电场
wind farm filter 风电场过滤器
wind farm site 风电场址；风电场工地
wind following 风向跟踪，对风
wind force 风力
wind force scale 风力级
wind frequency curve 风频数曲线
wind furnace ['fɜːnɪs] 风力加热器
wind gap 风口；风隙
wind gauge [geɪdʒ] 风速计
wind gradient ['greɪdɪənt] 风梯度
wind gust 阵风
wind induced load [ɪn'djuːst] 风致荷载
wind induced pressure 风压
wind induced vibration [vaɪ'breɪʃ(ə)n] 风致振动
wind information 风资料
wind intensity [ɪn'tensɪtɪ] 风强度
wind load 风荷载
wind load spectrum ['spektrəm] 风载谱
wind loading design criterion [kraɪ'tɪərɪən] 风载设计准则
wind loading standard ['stændəd] 风载标准
wind map 风图
wind mast 测风塔
wind meter 风速计
wind microzonation 风力微区划
wind noise 风噪声
wind nut 风机用螺母
wind porch [pɔːtʃ] 风廊
wind power density 风功率密度
wind power duration curve [djʊ'reɪʃ(ə)n] 风能持续时间曲线
wind power capture 风能捕获
wind power plants 风力发电厂
wind power potential [pə(ʊ)'tenʃ(ə)l] 风能潜力
wind power profile ['prəʊfaɪl] 风能廓线
wind power station 风电场
wind power system 风能转换系统
wind powered irrigation system [ˌɪrə'geʃən] 风力提灌系统
wind powered machinery [mə'ʃiːn(ə)rɪ] 风力机械
wind pressure 风压
wind pressure tap 风压测量孔
wind profile 风廓线
wind profile wind shear law ['prəʊfaɪl] 风廓线风切变律
wind proof performance [pə'fɔːm(ə)ns] 耐风性能
wind propulsion system [prə'pʌlʃ(ə)n] 风力助航系统

wind pump 风力泵
wind reduction efficiency 减风效率
wind regime [reɪ'ʒiːm] 风况
wind resistance [rɪ'zɪst(ə)ns] 抗风；风阻（力）
wind resistant design 抗风设计
wind resource 风力资源
wind resource assessment 风力资源评价
wind resource estimation [ˌestɪ'meɪʃən] 风能资源估计
wind rose 风玫瑰（图）
wind rose log 风向玫瑰图日志（记录）
wind rotor 风轮
wind run 风程
wind scale 风（力）级
wind sector 风区
wind sensor ['sensə] 风感受器
wind shade of the tower 塔的背风侧
wind shadow 风影；背风区
wind shaft 风矢杆；风轮转轴
wind shear 风切变
wind shear exponent [ɪk'spəunənt] 风切变指数
wind shift 风向转变
wind simulation 风模拟
wind site 风场
wind spectrum ['spektrəm] 风谱
wind speed 风速
wind speed bin 风速组
wind speed distribution 风速分布
wind speed duration curve [djʊ'reɪʃ(ə)n] 风速持续时间曲线
wind speed frequency 风速频数
wind speed frequency curve 风速频数曲线
wind speed measurement 风速测量
wind speed profile 风速廓线
wind speed Reynolds number ['renəldz] 风速雷诺数
wind speed scale 风速级
wind speed spectrum ['spektrəm] 风速谱
wind spilling [spɪlɪŋ] 风能溢出（限速法）
wind spilling type of governor ['gʌv(ə)nə] 风能溢出式限速装置
wind spun vortex [spʌn] ['vɔːteks] 风动涡旋
wind statistics [stə'tɪstɪks] 风统计（学）
wind stream power density 风流能密度
wind strength 风力
wind stress 风力
wind suction 风吸力；风抽吸
wind survey 测风
wind swept area 受风区；风扫掠面积
wind tower 测风塔
wind tower shop inspection 监造塔筒
wind transducer [trænz'djuːsə] 风向传感器
wind truss [trʌs] 抗风桁架
wind tunnel fetch 风洞吹程
wind tunnel ['tʌnl] 风洞

wind tunnel axes system [ˈæksiːz] 风洞轴系
wind tunnel centerline [ˈsentəlaɪn] 风洞轴线
wind tunnel choking 风洞壅塞
wind tunnel circuit [ˈsɜːkɪt] 风洞回流道
wind tunnel contraction [kənˈtrækʃ(ə)n] 风洞收缩段；风洞收缩比
wind tunnel data 风洞实验数据
wind tunnel diffuser [dɪˈfjuːzə] 风洞扩散段
wind tunnel efficiency 风洞效率
wind tunnel energy ratio [ˈreɪʃɪəʊ] 风洞能量比
wind tunnel geometry [dʒɪˈɒmɪtrɪ] 风洞截面几何形状
wind tunnel laboratory [ˈlæb(ə)rət(ə)rɪ] 风洞实验室
wind tunnel model 风洞模型
wind tunnel modelling technique [tekˈniːk] 风洞模拟技术
wind tunnel noise 风洞噪声
wind tunnel Reynolds number [ˈrenəldz] 风洞雷诺数
wind tunnel roof 风洞顶壁
wind tunnel speed 风洞风速
wind tunnel test section 风洞试验段
wind tunnel tests 风洞实验
wind tunnel time 风洞吹风时间；风洞占用时间
wind tunnel turbulence [ˈtɜːbjʊl(ə)ns] 风洞湍流（度）
wind tunnel turning vane 风洞导流片
wind tunnel wall effect 风洞洞壁效应
wind tunnel wall interference [ɪntəˈfɪər(ə)ns] 风洞洞壁干扰
wind tunnel window 风洞（试验段）观察窗
wind tunnel working section 风洞试验段
wind turbine 风力机
wind turbine furling 风轮机停机
wind turbine generator system (WTGS) 风力发电机组
wind turbine performance 风机性能
wind turbine stalling 风轮机失速
wind turbine terminals [ˈtɜːmɪnlz] 风力发电机端口
wind turbine test bench 整机出厂试验台；风轮机试验台
wind turbulence [ˈtɜːbjʊl(ə)ns] 风湍流（度）
wind vane 风向标；对风尾舵
wind variability [ˌveərɪəˈbɪlətɪ] 风变性
wind vector 风矢量
wind veering [ˈvɪərɪŋ] 风向转变
wind velocity 风速
wind velocity gradient [vɪˈlɒsətɪ] [ˈɡreɪdɪənt] 风速梯度
wind velocity profile 风速廓线
wind velocity spiral [ˈspaɪr(ə)l] （埃克曼）风速螺线
wind yaw angle [ˈæŋɡ(ə)l] 风侧滑角，风横偏角
windage 风阻；风力影响；船身挡风面

windage loss 风阻损失
windage resistance [rɪ'zɪst(ə)ns] 风阻
wind-approach angle 来流风迎角；风接近角
wind-axis wind machine 风轴风力机
wind-drift sand 流沙；风沙
windiness 多风性
winding ['waɪndɪŋ] 线圈；绕组
winding factor 绕组系数
windmill anemometer ['wɪn(d)mɪl] [ˌænɪ'mɒmɪtə] 风车式风速计
windmill brake state ['wɪndmɪl] 风车制动状态
windmills 风车
wind-off test 无风试验
wind-on test 有风试验
wind-power installation [ˌɪnstə'leɪʃən] 风能转换装置
wind-powered heat pump 风力热泵
wind-proof design [pruːf] 耐风设计
wind-resistant feature 抗风特性
windscreen 挡风玻璃；风挡
windscreen wiper 风挡刮水器；风挡雨雪刷
windseeking 调向，对风
wind-turbine noise 风力机噪音
windward ['wɪndwəd] 上风面
windward side 迎风面
windward slope 迎风坡
windward wall 迎风壁
windward (upwind) 上风向的，迎风向的
wing nut 蝶形螺母
wipe off 擦掉；抹掉
wire brush ['waɪə] 电刷
wire frames 线框
wire wool 线圈
wiredrawing 拉丝；拔丝；使延长
wire-stripping pliers ['plaɪəz] 剥线钳
wiring 布线；接线
withstand test 耐压试验
withstand voltage 耐受电压
wood-epoxy laminates 树脂层板
workbench 工作台
working earthing 工作接地
working flank 工作齿面
workshop 车间；工厂
worm [wɜːm] 蜗杆
worm gear 蜗杆螺杆；蜗杆齿
worm wheel 蜗轮
wound rotor 绕线转子
wound rotor induction generators 绕线式转子感应发电机
wrench 扳手；扳子

X

X-axis X 轴；横坐标轴
XLPE cable 交链聚乙烯电缆
XLPE (Cross Linked Polyethylene) [ˌpɒliː'eθəˌliːn] 交链聚乙烯(电缆)

Y

yaw [jɔː] 偏航
yaw acceleration [əkseləˈreɪʃ(ə)n] 偏航加速度；侧滑加速度；横摆加速度
yaw angle 偏航角
yaw bearing 偏航轴承
yaw brake 偏航制动；偏航闸
yaw control 偏航控制
yaw control mechanism [ˈmek(ə)nɪz(ə)m] 对风控制机构
yaw drive 偏航驱动
yaw drive dead band 调向死区
yaw error 偏航误差
yaw inertia [ɪˈnɜːʃə] 偏航惯性；对风惯性
yaw mechanism [ˈmek(ə)nɪz(ə)m] 调向机构；对风机构；偏航机构
yaw model 偏航模式
yaw moments 偏航力矩
yaw offset [ˌɒfˈset] 偏航偏移
yaw orientation [ɔːrɪənˈteɪʃ(ə)n] 调向，对风
yaw pinion 偏航齿轮
yaw probe [prəʊb] 方向探头
yaw sphere anemometer [sfɪə] [ˌænɪˈmɒmɪtə] 偏球风速计
yaw vane 调向尾舵，对风尾舵
yaw-active rotor 主动调向风轮
yaw-fixed rotor 定向风轮
yawing angle of rotor shaft 风轮轴偏航角
yawing device 调向装置，对风装置
yawing drive 偏航系统；偏航驱动
yawing flow 侧滑流
yawing moment 偏航力矩；[车]横摆力矩
yawing moment coefficient 偏航力矩系数
yawing moment of inertia [ɪˈnɜːʃə] 横摆惯性矩；偏航惯性矩
yaw-passive rotor [ˈpæsɪv] 被动调向风轮，被动对风风轮
Y-axis 纵坐标；Y轴
Y-Darrieus rotor Y形达里厄式风轮
yearly (average) wind speed 年平均风速
yoke plate [jəʊk] 联板
Young's modulus 杨氏系数

Z

zero displacement height 零风面位移高度；有效地面高度
zero lift drag 零升阻力
zero plane displacement 零风面位移
zero sequence current [ˈsiːkwəns] 零序电流
zero-lift angle of attack 零升迎角

zero-lift chord 零升力弦；零升力线
zero-lift moment 零升力矩
zero-torque speed [tɔːk] 零速力矩
zero-upcrossing frequency ['friːkwənsɪ] 零上穿频率
zero-wind drag coefficient [ˌkəʊɪ'fɪʃ(ə)nt] 无风阻力系数
zinc coating 镀锌层
zinc oxide ['ɒksaɪd] 氧化锌
zink spray [zɪŋk] [spreɪ] 镀锌
zones of visual impact (ZVI) ['ɪmpækt] 区视觉冲击力
ZVRT (Zero Voltage Run Through) 零压穿越

太 阳 能 发 电

A

absolute pressure 绝对压力
absolute temperature 绝对温度
absorber [əbˈsɔːbə] 吸热体
absorber area 吸热体面积
absorptance [əbˈsɔːptəns] 吸收率
absorption peaks [əbˈsɔːpʃən] [piːks] 吸收峰
absorptions coefficient [əbˈzɔːpʃ(ə)n] 吸收强度
accuracy 精确度
active loss 有功损耗
active power of PV power station 光伏系统有功功率
active solar (energy) system [ˈsəʊlə(r)] 主动式太阳能系统
actual pressure 实际压力
adding into 添加
air intake valve 进气阀
air mass (AM) 大气质量，空气质量
air preheater 空气预热器
air pressure 风压
air type collector 空气（型）集热器
allowable limit [əˈlaʊəbl] 允许限值
aluminum alloy casing [əˈluːmɪnəm] [ˈælɒɪ] 铝外壳
ambient(air)temperature(t_a) [ˈæmbɪənt] 环境（空气）温度
ambient air [ˈæmbɪənt] 环境空气
ambient pressure [ˈæmbɪənt] 环境压力
amorphous silicon solar cell [əˈmɔːfəs] 非晶硅太阳能电池
anneal 退火
anticorrosion [ˌæntɪkəˈrəʊʒən] 防腐蚀，防腐蚀的
aperture [ˈæpətʃə(r)] 采光口
aperture area (A) [ˈæpətʃə(r)] [ˈeərɪə] 采光面积
aperture diameter width 开口直径大小
aperture plane [ˈæpətʃə(r)] 采光平面
applying film to inner tank 内胆覆膜
applying film [əˈplaɪŋ] 覆膜
array (solar cell array) 方阵（太阳电池方阵）
as-built drawing 竣工图
assembling system 集成系统
assembly [əˈsemblɪ] 装配
atmospheric composition [ætməsˈferɪk] 大气成分，大气组成
attaching label and brand [ˈleɪbl] 粘贴标识
auto air vent 自动排气阀
automatic adjustment 自动调整

automatic control 自动控制
automatic track 自动跟踪
availability 可利用率
average daily efficiency 平均日效率
average temperature 平均温度
azimuth angle [ˈæzɪməθ] 方位角

B

back holder (inside) 后腿(内)
back holder (outside) 后腿(外)
back pressure 背压
background [ˈbækgraʊnd] 背景
ball cock valve 球阀
basic research [ˈbeɪsɪk] 基础研究
bent [bent] 弯头
biomass 生物质
blocking diode 隔离二极管
blow out opening 试压口
body composition 身体组成
boiler feed pump 给水泵
boring holes on main copper tube 主铜管钻孔
bottom frame 拉铝尾条
bracket 支架
Building-integrated PV (BIPV) 太阳能光伏建筑一体化
by-pass diode [ˈdaɪəʊd] 旁路二极管
bypass valve 旁路阀

C

capacity 容量
carbonate salt [ˈkɑːbəneɪt] 碳酸盐
cathode [ˈkæθəʊd] 阴极
CdTe solar cell 碲化镉太阳能电池
Central Control Unit (CCU) 中央控制器
Central Processing Module (CPM) 中央处理模块
centrifugal type pumps [senˈtrɪfjʊg(ə)l] 离心式泵
chamber 室,房间
characteristics of solar energy [ˌkærəktəˈrɪstɪks] 太阳能的特点
charge 充电
charge controller 充电控制器
check valve 检修口
Chemical Vapor Deposition (CVD) 化学气相沉积
chemical composition 化学成分;化学组成
chloride salt [ˈklɔːraɪd] 氯化盐
circuit breaker 断路开关
circulating water pipe (pump) [ˈsɜːkjuleɪtɪŋ] 循环水管
circulation pump [ˌsɜːkjʊˈleɪʃən] 循环泵
city planning 城市规划
closed system 封闭系统
coat 涂层
collector 集热器

collector array [ə'reɪ] 集热器列阵
collector cover plate 集热器盖板
collector efficiency [ɪ'fɪʃənsɪ] 集热器效率
collector efficient factor (F) ['fæktə(r)] 集热器效率因子
collector flow factor 集热器流动因子
color composition [kɒmpə'zɪʃ(ə)n] 色彩构成；彩色合成
combiner box [kəm'baɪnə] 汇流箱
compact non-pressurized solar water heater system ['preʃəraɪzd] 一体非承压太阳能热水器系统
compact pre-heat pressurized solar water heater 一体预热承压太阳能热水器
compact pressurized solar water heater ['preʃəraɪzd] 一体承压太阳能热水器
component [kəm'pəunənt] 部件
composition of a picture 画面结构
compound parabolic (concentrating) collector ['kɒmpaʊnd] [ˌpærə'bɒlɪk] 复合抛物面集热器
compound semiconductor solar cell 化合物半导体太阳电池
concentrating collector ['kɒnsəntreɪtɪŋ] 聚光（型）集热器
concentrating solar power (CSP) 聚光式太阳能发电
concentration ratio 聚光比
concentrator ['kɒnsəntreɪtə] 聚光器或聚光镜
concentricity [kɒnsen'trɪsɪtɪ] 同心度
conductor [kən'dʌktə] 导体
connection for expansion vessel [ɪks'pænʃən] 膨胀罐接口
connection plate 外壳接板
Consumer Information System(CIS) 用户信息系统
control room 控制室
Control Room Management System (CRMS) 控制室管理系统
controller [kən'trəulə(r)] 控制器
convective heat transfer [kən'vektɪv] 对流传热
convective flow 对流
conversion efficiency [kən'vɜːʃ(ə)n] 转换效率
Coolant temperature gauge 冷却水温度表
cooling DSC curve 冷却 DSC 曲线
corrosion test [kə'rəʊʒ(ə)n] 腐蚀试验
criteria [kraɪ'tɪərɪə] 标准
critical pressure 临界压力
critical temperature 临界温度
cross braces 前横拉杆
cross holder 后斜拉杆
cross shearing ['ʃɪərɪŋ] 横剪
crystalline ['krɪst(ə)laɪn] 晶体的
crystallite ['krɪst(ə)laɪt] 微晶
curve correction coefficient 曲线修正系数
cutting stock 切割下料

D

dark characteristic curve 暗特性曲线

dark current 暗电流

date of delivery [dɪˈɪlɪvərɪ] 交货期

DC/AC converter (inverter) 直流/交流电压变换器

DC/DC converter (inverter) 直流/直流电压变换器

deaerator [diːˈeəreɪtə] 除氧器

degree of finish 光洁度

density [ˈdensətɪ] 密度

diagram [ˈdaɪəgræm] 图表；图解

differential pressure [ˌdɪfəˈrenʃ(ə)l] 差压

differential temperature controller [kənˈtrəʊlə(r)] 温差控制器

diffuse irradiation (diffuse insolation) 散射辐照（散射太阳辐照）量

diffusion [dɪˈfjuːʒən] 扩散，漫射

direct flow solar tube collector [kənˈtrəʊlə(r)] 直流管集热器

direct irradiation (direct insolation) 直射辐照

Direct Normal Irradiance (DNI) 直接光照强度

Direct Steam Generation (DSG) 直接产生蒸汽系统

direct system [dɪˈrekt] 直接系统

distribution box 配电箱

distribution subsystem [ˌdɪstrɪˈbjuːʃn] [ˈsʌbˈsɪstəm] 输送系统

down time 故障时间

drain valve 排污阀

drainback system 回流系统

draindown system 排放系统

draw-off temperature 取水温度

drying oven [ˈdraɪɪŋ] [ˈʌvən] 干燥箱

E

efficiency [ɪˈfɪʃənsɪ] 效率，效能，功效

electric heater 电加热

electrode [ɪˈlektrəʊd] 电极

electromagnetism valve [ɪˌlektrəʊˈmægnətɪzəm] 电磁阀

element part [ˈelɪmənt] 器件

elemental composition 构成的化学元素

emittance [ɪˈmɪtəns] 发射率

energetic amortization period [əˌmɔːtaɪˈzeɪʃən] 能量偿还期

energy audit [ˈɔːdɪt] 能量审核

energy yield 能量输出

enhanced heat transfer [ɪnˈhaːnst] 强化传热

enthalpy [ˈenθ(ə)lpɪ] 焓

equivalent length [ɪˈkwɪvələnt] 当

量长度

eutectic point [juːˈtektɪk] 共晶点

evacuated heat pipe vacuum tube [ɪˈvækjʊeɪtɪd] 全集热真空管

evacuated tube collector 真空管集热器

evaporation [ɪˌvæpəˈreɪʃən] 蒸发

exit temperature 出口温度

expansion vessel [ɪksˈpænʃən] [ˈvesəl] 膨胀罐

experimental platform [ekˌsperɪˈmentəl] [ˈplætfɔːm] 实验平台

experimental system 实验系统

F

fabrication [ˌfæbrɪˈkeɪʃən] 制作，构成

facade system [fəˈsɑːd] 正面系统

faceted collector 多反射平面集热器

faceted stretched-membrane[ˈmembreɪn] 多膜式镜片

feed water flow 给水流量

feed water tank 供水箱

feed water valve 给水阀

feed-in meter 输入计

field system 野外系统

fill factor (curve factor) 填充因子

filling and rinsing cocks [ˈrɪnsɪŋ] 注液清洗阀

filling reagent to both sides filtrate valve [ˈfɪltreɪt] 两端补料

flaming 烧结

flanging holes on main copper tube [ˈflændʒɪŋ] 主铜管翻边

flanging side of inner tank 内胆翻边

flanging [ˈflændʒɪŋ] 翻边

flat plate (solar) collector 平板(型太阳)集热器

flatness 平滑性

flat-roof system 平台屋顶系统

flexible connector [ˈfleksəbl] 波纹管

flow condition 流动状态

flow meter [ˈmiːtə(r)] 流量计

flow switch 流量开关

fluid inlet temperature (t_i) 工质进口温度

fluid outlet temperature(t_0) [ˈfluːɪd] 工质出口温度

fluoride salt [ˈflu(ː)əraɪd] 氟化盐

flux concentrating ratio [ˈkɒnsəntreɪtɪŋ] [ˈreɪʃɪəʊ] 通量聚光比

foaming process [ˈfəʊmɪŋ] [ˈprəuses] 发泡工序

focal lengths 焦距

foil 金属薄片

foot holder 尾托

foot pad [pæd] 铁鞋

fossil [ˈfɒsəl] 化石

fossil fuels [ˈfɒs(ə)l] 矿物燃料

fractional energy savings [ˈfrækʃənəl] 节能率

frame for flat roof 平屋顶支架

frame for pitched roof [pɪtʃt] 斜屋顶支架
Fresnel collector [freɪ'nel][kəlektə] 菲涅尔集热器
Fresnel lens 菲涅尔透镜
friction losses 磨擦损失
frictional resistance ['frɪkʃənəl] [rɪ'zɪstəns] 摩擦阻力
furnace ['fɜːnɪs] 火炉，熔炉，马弗炉

G

gas composition 气体组分，气体成分
gasket 密封垫
general behavior [bɪ'heɪvjə] 常规性能
geometrica concentrator ratio [dʒɪə'metrɪkəl] 几何聚光比
girth welding [weldɪŋ] 环缝焊接
global irradiance (solar global irradiance) 总辐射度（太阳辐照度）
global radiation [reɪdɪ'eɪʃ(ə)n] 总辐射
greenhouse effect 温室效应
greenhouse gases 温室气体
grid-connected PV system 并网光伏系统
Grid-Connected PV system 并网太阳能光伏发电系统
gross aperture area (Aa)（聚光集热器）总采光面积
gross collector area (Ag) 集热器总面积
Group cover tatio (GCR) 地面覆盖率

H

header 集（流）管
heat conduction [kən'dʌkʃən] 热传导
heat exchange coil [kɒɪl] 换热盘管
heat exchanger 热交换器
heat pipe solar tube collector 热管集热器
heat preservation tank [ˌprezə 'veɪʃn] 恒温水箱
(net) heat rate（净）热耗率
heat transfer [træns'fɜː] 传热
heat transfer and thermal storage materials 熔融盐传热蓄热材料
heat transfer fluid [træns'fə] [fluːɪd] 传热工质；有机热载体（导热油）
heat-conducting blind pipe 导热盲管
heating surface 受热面
heliostats ['hiːlɪə(ʊ)stæt] 定日镜
Hertz (Hz) 赫兹
high-efficiency heat transfer and

thermal storage technology 高效传热蓄热技术

HVAC， Heating/Ventilation/Air/Condition 暖通

hybridization [ˌhaɪbrɪdaɪˈzeɪʃən] 混合技术

hydrant [ˈhaɪdrənt] 水栓

hydraulic calculation [haɪˈdrɔːlɪk] [ˌkælkjəˈleɪʃən] 水力计算

hydrogen [ˈhaɪdrədʒ(ə)n] 氢

I

IED (Intelligent Electronic Devices) 智能设备

incidence waves 入射波

incident angle modifier (K_0) [ˈɪnsɪdənt] [ˈmɒdɪfaɪə(r)] 入射角修正系数

inclined holder [ɪnˈklaɪnd] 斜拉杆

indicator [ˈɪndɪkeɪtə] 指示仪表

indirect system [ˈɪndɪˈrekt] 间接系统

indoor temperature 室内温度

infrastructure [ˈɪnfrəˌstrʌktʃə] 基础设施

injecting foaming reagent [ɪnˈdʒektɪŋ] [rɪˈeɪdʒənt] 发泡

inlet 进口

inlet of cold water 冷水入口

inlet pressure 入口压力

inlet temperature [ˈɪnlet] 入口温度，进口温度

inlet/outlet connectors [kəˈnektəz] 等径卡套

inner tank [ˈɪnə] [tæŋk] 内胆

inner tank lid 内胆盖

input power 输入功率

in-roof installation [ɪnstəˈleɪʃ(ə)n] 镶嵌屋顶系统

inserting die [ɪnˈsɜːtɪŋ] 支模具

in-service condition 工作状态

installing rubber rings to holes on outer casing 上外壳胶圈

instantaneous (collector) efficiency (h) [ˌɪnstənˈteɪnɪəs] 集热器（瞬时）效率

instruction manual [ɪnˈstrʌkʃən] [ˈmænjuəl] 说明书

insulator [ˈɪnsjʊleɪtə(r)] 隔热体

integral collector storage solar water heater [ˈɪntɪgrəl] 整体式太阳热水器

integret 集成

intensity [ɪnˈtensɪtɪ] 光照强度

inverter [ɪnˈvɜːtə] 逆变器

inverter efficiency 逆变器变换效率

irradiance [ɪˈreɪdɪəns] 日照强度；辐照度

island system 独立系统

islanding 孤岛效应

I－V characteristic curve of solar cell 太阳电池的伏安特性曲线

L

latent heat changed [ˈleɪtənt] 相变潜热

lever valve [ˈliːvə] 球阀

lidding up both sides 两端扣盖

line focus collector 线聚焦（型）集热器

liquid type collector [ˈlɪkwɪd] 液体（型）集热器

load（L）日负荷

local resistance [ˈləukəl] [rɪˈzɪstəns] 局部阻力

longitudinal seaming [ləndʒɪˈtjuːdɪnl] 直缝焊

losses 损失 损耗

low cost 低成本

low pressure [ˈpreʃə] 低压力

Lower Temp 降低熔点

M

manual intervention [ɪntəˈvenʃ(ə)n] 人工干预

market prediction [prɪˈdɪkʃən] 市场预测

maximum power (P_m) 最大功率

maximum power point 最大功率点

mean fluid temperature (t_m) [ˈtempərɪtʃə] 工质平均温度

measured value 测量值

measuring instrument [ˈɪnstrum(ə)nt] 仪表

mechanical composition [mɪˈkænɪk(ə)l] 机械组成

mechanism [ˈmekənɪzəm] 机制；原理；进程；机械装置

Mega Watt（MW）兆瓦

melting point [ˈmeltɪŋ] [pɒɪnt] 熔点

mineral composition [ˈmɪn(ə)r(ə)l] 矿物组成，矿物成分

modulate [ˈmɒdjʊleɪt] 调整，调节

module 模块

module efficiency 组件效率

module（solar cell module）太阳电池组件

molten salt [ˈməultɜːn] 熔融盐

molten salt mixture [ˈmɪkstʃɜː] 混合熔盐

mono-crystalline silicon solar cell [ˈkrɪst(ə)laɪn] [ˈsɪlɪk(ə)n] 单晶硅太阳能系统

monolithic stretched-membrane [mɒnəˈlɪθɪk] 单膜式镜片

motor-driven hoist 电动葫芦

mounted 安装的，已建立的

MS-Main Steam 主蒸汽

multi-crystalline silicon solar cell 多晶硅太阳能电池

multijunction solar cell 多结太阳电池

multi-molten salts [ˈmʌltɪ] 多元熔盐

N

National Electrical Code (NEC) 国家电气代码

National Electrical Manufacturers Association 国家电力生产商协会

National Renewable Energy Laboratory (NREL) 国家可再生能源实验室

natural cycle solar water heater 自然循环式（太阳）热水器

necking down [ˈnekɪŋ] 下料缩口

necking down main copper tube 主铜管缩口

necking down of outer tank [ˈnekɪŋ] 外桶缩口

necking up heat-conducting blind pipe [ˈnekɪŋ] 导热盲管括口

net aperture area (Ae) [ˈæpətʃə] 净采光面积（聚光集热器）

nipple [ˈnɪpl] 锥丝

nitrate salt [ˈnaɪtreɪt] 硝酸盐

nominal operating cell temperature (NOCT) 组件的电池额定工作温度

nominal working pressure [ˈnɒmɪnl] 额定工作压力

non-concentrating collector [ˈkɔnsəntreɪtɪŋ] 非聚光（型）集热器

non-eutectic point [juːˈtektɪk] 非共晶点

non-imaging collector 非成像集热器

non-pressurized solar tube collector [preˈʃəraɪzd] 非承压太阳能热管集热器

non-return valve [vælv] 止回阀

non-selective surface [sɪlˈektɪv] 非选择性表面

O

open circuit voltage (V_{oc}) 开路电压

open system 敞开系统

optimum operating current (I_n) 最佳工作点电流

optimum operating voltage (V_n) 最佳工作点电压

other spare parts 其他零配件

outdoor temperature 室外温度

outer casing 外壳

outlet of hot water 热水出口

outlet pressure 出口压力

outlet temperature 出口温度

overall heat loss coefficient (UL) [ˌkəʊɪˈfɪʃnt] 总热损系数

overheating steam [ˌəʊvərˈhiːtɪŋ] 过热蒸汽

P

packaging [ˈpækɪdʒɪŋ] 包装
parabolic [ˌpærəˈbɒlɪk] 抛物线的
parabolic dish collector [ˌpærəˈbɒlɪk] 旋转抛物面集热器
parabolic trough 抛物槽
parabolic trough collector [ˌpærəˈbɒlɪk] 槽形抛物面集热器
partial flow [ˈpɑːʃ(ə)l] 部分流量
passive solar (energy) system [ˈpæsɪv] 被动式太阳能系统
peak current 峰值电流
peak load 最大负荷
peak power 峰值功率
peak voltage 峰值电压
performance guarantee 性能质保
performance prediction [pəˈfɔːm(ə)ns] [prɪˈdɪkʃ(ə)n] 性能预测
performance tolerance [ˈtɒl(ə)r(ə)ns] 性能公差
phase change solar system 相变太阳能系统
phase [feɪz] 相位，相阶段
phase-change media [feɪz] 相变介质
photo-electron 光电子
photovoltaic(PV) [ˌfəʊtəʊvɒlˈteɪɪk] 光伏
photovoltaic array 光伏阵列
photovoltaic cell 光伏电池（格）
photovoltaic module [ˈmɒdjuːl] 光伏组件
photovoltaic system 光伏系统
pipe taps 管接头
pipeline [ˈpaɪpˌlaɪn] 管道
plastic end lid 塑料端盖
plumbing [ˈplʌmɪŋ] 给排水
point of common coupling 公共连接点
pollutant [pəˈluːtnt] 污染物
poly-crystalline solar cell 聚晶硅太阳能电池
polymer film [ˈpɒlɪmə] 高分子聚合薄膜
porcelain enamel coating [ˈpɔːsəlɪn] [ɪˈnæməl] 陶瓷内胆
power block 发电机组
power factor of PV power station 光伏系统功率因数
power generation [ˌdʒenəˈreɪʒən] 发电量；发电
power-factor 功率因数
preparation of molten salt [ˌprepəˈreɪʃən] 熔盐制备
pressing and ribbing [ˈrɪbɪŋ] 起筋压平
pressure gauge with mounting valve [geɪdʒ] 压力表
pressure switch 压力开关
pressure testing 试压
pressure testing for leaks 试压检漏
pressure-temperature relief device 压力温度安全器
product composition 产品构成
proprietary [prəʊˈpraɪətərɪ] 专利

protection device for grid 电网保护装置
pump station [ˈsteɪʃən] 泵站
punching holes [ˈpʌntʃɪŋ] 冲孔
punching holes and flanging [ˈflændʒɪŋ] 冲孔成型

R

radiation 辐射
radiometer 辐射计
rapping apparatus [ˌæpəˈreɪtəs] 振打装置
rated power 额定功率
reactive loss 无功损耗
reactive power of PV power station 光伏系统无功功率
receiver [rɪˈsiːvə] (热管) 接收器；
recorder 记录仪表
rectifier [ˈrektɪfaɪə] 整流器
reflectance 反射率
reflection coefficient(R) 反射系数
reflection film technology (ReflecTech) 反射薄膜技术
reflective surface 反射表面
refracte 折射
reliability [rɪˈlaɪəˈbɪlətɪ] 可靠性
remote storage system 分离式（太阳热水）系统
removing die 卸模
renewable energy [rɪˈnjʊː əbl] 再生性能源
repair welding [weldɪŋ] 补焊
research progress [ˈprəʊgres] 研究进展
return 回流口
rework 返工
robust [rə(ʊ)ˈbʌst] 稳健的
rock wall 岩棉
rolling rebar [ˈrəʊlɪŋ] [rɪˈbɑː] 滚筋
roof inclination [ɪnklɪˈneɪʃ(ə)n] 屋顶倾斜度
rubber plugs [ˈrʌbə(r)] [plʌgz] 尾架橡胶件
rubber rings [ˈrʌbə(r)rɪŋz] 防尘圈

S

safety relief valve [rɪˈliːf] 安全卸压阀
saturation steam [ˌsætʃəˈreɪʃən] 饱和蒸汽
scaffold [ˈskæfəld] 脚手架
sealing arrangement 密封装置
secondary concentrator [ˈkɒnsəntreɪtə] 二次聚光器
selective surface [sɪˈlektɪv] 选择性表面
sensitivity [sensɪˈtɪvɪtɪ] 灵敏度
sequence number [ˈsiːkwəns] 序号
series resistance 串联电阻
series-connected system；once-through system 直流式（太阳热水）系统

setpoint tracing 照射方向自动跟踪
short-circuit current（I_{sc}） 短路电流
shunt resistance 并联电阻
shut off cock 碟阀
side frame 前腿
signal pipe [ˈsɪgnəl] 信号管
silicon 化学元素，硅
silicon paste tube [ˈsɪlɪkən] 导热硅脂
silicon ribbon solar cell 带硅太阳电池
single crystalline silicon solar cell 单晶硅太阳电池
sleeve 套管
slitting [ˈslɪtɪŋ] 纵剪
smooth tube [smuːð] 光管
solar (energy) system 太阳能系统
Solar Advisor Model（SAM） 太阳能设计模型
solar cell 太阳电池
solar cell temperature 太阳电池温度
Solar Collector Assembly（小 SCA） 太阳能集热器组件
Solar Collector Elements（SCE） 聚光集热器单元
solar constant 太阳（照射）常数
solar controller 太阳能控制器
solar cylinder [ˈsɪlɪndə(r)] 太阳能水箱
Solar Electric Generation Systems (SEGS) 太阳能发电系统
solar energy parts 太阳能组件
solar field 太阳能镜场
solar fraction (f) [ˈfrækʃn] 太阳能保证率
solar heating system 太阳（能）加热系统
solar independent system 太阳能独立系统
solar module [ˈmɒdjuːl] 太阳能电池模块，光伏模块
solar multiple 太阳能倍数
solar photovoltaic（PV）system 太阳能光伏系统
solar radiation [ˌreɪdɪˈeɪʃən] 太阳辐射
solar thermal collector [ˈθɜːməl] (solar) 集热器（太阳）
Solar Thermal Power（STP） 光热太阳能发电
solar transfer fluid [trænsˈfɜː(r)] [ˈfluːɪd] 太阳能循环液
solar water heating system efficiency [ɪˈfɪʃnsɪ] 太阳热水系统效率
solenoid [ˈsəulənɒɪd] 螺线管
solidification [ˌsəlɪdɪfɪˈkeɪʃən] 凝固；团结；浓缩
spatial-time [ˈspeɪʃəl] 时空
specific energy yield [jiːld] 能量生产率（比能率）
spectral response (spectral sensitivity) 光谱响应
Split/Separate Pressurized Solar Water Heater [ˈpreʃəraɪzd] 分体承压热水器
stability [stəˈbɪlɪtɪ] 稳定性
stagnation irradiation [stægˈneɪʃən] [ɪˌreɪdɪˈeɪʃn] 闷晒
stagnation temperature [ˈtempərɪtʃə] 滞止温度

stainless steel [ˈsteɪnlɪs] [stiːl] 不锈钢

stand alone PV system 独立太阳能光伏发电系统

standard atmosphere [ˈætməsfɪə] 标准大气压力

Standard Test Conditions (STC) 标准测试条件

standby 备用

stand-by condition 备用状态

steady-state [sˈtedɪ] 稳态

steam flow 蒸汽流量

steam generator 汽发生器

steam pressure 汽压

steam purge system 蒸汽吹扫系统

steam turbine with alternator[ˈɔːltəneɪtə] 蒸汽汽轮发电机组

stiffness [ˈstɪfnɪs] 硬度、刚度

stop cock 停止阀

storage capacity [ˈstɔːrɪdʒ][kəˈpæsɪtɪ] 储热容

storage medium [ˈmiːdɪəm] 储热介质

storage tank [ˈstɔːrɪdʒ] 储水箱

strapes [stræps] 拉条

stress-relieved [rɪˈliːvd] 应力消除

stretched-membrane 拉伸膜

structure [ˈstrʌktʃə(r)] 结构

sub-array (solar cell sub-array) 子方阵

subcritical pressure [sʌbˈkrɪtɪk(ə)l] 亚临界压力

sun hours 太阳小时

sun spectrum [ˈspektrəm] 光谱

sun-tracking controller 太阳跟踪控制器

super conduction metal heat pipe vacuum tube 超导金属热管空管

supercritical pressure [suːpəˈkrɪtɪk(ə)l] 超临界压力

superheater 过热器

supply meter 电能表

surface condenser [kənˈdensə] 表面凝汽

surface heat exchanger 表面式换热器

surface temperature 表面温度

surge withstand capability [sɜːdʒ] [keɪpəˈbɪlɪtɪ] 抗冲击能力

surrounding air speed 环境风速

switch 开关

switchgear 开关设备

synthetic [sɪnˈθetɪk] 合成的，人造的

system efficiency 系统的效率

system of making autonomous electricity [ɔːˈtɒnəməs] 独立发电系统

system of making common electricity 并网发电系统

T

T/P valve [vælv] 降温解压阀

tank 桶，箱，罐

tank capacity (V) [kəˈpæsɪtɪ] 储水（容）量

taps 接头
target load [ˈtɑːgɪt] 目标负荷
tee [tiː] 三通
temperature coefficient 温度系数
temperature coefficients of I_{sc} 短路电流的温度系数
temperature coefficients of P_m 峰值功率的温度系数
temperature coefficients of V_{oc} 开路电压的温度系数
temperature compensation [kɒmpenˈseɪʃ(ə)n] 温度补偿
temperature difference [ˈdɪf(ə)r(ə)ns] 温差
temperature drops 温降
temperature sensors [ˈtempərɪtʃə] [ˈsensəz] 温度传感器
temperature switch 温度开关
test wells 测点插孔
theoretical analysis [ˌθɪəˈretɪkəl] [əˈnæləsɪs] 理论分析
theoretical efficiency [ɪˈfɪʃənsɪ] 理论效率
thermal cycle [ˈθɜːm(ə)l] 热力循环
thermal efficiency 热效率
thermal efficiency curve [ɪˈfɪʃənsɪ] 热效率曲线
thermal properties [ˈprɒpətɪs] 热特性
thermal shock tests [ʃɒk] 热冲击实验
thermal stability [stəˈbɪlɪtɪ] 热稳定性
thermal storage [ˈstɒrɪdʒ] 蓄热
thermal storage device [ˈstɔːrɪdʒ] [dɪˈvaɪs] 储热器
thermal behavior [bɪˈheɪvjə] 热性能
thermometer [θəˈmɒmɪtə] 恒温计
thermophysical property [ˌθɜːməʊˈfɪzɪkəl] [ˈprɒpətɪ] 热物理特性
thickness losses [ˈθɪknɪs] 厚度损耗
thin-film solar cell 薄膜太阳能电池
three-phase voltage control 三相电压控制器
tidying up appearance [əˈpɪərəns] 外观清理
tilt angle 倾斜角
time constant (T_c) 时间常数
total flow 总流量
total generation 总发电量
toughened glass 钢化玻璃
tracing 底座
tracking [ˈtrækɪŋ] 跟踪
tracking collector 跟踪集热器
tracking error 跟踪误差
transmission coefficient (T) 透射系数
transmittance [trænzˈmɪt(ə)ns] 透射率
trickle collector [ˈtrɪkl] 涓流集热器
trimming and cleaning-up 削料清理
tube bank 排管
tubular [ˈtjuːbjʊlə] 管状的
turbine room 汽轮机室
turbulent [ˈtɜːbjʊl(ə)nt] 湍流
turn-key solution 交钥匙工程

type of parts 组件的规格

U

up time 工作时间

utility interface 并网接口

V

vent cock with hand wheel 手动排气阀

vent pot 排空罐

vented system [ventɪd] 开口系统

viscosity [vɪˈskɒsətɪ] 黏度

W

wall temperature 壁温

warehousing [ˈweəhaʊzɪŋ] 入库

water draw rate 取水流量

water level sensor [ˈsensə(r)] 水位探测器

water nozzle pipe 水嘴管

water preheater 水预热器

watts peak 峰瓦

welding water nozzle 水嘴焊接

whole machine testing 整机检验

wind speed (v) 风速

working pressure 工作压力

working temperature 工作温度

核 能 发 电

132kV electrical power system 132kV 电气系统

220kV & 132kV indoor switchgear building HVAC system 220kV 和 132kV 室内配电装置 HVAC 系统

220kV electrical power system 220kV 电气系统

220V AC computer power supply system 220V 交流计算机电源系统

220V AC important instrumentation power system [ˌɪnstrumenˈteɪʃ(ə)n] 220V 交流重要仪表电源系统

380V AC distribution and control system 380V 交流配电和控制系统

6kV normal auxiliary power system [ɔːgˈzɪlɪərɪ] 6kV 厂用电源系统

β/α ratemeter [ˈreɪtˌmiːtə] β/α 比值仪

A

abnormal condition [əbˈnɔːm(ə)l] 异常工况

absorbed dose rate 吸收剂量

absorber chiller 吸收式制冷机

absorber rod 吸收棒

absorption coefficient [əbˈzɔːpʃ(ə)n] 吸收系数

absorption cross section 吸收截面

absorption ratio 吸收比

acceleration pressure drop [əkseləˈreɪʃ(ə)n] 加速度压降

acceleration pressure loss 加速度损失

acceptable daily intake 日允许摄入量

acceptance criterion [kraɪˈtɪərɪən] 验收准则

acceptance limit 可接受限值

acceptance report 验收报告

acceptance standard 验收标准

acceptance test 验收实验

access [ˈækses] 通道，入口

accident analysis 事故分析

accident conditions 事故工况

accident interlocking module [ˌɪntəˈlɒkɪŋ] 事故联锁组件

accident management [ˈæksɪdənt] 事故处理

accident mitigation [mɪtɪˈgeɪʃ(ə)n] 事故缓解

accident prevention [prɪˈvenʃn] 事故预防

accident shutdown [ˈʃʌtdaʊn] 事故停堆

accident source [sɔːs] 事故源

accidental exposure [ɪkˈspəʊʒə] 事

故照射

accumulated dose [ə'kjʊmjəˌlet] 累积剂量

acid-proof tile ['æsɪd] 耐酸瓷砖

activation [ˌæktɪ'veɪʃən] 活化

active carbon filter 活性碳过滤器

active component [kəm'pəʊnənt] 能动部件

active core height 堆芯活性高度

active power 有用功率

activity 活度

activity concentration [kɒns(ə)n'treɪʃ(ə)n] 放射性浓度

activity level 放射性活度

actuate ['æktʃʊeɪt] 驱动，动作

adoption by equivalent [ə'dɒpʃ(ə)n] [ɪ'kwɪv(ə)l(ə)nt] 等同采用

aerial cable ['eərɪəl] 架空电缆

aerodynamic behavior [ˌeərəʊdaɪ'næmɪks] [bɪ'heɪvjə] 空气动力特性

aerosol ['eərəsɒl] 气溶胶

after filter (or post-filter) 后置过滤器

after-heat 剩余释热

after-power 剩余功率

air breaker 空气断路器

air change rate 换气率

air cooler 空气冷却器

air delivery pipe [dɪ'lɪv(ə)rɪ] 供气管道

air filter and absorber unit [əb'sɔːbə] 空气过滤吸附机组

air filter unit 空气过滤机组

air flow rate 空气流量

air handling unit 空气处理机组

air hose 空气软管

air intake 采风口，进风口

air pump [pʌmp] 抽气泵

air sampling device 空气取样设备

air self-cooling type 空气自冷式

air submersion dose [səb'mɜːʃən] 空气浸没剂量

airborne particulate sample [pɑː'tɪkjʊlət] 气载粒子取样器

airborne radioactivity ['eəbɔːn] [ˌreɪdɪəʊæk'tɪvɪtɪ] 气载放射性

alarm signal 报警信号

alarm window 报警窗

alarm (output) module ['mɒdjuːl] 报警（输出）组件

albedo [æl'biːdəʊ] 反射率

aligning pin 定位销

alphanumeric keyboard [ˌælfənjuː'merɪk] 字母数字键盘

American Concrete Institute 美国混凝土学会

American Institute of Steel Construction 美国钢铁结构协会

American National Standards Institute 美国国家标准学会

American Nuclear Society 美国核学会

American Society for Testing and Material 美国材料与试验学会

American Society of Civil Engineers 美国土木工程师学会

American Society of Mechanical Engineers 美国机械工程学会

ampacity [æm'pæsɪtɪ] 载流容量；安培流量
analog channel 模拟量通道
analog input 模拟量输入
analog output 模拟量输出
analog signals 模拟信号
analog/digital converter 模拟/数字转换器
analogue link ['ænəlɒg] 模拟接续线
anchor bolt ['æŋkə] 地角螺栓
anion bed ['ænaɪən] 阴离子床
anticipated operational occurrences [æn'tɪsɪpeɪt] [ə'kʌr(ə)ns] 预期运行事件
anticipated transient without scram (ATWS) 未能紧急停堆的预计瞬态
antifoam reagent ['æntɪˌfəʊm] 防泡剂
anti-rust agent 防锈漆
anti-seizer lubricant ['lu:brɪk(ə)nt] 防咬润滑剂
anti-vibration bar [vaɪ'breɪʃ(ə)n] 防振条
anti-whip device 管道防甩装置
apparent power 视在功率
application package 应用软件包
application program [ˌæplɪ'keɪʃ(ə)n] 应用程序
approval [ə'pru:v(ə)l] 批准，认可，审定
arc-lamp 弧光灯
area radiation monitoring system [reɪdɪ'eɪʃ(ə)n] 区域辐射监测系统
armored cable ['ɑ:məd] 铠装电缆
as low as reasonable achievable (ALARA) 合理可行尽量低
as-built drawing 竣工图
aseismic joint [eɪ'saɪzmɪk] 防震缝
assessment of exposure [ɪk'spəʊʒə] 照射评价
assessment of radiation protection [ə'sesmənt] [reɪdɪ'eɪʃ(ə)n] 辐射防护评价
asynchronous motor [ə'sɪŋkrənəs] 异步电动机
atmospheric corrosion [ætməs'ferɪk] [kə'rəʊʒ(ə)n] 大气腐蚀
atmospheric dispersion factor [dɪ'spɜ:ʃ(ə)n] 大气弥散因子
atmospheric overvoltage [ˌəʊvə'vəʊltɪdʒ] 大气过电压
atmospheric pressure 大气压力
atmospheric radiation monitoring apparatus [ætməs'ferɪk] [reɪdɪ'eɪʃ(ə)n] [æpə'reɪtəs]大气辐射监测装置
atmospheric stability 大气稳定度
atom absorption spectroscopy [əb'zɔ:pʃ(ə)n] [spek'trɒskəpɪ] 原子吸收谱
audible count rate signal ['ɔ:dɪb(ə)l] 计数率音响系统
audible signals 音响信号
audit ['ɔ:dɪt] 监查；审计
audit follow-up 监查后续行动
audit plan 监查计划
audit record 监查记录

C

circuit diagram [ˈsɜːkɪt] 电路图
circular railway 环形轨道
circulating cooler 循环冷风机
circulating cooling water drainage system 循环冷却水排放系统
circulating cooling water energy dispersion [dɪˈspɜːʃ(ə)n] 循环冷却水排放口消能系统
circulating cooling water intake system 循环冷却水吸入系统
circulating cooling water pumping station system 循环冷却水泵房系统
circulating cooling water treatment system 循环冷却水处理系统
circulating cooling water yard pipe network system 循环冷却水厂区管路系统
circulating water pump 循环水泵
circulation ratio [sɜːkjʊˈleɪʃ(ə)n] 循环倍率
circumferential ridge [səˌkʌmfəˈrenʃəl] 周向环脊
civil architecture 民用建筑
civil construction/engineering 土木建筑/工程
civil service 行政事务
civil works 土建工程
clad pellet gap [klæd] 包壳与芯块间的间隙
cladding 包壳
cladding creep 包壳蠕变
cladding flattening 包壳压扁
cladding strain 包壳应变
clarification [ˌklærɪfɪˈkeɪʃ(ə)n] 澄清，净化
clarifier [ˈklærɪfaɪə] 澄清槽
classification of records [ˌklæsɪfɪˈkeɪʃ(ə)n] 记录分类
clear-water reservation 清水池
clevis insert [ˈklevɪs] 镶块
climatology [klaɪməˈtɒlədʒɪ] 气候学；风土学
clock system 时钟系统
clockwise [ˈklɒkwaɪz] 顺时针方向
closed circuit TV system 闭路电视系统
closed cycle cooling water system 闭式循环冷却水系统
closed-circulating cooling water 闭路循环冷却水
closing (circuit breaker) 合闸（断路器）
closure head 顶盖
coagulant [kəʊˈægjʊlənt] 凝聚剂
coarse sand [kɔːs] 粗砂
coast down flow 惯性流量
coated electrode 药皮焊条
coaxial cable [kəʊˈæksɪəl] 同轴电缆
cobalt-base alloy 钴基合金
code case 规范案例
code controller 编码控制器
code of federal regulation (CFR)

['fed(ə)r(ə)l] 联邦管理法规
coefficient of thermal expansion [ˌkəʊɪ'fɪʃ(ə)nt] ['θɜːm(ə)l] [ɪk'spænʃ(ə)n] 热膨胀系数
coil stack assembly [ə'semblɪ] 线圈组件
cold junction box ['dʒʌŋ(k)ʃ(ə)n] 冷端补偿箱
cold performance test 冷态性能试验
cold shutdown 冷停堆
cold startup [stɑːtʌp] 冷态启动
cold test 冷态试验
cold void volume 冷空腔
cold-wall effect 冷壁效应
collapse [kə'læps] 压扁
collapse level 坍塌水位
collapse-fissure [kə'læps] ['fɪʃə] 塌陷裂缝
collective dose equivalent [ɪ'kwɪv(ə)l(ə)nt] 体剂量当量
color code 色标
commissioning 调试
committed dose equivalent [ɪ'kwɪv(ə)l(ə)nt] 积剂量当量
common base plate 公用底板
common point 公共点
common switchboard 公共配电盘
communication link 通信连接
communication system 通信系统
company standard 企业标准
compartment [kəm'pɑːtm(ə)nt] 隔间，小室
compensated ionization chamber [ˌaɪənaɪ'zeɪʃən] 补偿电离室
compensation [kɒmpen'seɪʃ(ə)n] 补偿
compensation cable 补偿电缆
compiler [kəm'paɪlə] 编译程序
complete logbook 全日志
completion certificate 完工证书
completion of erection release [ɪ'rekʃ(ə)n] 安装完工证书
compliance with requirements [kəm'plaɪəns] 符合要求
component cooling water [kəm'pəʊnənt] 设备冷却水
component cooling water heat exchanger 设备冷却水热交换器
component cooling water pump 设备冷却水泵
component cooling water surge tank 设备冷却水波动箱
component cooling water system 设备冷却水系统
composite structure ['strʌktʃə] 混合结构
compressed air system 压缩空气系统
compressible flow [kəm'presəbl] 可压缩流
computer network 计算机网络
concentrate storage tank ['kɒns(ə)ntreɪt] 浓缩液储存箱
conceptual design [kən'septjʊəl] 方案设计
condensate ['kɒnd(ə)nseɪt] 凝结水
condensate cooler 凝结水冷却器
condensate polishing system 凝结

水净化系统
condensate pump 凝结水泵
condensate transfer pump 凝结水输送泵
condenser [kən'densə] 冷凝器
condenser vacuum system ['vækjʊəm] 冷凝器抽真空系统
condition adverse to quality 不利于质量条件
conducing wire 引出线
conductive integral [kən'dʌktɪv] ['ɪntɪgr(ə)l] 积分热导率
confidence ['kɒnfɪd(ə)ns] 置信度
confined point 约束点
conical head ['kɒnɪk(ə)l] 锥形封头
connection box 接线盒
connection diagram 接线图
connection point 连接点
connector [kə'nektə(r)] 连接器
conservation of energy [kɒnsə'veɪʃ(ə)n] 能量守衡
conservation of mass 质量守恒
conservation of momentum 动量守衡
console [kən'səʊl] 控制台
constant current power supply 恒流电源
constant voltage power supply 恒压电源
construction manager [kən'strʌkʃ(ə)n] 工程经理
construction permit 建造许可证
construction quality 施工质量
contact conductance [kən'dʌkt(ə)ns] 接触热导
contained source 内在源
containment air cleanup system 安全壳空气净化系统
containment hydrogen mixing system ['haɪdrədʒ(ə)n] 安全壳氢气混合系统
containment hydrogen recombine system [ˌriːkəm'baɪn] 安全壳消氢系统
containment hydrogen ventilation system [ˌventɪ'leɪʃ(ə)n] 安全壳氢气排风系统
containment instrumentation system [ˌɪnstrʊmen'teɪʃ(ə)n] 安全壳仪表系统
containment isolation signal [kən'teɪnm(ə)nt] [aɪsə'leɪʃ(ə)n] 安全壳隔离信号
containment penetration [penɪ'treɪʃ(ə)n] 安全壳贯穿件
containment purge ventilation system 安全壳清洗通风系统
containment reactor coolant drain system ['kuːl(ə)nt] 安全壳疏排系统
containment spray heat exchanger [spreɪ] 安全壳喷淋热交换器
containment spray pump 安全壳喷淋泵
containment spray signal 安全壳喷淋信号
containment spray system [kən'teɪnm(ə)nt] 安全壳喷淋系统

containment spray system heat exchanger 安全壳喷淋系统热交换器
containment sump 安全壳地坑
content by weight 重量含量
continuous welding 连接焊
contraction coefficient 收缩系数
contraction pressure loss [kən'trækʃ(ə)n] 收缩压力损失
control (rod) assembly 控制棒组件
control (rod) bank/group 调节棒组
control (rod) calibration [kælɪ'breɪʃ(ə)n] 控制棒刻度
control (rod) cluster 控制棒组
control board 控制盘
control cable 控制电缆
control damper ['dæmpə] 控制风阀
control level of surface contamination [kənˌtæmɪ'neɪʃən] 表面污染控制水平
control logic cabinet ['kæbɪnɪt] 控制逻辑柜
control of items 物项管理
control of purchased items 已购物项管理
control rod 控制棒
control rod control system 控制棒控制系统
control rod drive line 控制棒驱动线
control rod drive line alignment [ə'laɪnm(ə)nt] 控制线对中
control rod drive mechanism (CRDM) 控制棒驱动机构
control rod dropping time 控制棒落棒时间
control rod guide thimble ['θɪmb(ə)l] 控制棒导向管
control rod movement speed 控制棒移动速度
control rod position indication [ɪndɪ'keɪʃ(ə)n] 控制棒棒位指示
control rod withdrawal 控制棒提升
control switch [swɪtʃ] 控制开关
control volume 控制容积
controlled area 控制区
convection boiling [kən'vekʃ(ə)n] 对流沸腾
conventional island DC&AC UPS system 常规岛直流和交流电源系统
conventional island (CI) [kən'venʃ(ə)n(ə)l] 常规岛
converter [kən'vɜːtə] 转换装置
conveyer belt [kən'veɪə] 传送带
conveyer car 运输小车
coolant chemistry ['kemɪstrɪ] 冷却剂水化学
coolant flow rate 冷却剂流速
coolant mass flow rate 冷却剂质量流量
coolant mixing 冷却剂交混
coolant specific enthalpy [spɪ'sɪfɪk] 冷却剂比焓
cooling coil 冷却盘管
cooling tower 冷却塔

cooling water 冷却水
core area 堆芯区域
core average heat flux (density) 堆芯平均热通量
core barrel cylinder [ˈsɪlɪndə] 堆芯吊篮组件
core barrel flange 吊篮筒体法兰
core configuration [kənˌfɪgəˈreɪʃ(ə)n] 堆芯布置
core flow subfactor 堆芯流量分因子
core integrity [ɪnˈtegrɪtɪ] 堆芯完整性
core maximum heat flux 堆芯最大热通量
core mixing subfactor 堆芯交混分因子
core physics 堆芯物理
core reflooding 堆芯再淹没
core shell 筒体段，堆芯筒体
core support pad 堆芯支承块
core thermal design [ˈθɜːm(ə)l] 堆芯热工设计
core uncover 堆芯裸露
core unloading 堆芯卸料
core void fraction [ˈfrækʃ(ə)n] 芯空泡份额
corrective action 纠正措施
corrosion fatigue [kəˈrəʊʒ(ə)n] [fəˈtiːg] 腐蚀疲劳
corrosion inhibitor [ɪnˈhɪbɪtə] 缓蚀剂
corrosion inhibitor addition tank [ɪnˈhɪbɪtə] 缓蚀剂添加箱
corrosion medium 腐蚀介质
corrosion product 腐蚀产物
corrosion rate 腐蚀速率
corrosion resistance 耐腐蚀性
cosmic radiation [ˈkɒzmɪk] 宇宙射线
cost benefit analysis 成本效益分析
counter area monitor [ˈkaʊntə] 计数管区域监测仪
counter flow 反向流
coupling breaker 耦合开关
coupon 试块
CRDM adaptor 驱动机构管座
CRDM cooling system 控制棒驱动机构冷却通风系统
CRDM power supply system 棒电源系统
CRDM ventilation shroud [ˌventɪˈleɪʃ(ə)n] 驱动机构通风管座
CRDM ventilation system 控制棒驱动机构通风系统
creep 蠕变
creepage distance 表面漏电距离
crevice corrosion [ˈkrevɪs] [kəˈrəʊʒ(ə)n] 缝隙腐蚀
criterion 准则
critical concentration [kɒns(ə)nˈtreɪʃ(ə)n] 临界浓度
critical experiment [ɪkˈsperɪm(ə)nt] 临界实验
critical exposure pathway [ɪkˈspəʊʒə] 关键照射途径
critical flow 临界流
critical heat flux modified coldwall factor 临界热流密度冷壁修正

因子

critical heat flux modified shape factor 临界热流密度形状修正因子

critical heat flux modified spacer factor 临界热流密度格架修正因子

critical heat flux（CHF） 临界热流密度

critical load 危险载荷

critical mass 临界质量

critical mass flux 临界质量流量

critical nuclide [ˈnjuːklaɪd] 关键核素

critical size 临界尺寸

critical transfer pathway 临界键转移途径

critical volume 临界体积

criticality accident [ˈæksɪdənt] 临界事故

criticality alarm system 临界报警系统

criticality approach curve 接近临界曲线

criticality safety 临界安全

cross flow 横向流

cross section 截面

crossover leg U形管段

culvert 涵洞

current curve versus time [ˈvɜːsəs] 电流波形图

current density 电流密度

curvature [ˈkɜːvətʃə] 曲率

cushion [ˈkʊʃ(ə)n] 垫层

cut-off energy 切割能

cut-off valve 断流阀

cylindrical shell [sɪˈlɪndrɪkəl] 筒形壳体

D

daily variation coefficient [veərɪˈeɪʃ(ə)n] 日变化系数

damper 风阀

dashpot action 缓冲作用

dashpot characteristics [ˌkærəktəˈrɪstɪks] 缓冲特性

dashpot drop time 缓冲落棒时间

dashpot region 缓冲区

dashpot section 缓冲段

data acquisition [ˌækwɪˈzɪʃ(ə)n] 数据采集

data acquisition and processing system 数据采集和处理系统

data base 数据基础

data bus 数据总线

data channel 数据通道

data communication system [kəmjuːnɪˈkeɪʃ(ə)n] 数据传输系统

data file 数据文件

data organization 数据结构

data processing module 数据处理模块

date of standard implementation [ˌɪmplɪmenˈteɪʃən] 标准实施日期

daughter product 子体产物
DC hold cabinet [ˈkæbɪnɪt] 直流保持柜
DC stabilized power supply 直流稳压电源
DC&AC UPS system 直流和交流不停电电源系统
dead band 死区
deborating demineralizer [diːˈmɪnərəlaɪzə] 除硼床
debug 查错
decay chain 衰变链
decay constant 衰变常数
decay heat 衰变热
decay tank 衰变箱
decay [dɪˈkeɪ] 衰变
decommissioning [ˌdiːkəˈmɪʃən] 退役
decontamination [ˈdiːkənˌtæmɪˈneɪʃən] 去污
decontamination factor 去污因子
decontamination system 去污系统
deformation [ˌdiːfɔːˈmeɪʃ(ə)n] 畸变
degradation 恶劣
degree of enrichment [ɪnˈrɪtʃmənt] 富集度
degree of subcooling 过冷度
degree of superheat 过热度
delay circuit module 延迟电路组件
delayed critical [ˈkrɪtɪk(ə)l] 缓发临界
delayed neutron fraction 缓发中子比
delayed neutron [ˈnjuːtrɒn] 缓发中子
delithium demineralizer [diːˈmɪnərəlaɪzə] 除锂床
delta connection 三角形连接
demineralized pre-filter 床前过滤器
demineralized water 除盐水
demineralizer 除盐装置
demineralizer after-filter 床后过滤器
demography [diːˈmɒgrəfɪ] 人口统计学
densification [ˌdensɪfiˈkeɪʃən] 密实化
densimeter [denˈsɪmɪtə] 密度计
density wave 密度波
denting 凹陷
deoxygenated water cooler [diːˈɒksɪdʒɪneɪtɪd] 除氧水冷却器
deoxygenated water pump 除氧水泵
deoxygenated water tank [diːˈɒksɪdʒəneɪt] 除氧水箱
deoxygenated water [diːˈɒksɪdʒɪneɪtɪd] 除氧水
departure from film boiling DFB 偏离膜态沸腾
departure from nuclear boiling ratio (DNBR) 偏离核态沸腾率
departure from nuclear boiling (DNB) 偏离核态沸腾
depleted fuel 乏燃料
depletion [dɪˈpliːʃn] 燃耗，烧乏
deposit corrosion 沉积腐蚀
deposited corrosion product

[kəˈrəʊʒ(ə)n] 沉积腐蚀产物

deposited metal 熔敷金属

deposition [ˌdepəˈzɪʃ(ə)n] 沉积量

depressurization [diːˌpreʃəraɪˈzeɪʃən] 卸压

depth of corrosion 腐蚀深度

depth of foundation 基础埋置深度

derived air concentration [kɒns(ə)nˈtreɪʃ(ə)n] 导出空气浓度

derived limit [dɪˈraɪvd] 导出极限

description for drawing up standard [dɪˈskrɪpʃ(ə)n] 标准编制说明

design assumption [əˈsʌm(p)ʃ(ə)n] 设计假设

design basis 设计基准

design basis accident (DBA) [ˈæksɪdənt] 设计基准事故

design basis event 设计基准事件

design basis source terms 设计基准源项

design change 设计变更

design code 设计规范

design condition 设计工况

design control 设计管理

design for reactor core safety in NPP 核电厂堆芯安全设计

design interface control 设计接口管理

design objective limit 设计目标限值

design pressure 设计压力

design report 设计报告

design review 设计审查

design specification [ˌspesɪfɪˈkeɪʃ(ə)n] 设计规范书

design temperature 设计温度

design verification [ˌverɪfɪˈkeɪʃən] 设计验证

designation of system 系统名称

desilting basin 沉砂池

detail design 施工设计

detection limit 探测限值

detector push-pull device [dɪˈtektə] 探测器推拉装置

detector storage tube 探测器存放管

detector well 探测器孔道

diagnostic program 诊断程序

dialog keyboard 对话键盘

diaphragm valve [ˈdeəfræm] 隔膜阀

dielectric strength 绝缘强度

diesel generator 柴油发电机

diesel generator set 柴油发电机组

differential pressure controller 差压调节器

differential pressure gauge [ˌdɪfəˈrenʃ(ə)l] 差压计

differential pressure transmitter 差压变送器

differential protection 差动保护

differential worth 微分价值

differentiator module [ˌdɪfəˈrenʃɪeɪtə] 微分器组件

diffuser plate 流量分配板

diffusion coefficient [ˌkəʊɪˈfɪʃənt] 扩散系数

diffusion theory [dɪˈfjuːʒ(ə)n] 扩散理论

digital input 数字输入
digital misalignment alarm module [mɪsəˈlaɪnmənt] 数字失步报警
digital voltmeter module [ˈvəʊltmiːtə] 数字电压表组件
dilution factor [daɪˈluːʃn] 稀释因子
dimineralized water pump 除盐水泵
direct burial [ˈberɪəl] 直埋
direct grounding 直接接地
direct inward dialing (DID) 直接内拨
direct outward dialing 直接外拨
direction of welding 焊接方向
disadvantage factor 不利因子
discharge 卸料
discharge burnup [ˈbɜːnʌp] 卸料燃耗
discharge pump 排放泵
disconnect 断连，断路
disconnect button 拆卸按钮
disconnect plug 可拆接头
disconnect rod 拆卸杆
dished head 碟形封头
dished pellet 碟形芯块
disinfection [ˌdɪsɪnˈfekʃən] 消毒
dispersed flow [dɪˈspɜːst] 弥散流
display frame 显示帧帽
display operation station 操作器（显示）
distributed computer system 分布式计算机系统
distributed network 分布式网络
distribution box 配电柜
distribution cabinet [ˈkæbɪnɪt] 分配柜
distribution list 分发清单
distribution system 配水管
document preparation 文件编制
document review 文件审查
domestic sewage 生活污水
domestic water 生活水
domestic water yard pipe network system 生活水厂区管网系统
Doppler coefficient 多普勒系数
Doppler effect 多普勒效应
dose assessment 剂量评价
dose commitment 剂量负担
dose equivalent 剂量当量
dose equivalent commitment 剂量当量负担
dose equivalent limit 剂量当量限值
dose rate conversion factor [kənˈvɜːʃ(ə)n] 剂量率转换因子
dosemeter reader [ˈdəʊsˌmiːtə] 剂量计读数器
double ended break 双端断裂
double fume chamber and hood 双位通风柜
double hook 双钩
double period 倍增周期
double-pole single-throw (DPST) switch 双极单掷开关
double-pulse generator 双脉冲发生器
double-reeling system 双卷绕系统
downhand welding 平焊
draft standard for approval 标准待批稿

draft standard for examination 标准送审稿
drain flash tank 扩容器
drain recovery tank 疏水回收箱
drain tank 疏水箱
drawing 图纸
drawing (volume) list 图册目录
drift flux 漂移流密度
drip pan 集油盘
drive shaft 驱动轴
drive shaft assembly 驱动轴部件
drive unit 驱动机构
drop time 落棒时间
droplet 液滴
droplet entrainment [ɪn'treɪnmənt] 液滴夹带
dry bulb temperature 干球温度
dry deposition 干沉降
drying oven 干燥箱，烘干箱，烘箱，干燥炉
dry-insulated transformer 干式绝缘变压器
dryout 干涸
dry-type transformer 干式变压器
dual trace oscilloscope 双迹示波器
dual-frequency simplex operation 共频单工操作
dual-tone multifrequency 双音多频
duckwork 管网
duct 风管
duplex transceiver 双工收发信机
duration time 持续时间
dye penetrant examination ['penətrənt] 着色渗透检查
dynamic analysis 动态分析
dynamic balance [daɪ'næmɪk] 动平衡动压头
dynamic equilibrium [ˌi:kwɪ'lɪbrɪəm] 动态平衡
dynamic head 动压头
dynamic instability [ɪnstə'bɪlɪtɪ] 动力学不稳定性
dynamic parameter [pə'ræmɪtə] 动态参数

E

earth leakage protection ['li:kɪdʒ] 接地保护
earthing system 接地系统
earthquake subsidence ['sʌbsɪd(ə)ns] 震陷
effective dose equivalent 有效剂量当量
effective flow rate 有效流量
effective full power days 等效满功率天
effluent radiation monitoring system 排出流辐射监测系统
egg crate grid 蛋篓型格架
ejector 抽气器；喷射器
elastic scattering 弹性散射
electret microphone ['maɪkrəfəʊn] 驻极体传声器
electric building 电气厂房
electric building cooler system 电气厂房循环冷风机组系统

electric building smoker exhaust system [ɪɡ'zɔːst] 电气厂房排烟系统
electric building ventilation system [ˌventɪ'leɪʃ(ə)n] 电气厂房通风系统
electric building water supply and drainage system 电气厂房给排水系统
electric dispatch network [dɪs'pætʃ] 电力调度网
electric furnace ['fɜːnɪs] 电炉
electric heater 电加热器
electric horn 电喇叭
electric monorail hoist ['mɒnə(ʊ)reɪl] 单轨电动葫芦
electrical conduit ['kɒndjʊɪt] 引出线管
electrical panel 配电屏
electrical penetration [penɪ'treɪʃ(ə)n] 电气贯穿件
electric-hydraulic actuating brake 电磁液压制动器
electroacoustic component [ɪˌlektrəvə'kuːstɪk] 电声元件
electromagnetic braker ['breɪkə] 电磁制动器
electromagnetic flowmeter 电磁流量计
electronic pocket dosimeter 电子式袖珍剂量计
electro-slag remelting 电渣重熔
electro-sound telephone system 检修电话系统
electrotechnical box [ɪˌlektrəu'teknɪkəl] 电气箱
ellipsoidal head 椭圆形封头
embedded part 预埋件
emergency announcement [ə'naʊnsm(ə)nt] 紧急通知
emergency boration [ɪ'mɜːdʒ(ə)nsɪ] 紧急硼化
emergency control room system 应急控制室
emergency core cooling 应急堆芯冷却
emergency diesel generation system 应急柴油发电机系统
emergency diesel generator room ventilation system 应急柴油发电机通风系统
emergency disk braker 应急盘式制动器
emergency drain 应急疏水
emergency exposure 应急照射
emergency feedwater tank 应急给水箱
emergency oil pump 应急油泵
emergency outlet 事故排出口
emergency power source (or supply) 应急电源
emergency shutdown 紧急停堆
emergency telephone system 应急自动电话系统
emergency transfer equipment 应急电源转换设备
emergency trip time 紧急停堆时间
enclosed motor 封闭式电机，防水电动机
end of cycle 循环末

end of life（EOL）寿期末
end plug 端塞
end truck 端梁
end-of erection state report 安装完工报告
end-of travel limit switch 行程终端限位开关
end-window counter 钟罩形计数管
energization 通电
energy group 能群
energy level 能级
energy resolution 能量分辨率
energy response 能量响应
energy spectrum [ˈspektrəm] 能谱
engineer station 工程师站
engineering company 工程公司
engineering enthalpy rise factor [ˈenθ(ə)lpɪ] 工程焓升因子
engineering factor 工程因子
engineering heat flux hot channel factor 工程热通量热管因子
engineering heat flux hot spot factor 工程热通量热点因子
engineering manual [ˈmænjʊ(ə)l] 工程手册
enriched fuel [ɪnˈrɪtʃt] 富集燃料
enriched uranium 浓缩铀
enriched uranium reactor 浓缩铀反应堆
enrichment 浓缩度，富集度
enthalpy rise factor 焓升因子
enthalpyrise uncertainty factor [enˈθælpɪ] 焓升不确定因子
enthapy rise engineering hot channel factor 焓升工程热管因子
entrainment 夹带
environment monitoring vehicle 环境监测车
environment qualification 环境鉴定
environmental assessment（EA）环境评价
environmental monitoring [ɪnvaɪrənˈment(ə)l] 环境监测
environmental quality（EQ）环境质量
epicenter 震中
epithermal neutron [ˌepɪˈθɜːməl] [ˈnjuːtrɒn] 超热中子
equilibrium enrichment [ɪnˈrɪtʃmənt] 平衡富集度
equilibrium poisoning 平衡中毒
equipment hatch 设备闸门
equipment list 设备明细表
equipotential connection [ˌiːkwɪpəˈtenʃ(ə)l] 等电压连接
equivalent core diameter [ɪˈkwɪv(ə)l(ə)nt] 堆芯等效直径
equivalent diameter 当量直径
erosion corrosion 磨损腐蚀
escape 泄漏，逃逸
escape rate coefficient 逃脱率系数
essential chilled water system 重要冷冻水系统
essential service water intake filteration system 重要厂用水取水过滤系统
essential service water pump 重要厂用水泵
essential service water pumping

station system 重要厂用水泵房系统
essential service water yard pipe network 重要厂用水厂区管路系统
evaporative air cooling (EAV) unit 蒸发式空气冷却机组
evaporator 蒸发器
excess letdown 过剩下泄
excess letdown heat exchanger 过剩下泄热交换器
excess power 超功率
excess reactivity 过剩反应性
exclusion area boundary [ɪk'skluːʒ(ə)n] ['baʊnd(ə)rɪ] 禁区边界
ex-core instrumentation [ˌɪnstrʊmen'teɪʃ(ə)n] 堆外测量
exhaust cleanup unit (ECU) 排风净化机组
exhaust fan [ɪg'zɔːst] 排风机
exhaust hole 排气孔
exhausting smoke fan 排烟风机
expansion joint 膨胀结合，伸缩缝
expansion ring 膨胀环
expansion tank 膨胀水箱
expert's opinion 专家意见
expert's report 专家报告
external contamination 外污染
external exposure 外照射
external interface 外部接口
external load 外负荷
external pressure 外压

F

face of weld 焊缝正面
face plate 面板
failed fuel assembly location detecting system 破损燃料足见定位检测系统
failed fuel fraction 破损燃料份额
failure detector module ['mɒdjuːl] 故障检测元件
fan 风机
fast connector 快拆接头
fast fission factor 快裂变因子
fast neutron fluence 快中子通量
fast neutron flux 快中子通量
feasibility study 可行性研究
feed water 给水
feed water heater 给水加热器
feed water pump 给水泵
feed water storage tank 给水箱
feeder ring 给水环管
feedwater inlet nozzle ['nɒz(ə)l] 给水进口管嘴
fiberglass 玻璃纤维
field change 现场修改
field inspection [ɪn'spekʃn] 巡回检查
filler metal 填充金属
film badge 胶片剂量计
film boiling 膜态沸腾
filter 过滤器；滤池
fin 散热片
final acceptance 最终验收
final acceptance report 最终验收

报告

final safety analysis report (FSAR) [ə'nælɪsɪs] 最终安全分析报告

final storage of document 文件最终储存

fine sand 细砂

fire alarm 失火报警

fire annihilation protecting equipment 熄火保护装置

fire break 防火墙

fire damper 防火阀

fire extinction 灭火

fire extinction agent 灭火剂（器）

fire extinction pump [pʌmp] 消防泵

fire fighting foam system 泡沫消防系统

fire hose 消防水龙带

fire hydrant 消火栓

fire proof installation [ɪnstə'leɪʃ(ə)n] 防火设施

fire protection 消防

fire protection water system 消防水系统

fire rating 耐火等级

fire resistant 耐火的

fire stop 阻火件

fire stopping wall 挡火墙

fire-fighting water 消防水

first collision probability [kə'lɪʒ(ə)n] 首次碰撞几率

first core 第一堆芯

first out alarm 首发报警

fissile nucleus 易裂变核

fission and corrosion product activity 裂变和腐蚀产物活度

fission chain [tʃeɪn] 裂变链

fission chamber (miniature) 裂变室（微型）

fission cross section 裂变截面

fission energy 裂变能

fission fragment 裂变碎片

fission gas 裂变气体

fission neutron 裂变中子

fission product 裂变产物

fission rate 裂变率

fission source 裂变源

fission spectrum 裂变谱

fission yield 裂变产额

fissionable 可裂变的

fixed point 固定点

fixed radio station 无线固定台

fixed tube 固定套管

flame arrester 阻火器

flame resistant 耐火焰的

flammable 易燃的

flange welding 卷边焊

flanged connection 法兰连接

flare groove welding 喇叭型坡口焊

flare test 扩口试验（传热管）

flashing 闪蒸

flashing alarm 闪光报警

flask/cask 屏蔽容器

flat head 平封头

flexible connection 柔性连接

flexible cord ['fleksɪb(ə)l] 柔线

flexible coupling 挠性联轴器，挠性接头

float charge 浮充电

float type flowmeter 浮子流量计

floatation [fləu'teɪʃən] 气浮池
floating point 浮点
flood light 泛光灯，探照灯
floor contamination monitor [kən,tæmɪ'neɪʃən] 地面污染监测仪
floor drain 楼面疏水
floor drain tank 楼面疏水箱
flow coastdown ['kəustdaun] 惯性流量下降
flow diagram 流程图（工艺）
flow distribution 流量分配
flow distribution plate 流量分配板
flow induced vibration [vaɪ'breɪʃ(ə)n] 流致振动
flow instability [ɪnstə'bɪlɪtɪ] 流动不稳定性
flow lift force 流动升力
flow restrictor [rɪs'trɪktə] 限流器
flow transmitter [trænz'mɪtə] 流量变送器
flow-head cure 流量-扬程曲线
fluorescent lamp 荧光灯
flush pump 冲洗泵
flushing water 冲洗水
flush-mounted 镶嵌，暗装
flux density 通量密度
flux distribution 通量分布
flux flattening 通量展平
flux gradient ['greɪdɪənt] 通量梯度
flux instrumentation guide tube [ˌɪnstrumen'teɪʃ(ə)n] 通量测量导向管
flux map 通量分布图
follow-up action 后续行动
forced circulation 强迫循环
foreground 前台
foreign material; impurity [ɪm'pjuərətɪ] 杂质
former 辐板（堆芯围板的径向支撑板）
fouling coefficient [ˌkəuɪ'fɪʃ(ə)nt] 污垢系数
foundation ring [faun'deɪʃ(ə)n] 基础环
fracture mechanics ['fræktʃə] 断裂力学
framework 框架，构架
free-standing cladding 自立式包壳
frequency analysis [ə'nælɪsɪs] 频谱分析
frequency meter 频率计，周波表
fresh water 淡水
fretting corrosion 激振腐蚀
fretting wear 微振磨损
friction coefficient [ˌkəuɪ'fɪʃ(ə)nt] 摩擦系数
friction loss 摩擦损失
front computer 前置计算机
front panel 前面板
fuel assembly [ə'semblɪ] 燃料组件
fuel assembly sipping device 燃料组件啜吸装置
fuel assembly visual inspection equipment 燃料组件目视检查装置
fuel basket 燃料篮
fuel building 燃料厂房
fuel building bridge 燃料厂房桥架
fuel building crane 燃料厂房起

重机

fuel building ventilation system 燃料厂房通风系统

fuel building water supply and drainage system 燃料厂房给排水系统

fuel burnup [ˈbɜːnʌp] 燃料燃耗

fuel cluster 燃料棒束

fuel column 燃料柱

fuel cycle 燃料循环

fuel cycle lifetime 燃料循环寿期

fuel element 燃料元件

fuel element failure accident 燃料元件破碎事故

fuel handling system 燃料装卸系统

fuel inventory [ˈɪnv(ə)nt(ə)rɪ] 燃料装量

fuel loading and unloading 装料与卸料

fuel management 燃料管理

fuel mispositioning accident 错装料事故

fuel pellet 燃料芯块

fuel pool bridge 燃料抓取机（燃料池桥架）

fuel rod 燃料棒

fuel shipping container rig 燃料运输容器吊具

fuel storage，transportation and inspection system 燃料储运和检验系统

fuel swelling 燃料肿胀

fuel transfer shielding 燃料运输屏蔽

fuel transfer system 燃料运输系统

fuel transfer tube 燃料运输通道

fuel tube 燃料包壳管

full penetration weld 全穿透焊

full penetration welding 全透焊缝

full power 满功率

full-load torque [tɔːk] 满载能力

function key 功能键

function test 功能试验

functional assignment 职责分工

fuse isolator [ˈaɪsəleɪtə] 熔断器闸刀

fuse switch 熔断器开关

fusion welding 熔焊

G

gantry [ˈgæntrɪ] 龙门架

gantry crane 龙门吊

gap conductance 间隙导热

gas accumulated flowmeter [əkˈjuːmjʊleɪtɪd] [ˈfləʊmiːtə] 气体累计流量计

gas chromatograph [ˈkrəʊmətəgrɑːf] 气体色谱

gas cooler [ˈkuːlə(r)] 气体冷却器

gas heater [ˈhiːtə(r)] 气体加热器

gas insulated switchgear（GIS）[ˈɪnsjʊleɪtɪd] [ˈswɪtʃgɪə] 气体绝缘开关装置

gas stripper [ˈstrɪpə(r)] 脱气塔

gas stripping [ˈstrɪpɪŋ] 脱气
gas stripping unit 脱气装置
gaseous sample [ˈgæsɪəs] 气体取样
gaseous space 气空间
gaseous waste treatment system [ˈgæsiːəs] 气体废物处理系统
gasket [ˈgæskɪt] 垫片
gate valve 闸阀
gear coupling [ˈkʌplɪŋ] 齿轮联轴器
gear reducer [riːˈdjuːsə] 齿轮减速器
Geiger-Muller counter [ˈkaʊntə(r)] 盖革/弥勒管
general arrangement [ˈdʒenərəl] [əˈreɪndʒmənt] 厂房总布置图
general corrosion [kəˈrəʊʒn] 全面腐蚀
general scientillation counter 通用闪烁计数器
general standard 通用标准
generator & main transformer system 发电机和主变压器系统
genetic effect [dʒɪˈnetɪk] 遗传效应
girder [ˈgɜːdə] 主梁
girth welding 环缝焊接
gland packing 压盖填料
glass dosimeter [dəʊˈsɪmɪtə] 玻璃剂量计
globe valve 截止阀
glove box 手套箱
go pulse [pʌls] 启动脉冲
governor valve [ˈgʌvənə] 调节阀
grab sampling [ˈsæmplɪŋ] 手选取样
grain-size 晶粒度
gravelly sand [ˈgrævəlɪ] 砾砂
Gray code [greɪ] [kəʊd] 葛莱码
Gray code shaper 葛莱码整形器
Gray/Gy [greɪ] 戈瑞
grid cell 定位格架栅元
grid spring 定位格架
grid telephone network 电网通话网络
gripper [ˈgrɪpə] 抓具
groove [gruːv] （焊接）坡口
groove angle [ˈæŋgl] 坡口角度
groove welding 坡口焊
ground fault 接地不良
ground water 地下水
grounded electrode [ɪˈlektrəʊd] 接地电极
grounding conductor [kənˈdʌktə(r)] 接地导体
grounding wire [ˈwaɪə(r)] 接地线
grounding/ earthing resistance [rɪˈzɪstəns] 接地电阻
groundwater level 地下水位
group display 成组显示
group selector [sɪˈlektə(r)] 组选择器
guide cylinder [ˈsɪlɪndə（r）] 导流筒
guide rail 导向轨道
guide tube（thimble）[ˈθɪmbl] 导向管
guide tube support plate 压紧顶帽（导向管支撑板）

H

hand and foot monitor [ˈmɒnɪtə(r)] 手足监测仪
hand hole 手孔
handling opening 吊装孔
hanger 支吊架
hard copy printer 硬拷贝打印机
harmful effect 有害影响
hazardous atmosphere [ˈhæzədəs] [ˈætməsfɪə(r)] 危险环境
head 封头
head block assembly [əˈsemblɪ] 定滑轮组
head dome insulation [ˌɪnsjʊˈleɪʃn] 顶盖球冠保温层
head flange insulation [flændʒ] [ˌɪnsjʊˈleɪʃn] 顶盖法兰保温层
head loss 水头损失
headset [ˈhedset] 头戴式送受话器
health physics program 保健物理大纲
health physics system 保健物理系统
heat affected zone 热影响区
heat balance [ˈbæləns] 热平衡
heat capacity [kəˈpæsətɪ] 热容
heat conductivity [ˌkɒndʌkˈtɪvɪtiː] 热导率
heat diffusion [dɪˈfjuːʒn] 热扩散
heat exchanger 热交换器
heat flow rate 热流量
heat flux [flʌks] 热通量
heat generation [ˌdʒenəˈreɪʃn] 释热
heat load 热负荷
heat tracing (electric) [ˈtreɪsɪŋ] 加热保温(电)
heat transfer coefficient [trænsˈfɜː] [ˌkəʊəˈfɪʃənt] 热交换系数
heat treatment 热处理
heat (or ladle) analysis [əˈnæləsɪs] 熔炼分析
heater 加热器
heating coil [kɒɪl] 加热盘管
heating ventilation and air conditioning [ˌventɪˈleɪʃn] 采暖通风及空调
heavy oil 重油
helical-wrap drive cable [ˈhelɪkl] 螺旋形驱动电缆
helium leak check [ˈhiːliːəm] 氦气检漏
hemispherical head [ˌhemɪˈsferɪkl] 半球形封头
hexadecimal [ˌheksəˈdesɪml] 十六进制的
high efficient filter [ˈfɪltə(r)] 高效过滤器
high energy pipe 高能管道
high neutron flux trip [ˈnuːˌtrɒn] 中子高通量紧急停堆
high range gamma radiation monitor for accident 事故用高量程辐射监测仪
high voltage power supply [ˈvəʊltɪdʒ] 高压电源

high-level noble gas gamma monitor ['gæmə 'mɒnɪtə(r)] 高放惰性气体 γ 监测仪

high-pressure ionization chamber [ˌaɪənaɪ'zeɪʃən] ['tʃeɪmbə] 高压电离室

high-purity germanium ['pjʊərɪtɪ] [dʒɜː'meɪnɪəm] 高纯锗

hoist speed [hɒɪst] 提升速度

hold down spring 压紧弹簧环

hold point 停工待检点

hold-down support assembly [ə'semblɪ] 压紧支撑结构

hold-up tank 暂存箱

horizontal acceleration [ækˌselə'reɪʃən] 水平加速度

horizontal guide wheel [ˌhɒrɪ'zɒntəl] 水平导向轨

horizontal pipe [ˌhɒrɪ'zɒntəl] 横管

horizontal position welding 横焊

horn loudspeaker [hɔːn] ['laʊdˌspiːkə] 号筒扬声器

hot cell 热室

hot channel 热通道

hot channel factor 热管因子

hot laundry ventilation system ['lɔːndrɪ] [ˌventɪ'leɪʃn] 洗衣房通风系统

hot leg 热段

hot performance test 热态性能试验

hot point factor 热点因子

hot shutdown 热停堆

hot sink; heat sink 热井

hot spot factor 热点因子

hot standby 热备用

hot startup ['stɑːtʌp] 热启动

hot test 热态试验

hot water production and distribution system 热水生产与分配系统

hot water system 热水系统

hourly variation coefficient [ˌveərɪ'eɪʃ(ə)n] [ˌkəʊə'fɪʃənt] 时变化系数

house load operation 带厂用电运行

housekeeping 场地管理

HP sampling cooler 高压取样冷却器

HP-heater & LP heater drain system 高-低压加热器疏水系统

humidifier [hjuː'mɪdɪfaɪə(r)] 加湿器

hydraulic press [haɪ'drɒːlɪk] 水压机

hydraulic resistance [haɪ'drɔːlɪk] [rɪ'zɪstəns] 水力阻力

hydrazine ['haɪdrəziːn] 联氨

hydrogen analyzer ['ænəlaɪzə] 氢分析仪

hydrogen embrittlement ['haɪdrədʒən] [em'brɪtlmənt] 氢脆

hydrogen induced cracking 氢致开裂

hydrogen measuring device ['meʒərɪŋ] 氢气测量装置

hydrogen pickup ['haɪdrədʒən] 吸氢

hydrogen recombiner ['riːkəm'baɪnə] 消氢器

hydrogen system 氢气系统

I

identification [aɪˌdentɪfɪˈkeɪʃn] 识别，标记
ignition [ɪgˈnɪʃn] 点燃，触发
illuminated pushbutton [ɪˈluːmɪneɪtɪd] [pʊʃˈbʌtn] 带灯按钮
illuminated switch [ɪˈljʊmɪneɪtɪd] 带灯开关
illumination [ɪˌluːmɪˈneɪʃn] 照明
impedance [ɪmˈpiːdns] 阻抗
impedance voltage (transformer) [ɪmˈpiːdəns] [ˈvəʊltɪdʒ] 短路电压，阻抗电压（变压器）
inaccessible equipment [ˌɪnækˈsesəbl] 不可接近的设备
incandescent lamp [ˌɪnkənˈdesənt] 白炽灯
incident of moderate frequency [ˈmɒdərət] [ˈfriːkwənsɪ] 中等频率事故
inclined positioning welding [ɪnˈklaɪnd] [pəˈzɪʃnɪŋ] 倾斜焊
inclusion [ɪnˈkluːʒn] 夹杂物
incoming breaker 线开关
inconel 718 alloy [ɪnˈkəʊnl] 因科镍 718 合金
inconel alloy [ˈælɒɪ] 因科镍合金
in-core 堆芯
in-core instrumentation [ˌɪnstrəmenˈteɪʃn] 堆芯测量
in-core instrumentation room [ˌɪnstrəmenˈteɪʃn] 堆芯仪表室
in-core instrumentation thimble [ˈθɪmbl] 堆芯测量管
in-core neutron flux monitoring system [ˈnuːˌtrɒn] 堆芯中子通量测量系统
in-core temperature monitoring system [ˈmɒnɪtərɪŋ] 堆芯温度测量系统
independence from cost and schedule 不受经费和进度的约束
independent review 独立检查
index file [ˈɪndeks] 索引文件
indicating lamp; indicator lamp (indicator) [ˈɪndɪkeɪtɪŋ] 指示灯，指示仪
indication of inspection status [ɪnˈspekʃn] [ˈsteɪtəs] 检查状态标识
induced radioactivity [ɪnˈdjʊːst] 感生放射性
industrial wastewater [ɪnˈdʌstrɪəl] [ˈweɪstwɒːtə] 工业废水
inelastic scattering [ɪnɪˈlæstɪk] [ˈskætərɪŋ] 非弹性散射
infinite lattice [ˈɪnfɪnɪt] [ˈlætɪs] 无限栅格
infrequent incident [ɪnˈfriːkwənt] 稀有事故
initial power [ɪˈnɪʃəl] 初始功率
inlet nozzle insulation [ˈnɒzəl] [ˌɪnsjuˈleɪʃn] 进口接管保温层
inlet temperature 进口温度
inner strap [ˈɪnə(r) stræp] （格架

内）条带

in-pile densification [densɪfɪ'keɪʃən] 堆内密实

input device 输入设备

input isolator module ['aɪsleɪtə] ['mɒdjuːl] 输入隔离组件

input/output (I/O) 输入/输出

insertion of reactivity [ɪn'sɜːʃn] 反应性引入

in-service inspection [ɪn'spekʃn] 在役检查

in-site controlled area environment radiation monitoring system 厂区环境辐射监测系统

inspection at delivery point 交接检查

inspection of work [ɪn'spekʃn] 工作检查

inspection plan 检查计划

inspection program [ɪn'spekʃn] 检查大纲

inspection record 检查记录

instantaneous relay [ˌɪnstən'teɪnjəs] 瞬时继电器

instruction [ɪn'strʌkʃn] 说明(书)，细则

instruction sheet 指令单

instrument room ['ɪnstrʊmənt] [ruːm] 仪表室

instrumental joint [ˌɪnstrʊ'mentl] 仪表接头

instrumentation (stop) valve [ˌɪnstrəmen'teɪʃn] 仪表（截止）阀

instrumentation tube 测量管

insulation level [ˌɪnsə'leɪʃən] 绝缘强度

insulator ['ɪnsjʊleɪtə(r)] 隔热片(块)

intact loop [ɪn'tækt] [luːp] 完好环路

intake 取水口，摄入量

intake structure ['strʌktʃə(r)] 取水构筑物

integrated flux [flʌks] 积分通量

integrated worth ['ɪntɪgreɪtɪd] 积分价值

intensity of back washing [ɪn'tensətɪ] 返洗强度

interaction [ˌɪntər'ækʃn] 交互作用

interactive mode [ˌɪntə'æktɪv] 交互方式

interbedding [ɪntɜːr'bedɪŋ] 互层

intercept valve [ˌɪntə'sept] 再热调节阀

intercom network 内部通信网

interface drawing 接口图

intergranular corrosion [ˌɪntə'grænjʊlə] [kə'rəʊʒən] 晶间腐蚀

intergranular stress corrosion cracking ['krækɪŋ] 晶间应力腐蚀开裂

interlock [ˌɪntə'lɒk] 联锁

intermediate range [ˌɪntə'miːdjət] 中间量程

intermittent welding [ˌɪntə'mɪtənt] 断续焊

internal contamination [kənˌtæmə'neɪʃən] 内污染

internal exposure [ɪks'pəʊʒə] 内照射

internal interface [ɪn'tɜːnəl] ['ɪntəfeɪs] 内部接口

internal pressure 内压

internals lifting support stand [ɪn'tɜːnlz] 堆内构件存放台

International Atomic Energy Agency (IAEA) 国际原子能机构

International Electrotechnical Commission [kə'mɪʃn] 国际电工委员会

International Organization for Standardization (ISO) 国际标准化组织

International Standardization Association (ISO) 国际标准化协会

interpass temperature [ɪntɜː'pɑːs] 层间温度

interrupting capacity [ˌɪntə'rʌptɪŋ] [kə'pæsətɪ] 遮断容量

intervention level [ˌɪntə'venʃən] 干预水平

intrinsically safe equipmen [ɪn'trɪnzɪklɪ] 固有安全设备

inverter [ɪn'vɜːtə] 逆变装置

iodine cartridge ['aɪədiːn] ['kɑːtrɪdʒ] 碘吸附盒

iodine removal efficiency ['aɪədiːn] [ɪ'fɪʃənsɪ] 除碘效率

iodine-131 monitor ['aɪədiːn] ['mɒnɪtə(r)] 碘-131 监测仪

ion chromatograph ['mætəgrɑːf] 离子色谱

ion exchange resin [ɪks'tʃeɪndʒ] ['rezɪn] 离子交换树脂

ion exchanger ['aɪən] 离子交换器

ionization chamber [ˌaɪənaɪ'zeɪʃən] ['tʃeɪmbə] 电离室

irradiated fuel [ɪ'reɪdiːeɪtɪd] ['fjuːəl] 已辐照燃料

irradiation [ɪˌreɪdɪ'eɪʃ(ə)n] 辐照

irradiation damage [ɪˌreɪdɪ'eɪʃn] 辐照损伤

irradiation embrittlement [em'brɪtəlmənt] 辐照脆化

irradiation induced growth [ɪˌreɪdɪ'eɪʃn] 辐照感生生长

irradiation surveillance capsule [sə'veɪl(ə)ns] ['kæpsjuːl] 辐照监督管

isokinetic sampling nozzle for stack [aɪsəʊkɪ'netɪk] 烟囱等速取样头

isolating amplifier ['aɪsəleɪtɪŋ] ['æmplɪfaɪə] 隔离放大器

isolating switch 隔离开关

isolating valve ['aɪsəleɪtɪŋ] 隔离阀

isolation damper ['dæmpə] 隔离(风)阀

isolator ['aɪsleɪtə] 绝缘子，绝缘体，隔离器

isoseisms [aɪ'səʊsaɪzəmz] 等震线

isotopic abundance [ˌaɪsəʊ'tɒpɪk] [ə'bʌnd(ə)ns] 同位素丰度

isotopic stripping fraction[aɪsəʊ'tɒpɪk] ['frækʃn] 同位素萃取份额

item important to safety 安全重要物项

K

kerma [ˈkɜːmə] 比释功能
key plan 索引图
key-locked switch 带钥匙开关
knife corrosion [kəˈrəʊʒn] 刀口腐蚀
knob [nɒb] 旋钮
knob-and-tube wiring 穿墙布线

L

lack of fusion [ˈfjuːʒ(ə)n] 未熔合
lack of penetration [penɪˈtreɪʃ(ə)n] 未焊透
large volume air sampler [ˈsɑːmplə] 大容积空气取样器
latch assembly [əˈsemblɪ] 钩爪组件
latch housing 密封盒
latching device [lætʃ] [dɪˈvaɪs] 闭锁装置
latching relay [ˈriːleɪ] 闭锁继电器
lateral movement [ˈlæt(ə)r(ə)l] 横向运动
lattice pith [ˈlætɪs] [pɪθ] 栅格间距
lead acid battery [ˈæsɪd] 铅酸蓄电池
lead auditor 主检查员
lead container [kənˈteɪnə] 铅室
lead/lag compensation module [kɒmpenˈseɪʃ(ə)n] [ˈmɒdjuːl] 超前/滞后补偿组件
leak detector [dɪˈtektə] 泄漏探测器
leakage radiation [reɪdɪˈeɪʃ(ə)n] 泄漏辐射
leak-off connection 引漏接头
leak-off line 引漏管线
leakproof 密封的，防漏的
legal metrological unit [ˌmetrəuˈlɒdʒɪkəl] 法定计量单位
let down 下泄
letdown heat exchanger 下泄热交换器
letdown orifice [ˈɒrɪfɪs] 下泄节流孔板
letdown path 下泄通道
licensing document 许可证申请文件
licensing procedure [prəˈsiːdʒə] 许可证批准程序
life test 寿命试验
lifetime 寿期
lift armature [ˈɑːmətʃə] 提升衔铁
lift coil [kɒɪl] 提升线圈
lift height 起重
light oil 轻油
light source 光源
limit switch 行程开关，限位开关

limited access area 监控区
limited approval [ə'pru:v(ə)l] 有限批准
limited condition 极限工况
limiting fault 极限事故
line of communication 联络渠道，通信线路
linear power density ['densɪtɪ] 线性功率密度
linkage editor 连接编辑
linking loader 连接装入程序
liquefaction potential [pə(ʊ)'tenʃ(ə)l] 液化势
liquid filter ['fɪltə] 液体过滤器
liquid level transmitter 液位变送器
liquid penetrant examination ['penɪtr(ə)nt] 液体渗透检验
liquid phase [feɪz] 液相
liquid space 液体空间
liquid waste receiver tank [tæŋk] 废液接收箱
liquid waste treatment system 废液处理系统
live part 带电部分
load block assembly [ə'semblɪ] 动滑轮组
load change 负载变化
load factor 负荷因子
load follow 负荷跟踪
load sensor ['sensə] 载荷传感器
load shedding ['ʃedɪŋ] 甩电荷
load surge [sɜ:dʒ] 负荷冲击
load-deformation curve [ˌdi:fɔ:'meɪʃ(ə)n] [kɜ:v] 应力—应变曲线
local boiling 局部沸腾
local cabinet ['kæbɪnɪt] 就地机柜
local control room system 就地控制室系统
local sampling 就地取样
local shielding ['ʃi:ldɪŋ] 局部屏蔽
localing tube [tju:b] 定位（距）管
localized corrosion [kə'rəʊʒ(ə)n] 局部腐蚀
lock nut 锁紧螺母
locking button 锁定钮
logbook 日志
logic error monitor 逻辑错误监测电路
logic isolator module ['aɪsəleɪtə] ['mɒdju:l] 逻辑隔离组件
logic operation [ɒpə'reɪʃ(ə)n] 逻辑运算
logic variable 逻辑变量
long ionization chamber [aɪənaɪ'zeɪʃən] ['tʃeɪmbə] 长电离室
longitudinal seam [ˌlɒn(d)ʒɪ'tju:dɪn(ə)l] 纵（焊）缝
longitudinal section 纵剖面
long-lived fission product ['fɪʃ(ə)n] 长寿命裂变产物
longtime rating ['reɪtɪŋ] 长时间额定值
loss of coolant accident ['ku:l(ə)nt] 失水事故
loss of power 失去电源
low activity α，γ detector [dɪ'tektə] 低放 α，γ 探测器
low alloy steel ['ælɒɪ] 低合金钢

low range ionization chamber area γ monitor 低量程电离室γ监测器

low-background β detector [dɪˈtektə] 低本底β探测装置

lower core plate [kɔː] [pleɪt] 堆芯下板

lower core support structure 堆内下部支撑

lower internals storage stand 吊兰部件存放架

lower support plate assembly [əˈsemblɪ] 下栅格板组件

low-level noble gas β monitor [ˈnəʊb(ə)l] [ˈmɒnɪtə] 低放惰性气体β监测仪

low-level waste liquid monitor 低放废水监测器

M

macroscopic cross section [ˌmækrə(ʊ)ˈskɒpɪk] [ˈsekʃ(ə)n] 宏观截面

magnetic disk unit [mægˈnetɪk] 磁盘机

magnetic particle examination [ˈpɑːtɪk(ə)l] 磁粉检验

magnetic tape 磁带

magnetic tape unit 磁带机

magnetic trip breaker 磁脱扣断路器

main condensate pump [ˈkɒnd(ə)nseɪt] 主凝结水泵

main control habitability ventilation system [ˌhæbɪtəˈbɪlətɪ] [ˌventɪˈleɪʃ(ə)n] 主控室可居留区排烟（通风）系统

main control room alarm processing system [əˈlɑːm] 主控室警报处理系统

main control room system 主控室系统

main distribution frame [dɪstrɪˈbjuːʃ(ə)n] 总配线架

main feed water system 主给水系统

main feedwater pump 主给水泵

main hoist [hɒɪst] 主起升机构

main steam 主蒸汽

main steam and turbine bypass system [ˈtɜːbaɪn] [ˈbaɪpɑːs] 主蒸汽及旁路蒸汽系统

main steam dump control system 主蒸汽排放控制系统

main stop valve 主汽阀

mains outage [ˈaʊtɪdʒ] 主电路故障

maintenance tool [ˈmeɪnt(ə)nəns] 维修工具

make up water 补给水

make up water system 补给水系统

make-up water pump 补给水泵

make-up water tank 补给水箱

mandatory standard [ˈmændət(ə)rɪ] 强制性标准

manhole 人孔，检查孔

manipulator crane [məˈnɪpjʊleɪtə(r)] [kreɪn] 装卸料机

man-machine interaction 人机对话
man-machine interface [ˈɪntəfeɪs] 人机接口
manometer 压力表
manual control [ˈmænjʊ(ə)l] 手动控制
manual emergency drive [ɪˈmɜːdʒ(ə)nsɪ] 应急手动操作机构
manual welding [ˈweldɪŋ] 手工焊
manual-electric changeover switch 手动—电动转换开关
marshalling box [ˈmɑːʃəlɪŋ] 编线盒
masonry structure [ˈmeɪs(ə)nrɪ] 砌体结构
mass flux [flʌks] 质量流密度
mass velocity [vɪˈlɒsɪtɪ] 质量速度
master clock 母钟
master cycler 主循环器
master cycler output module [ˈmɒdjuːl] 主循环器输出组件
mat foundation [mæt] 片筏基础
matrix material [ˈmeɪtrɪks] [məˈtɪərɪəl] 基体材料
maximum air concentration [kɒns(ə)nˈtreɪʃ(ə)n] 最大空气浓度
maximum allowable cooldown rate 最大允许冷却速率
maximum credible accident (MCA) 最大可信事故
maximum daily output 最大日供应量
maximum letdown flow 最大下泄流
MCR/ECR transfer switch 主控/应控切换开关
MCR/LL transfer switch 主控/就地切换开关
measured rod position [rɒd] 测量棒位
measurement box 测量箱
measuring tank [tæŋk] 计量槽
mechanical clutch [mɪˈkænɪk(ə)l] [klʌtʃ] 机械离合器
mechanical interlock [ɪntəˈlɒk] 机械连锁
mechanical load [mɪˈkænɪk(ə)l] 机械载荷
mechanical property [ˈprɒpətɪ] 机械性能
mechanical seal [siːl] 机械密封
mechanical strength 机械强度
medium sand 中砂
medium voltage [ˈvəʊltɪdʒ] 中间电压
medium-level noble gas γ monitor [ˈnəʊb(ə)l] 中放惰性气体γ监测仪
memory module [ˈmɒdjuːl] 记忆组件
mesh [meʃ] 网格，筛孔
metal clad [klæd] 金属包复，金属外壳
metallographic examination [mɪˌtæləʊˈgræfɪk] 金相检验
meteorological data [ˌmiːtɪərəˈlɒdʒɪkəl] 气象资料
meteorological observatory 气象观测台

meteorological station 气象站
microscopic cross section [maɪkrəˈskɒpɪk] 微观截面
mimic diagram [ˈdaɪəgræm] 模拟(流程)图
mimic panel [ˈmɪmɪk] [ˈpæn(ə)l] 模拟盘
miniature fission chamber [ˈmɪnɪtʃə] [ˈfɪʃ(ə)n] [ˈtʃeɪmbə] 小型裂变室
mini-flow circulation mode [sɜːkjʊˈleɪʃ(ə)n] 小流量循环方式
miniflow orifice [ˈɒrɪfɪs] 小流量孔板
minimum service head 最小使用压头
misalignment [mɪsəˈlaɪnmənt] 失步
miscellaneous BOP building ventilation system [ˌmɪsəˈleɪnɪəs] [ˌventɪˈleɪʃ(ə)n] 其他 BOP 厂房通风系统
mixed resin [ˈrezɪn] 混合树脂
mixing coefficient [ˌkəʊɪˈfɪʃ(ə)nt] 混合系数
mode of operation 运行方式
moderation [mɒdəˈreɪʃ(ə)n] 慢化
moderator [ˈmɒdəreɪtə] 慢化剂
moderator temperature coefficient [ˈmɒdəreɪtə] [ˌkəʊɪˈfɪʃ(ə)nt] 慢化剂温度系数
moderator-to-fuel-ratio [fjʊəl] [ˈreɪʃɪəʊ] 慢化剂—燃料比
modification observations [ˌmɒdɪfɪˈkeɪʃ(ə)n] 修正说明
modified checkerboard pattern [ˈmɒdɪfaɪd] [ˈtʃekəbɔːd] [ˈpæt(ə)n] 修正的棋盘式交叉布置
modified Mercalli intensity [ɪnˈtensɪtɪ] 修正的麦加利烈度(地震术语)
module [ˈmɒdjuːl] 组件
moisture separator [ˈmɒɪstʃə] 汽水分离器
moisture separator-reheater [ˈsepəreɪtə] 汽水分离再热器
momentary rating [ˈməʊm(ə)nt(ə)rɪ] [ˈreɪtɪŋ] 瞬时额定值
monitor tank [tæŋk] 检测槽
monitoring unit [ˈjuːnɪt] 检测装置
mono energetic [ˈmɒnəʊ] 单能的
monostable module [ˈmɒnəʊˌsteɪbl] 单稳态组件
motor control center 电动机控制中心
motor-driven single beam suspending crane [biːm] [səˈspend] [kreɪn] 电动单梁悬挂起重机
movable armature [ˈmuːvəb(ə)l] [ˈɑːmətʃə] 保持衔铁(可移动衔铁)
movable detector [dɪˈtektə] 可移动探测器
movable gripper coil [ˈgrɪpə] [kɒɪl] 保持线圈(移动钩爪线圈)
movable gripper latch [lætʃ] 保持钩爪(可移动钩爪)

movable miniature fission chamber 可移动小型裂变室

movable neutron detector [ˈnjuːtrɒn] [dɪˈtektə] 可移动中子探测器

movable part 可动部分

movable pole [pəʊl] 保持磁极(可移动磁极)

MP sampling cooler 中压取样冷却器

mud settler [mʌd] [ˈsetlə] 澄清槽

multichannel spectrum analyzer [ˈspektrəm] 多道能谱分析器

multi-layer welding [ˈmʌltɪ] [ˈweldɪŋ] 多层焊

multimeter [ˈmʌltɪmiːtə] 万用表

multi-pin socket [ˈsɒkɪt] 多芯插座

multiplication constant [ˌmʌltɪplɪˈkeɪʃ(ə)n] [ˈkɒnst(ə)nt] 增殖因子

multiplication factor 增殖因子

multizone core [ˈmʌltɪzəun] [kɔː] 多区堆芯

N

narrow-gap welding 窄焊缝

national standard 国家标准

national standard body 国家标准机构

natural background 天然本底

natural circulation 自然循环

natural draft [ˈnætʃ(ə)r(ə)l] [drɑːft] 自然通风

natural isotopic abundance [ˌaɪsəʊˈtɒpɪk] [əˈbʌnd(ə)ns] 同位素天然丰度

natural radioactivity 天然放射性

negative pressure 负压

negative reactivity [ˌrɪækˈtɪvətɪ] 负反应性

negative reactivity margin [ˈmɑːdʒɪn] 负反应性裕量

net positive suction head [ˈsʌkʃ(ə)n] 净正吸入压头

neutral point 中性点

neutralizer tank [ˈnjuːtrəlaɪzə] 中和槽

neutron absorber [əbˈsɔːbə] 中子吸收剂

neutron absorber material 中子吸收材料

neutron absorption [əbˈzɔːpʃ(ə)n] 中子吸收

neutron absorption cross section 中子吸收截面

neutron activated corrosion product [kəˈrəʊʒ(ə)n] 中子活化腐蚀产物

neutron albedo [ælˈbiːdəʊ] 中子反射率

neutron counting rate 中子计数率

neutron density 中子密度

neutron detector 中子探测器

neutron doserate-equivalent monitor [ɪˈkwɪv(ə)l(ə)nt] 中子剂量当量率监测仪

neutron flux [flʌks] 中子通量

neutron flux detector 中子通量探测器

neutron flux range [reɪn(d)ʒ] 中子通量量程

neutron generation time 中子代时间

neutron hardening [ˈhɑːdənɪŋ] 中子谱硬化

neutron irradiation [ɪˌreɪdɪˈeɪʃ(ə)n] 中子辐照

neutron pad 中子衬垫

neutron poison [ˈpɒɪz(ə)n] 中子毒物

neutron source 中子源

neutron source assembly [əˈsemblɪ] 中子源组件

neutron spectrum [ˈspektrəm] 中子谱

neutron temperature 中子温度

neutron thermalization [ˌθɜːməlaɪˈzeɪʃən] 中子热能化

neutron transport 中子输送

new fuel assembly handling tool 新燃料组件操作工具

new fuel assembly inspection operating [əˈsemblɪ] [ɪnˈspekʃn] [ˈɒpəreɪtɪŋ] 新燃料组件检查台架

new fuel assembly measuring handling tool 新燃料组件测量装置操作工具

new fuel elevator [ˈelɪveɪtə] 新燃料升降机

new fuel handling tool 新燃料操作工具

new fuel shipping container [ˈʃɪpɪŋ] [kənˈteɪnə] 新燃料运输容器

new fuel shipping container console [kənˈsəʊl] 新燃料运输容器支架

new fuel storage rack [ræk] 新燃料储存格架

NI fire protection system 核岛消防水系统

nickel base alloy [ˈnɪk(ə)l] [beɪs] [ˈælɒɪ] 镍基合金

nil-ductility transition temperature [nɪl] [dʌkˈtɪlətɪ] 无延性转变温度

nitrogen blanket [ˈnaɪtrədʒ(ə)n] [ˈblæŋkɪt] 氮气覆盖层

nitrogen system 氮气系统

noble gas [ˈnəʊb(ə)l] 惰性气体

noble gas sampler 惰性气体取样器

no-load loss 空载损失

nominal power [ˈnɒmɪn(ə)] 名义功率

non-condensable gas [kənˈdensəbəl] 不凝性气体

non-conformance review 不符合项审查

non-consumable electrode [kənˈsjuːməb(ə)l] [ɪˈlektrəʊd] 不熔化电极

non-destructive examination [dɪˈstrʌktɪv] 无损检测

non-essential chilled water cooling water system [ɪˈsenʃ(ə)l] [tʃɪld] 非重要冷冻水冷却水系统

non-essential chilled water system [ɪˈsenʃ(ə)l] 非重要冷冻水系统

non-fission capture [ˈfɪʃ(ə)n] [ˈkæptʃə] 非裂变俘获
non-metallic inclusion [mɪˈtælɪk] 非金属夹杂物
non-polluted industrial wastewater [ɪnˈdʌstrɪəl] 生产废水（未污染的工业废水）
nonsafety-related 非安全相关的
non-stochastic effect [stəˈkæstɪk] 非随机效应
normal condition 正常条件，正常状态
normal operation 正常运行
normal power supply 正常电源
normal telephone system 自动电话系统
normally-closed contact [ˈkɒntækt] 常闭触点
normally-open contact 常开触点
nozzle belt 接管段
nuclear aux building lighting system [ˌɪnstrʊmenˈteɪʃ(ə)n] 核辅助厂房照明系统
nuclear boiling 泡核沸腾
nuclear data link 核数据通信线
nuclear enthalpy rise factor [enˈθælpɪ] 核焓升因子
nuclear instrumentation system [ˌɪnstrʊmenˈteɪʃ(ə)n] 核测仪表系统
nuclear island 核岛
nuclear island 220V AC instrumentation PSS 核岛 220V 交流仪表电源系统
nuclear island 220V DC power supply system 核岛 220V 直流电源系统
nuclear island 24V DC power supply system 核岛 24V 直流电源系统
nuclear island 380V AC power system 核岛 380V 交流电源系统
nuclear island and conventional island fire protection water distribution system 核岛、常规岛消防水分配系统
nuclear island building stair positive pressure supply system [ˈpɒzɪtɪv] 核岛厂房楼梯间正压送风系统
nuclear power plant 核电厂
Nuclear Regulatory Commission (NRC) [ˈregjulətərɪ] [kəˈmɪʃ(ə)n] 核管理委员会
nuclear safety inspection 核安全检查
nuclear uncertainty factor 核不确定因子

O

object program 目标程序
objective evidence 客观证据
occupancy factor 居留因子
occupation standard 行业标准
occupational exposure [ɪkˈspəʊʒə] 职业照射

occupational radiation worker 职业性辐射工作人员
off-line 离线
offset 残余变形，偏移，失调
offsite power source 厂外电源
oil injection [ɪn'dʒɛkʃən] 注油
oil pit 油槽，油坑
oil reservoir ['rezəvwɑː(r)] 油箱
oil tank 油箱
omega seal ['əumɪgə] Ω 密封
on (-the) -job training 岗位培训
on/off input 开关量输入
one breaker and a half type 一个半断路器
on-position 工作状态
onsite source 厂内电源
open wiring 明配线
operating license 运行许可证
operating organization 营运单位
operating overvoltage ['əuvə'vəultɪdʒ] 操作过电压
operating platform 操作平台
operation basis earthquake (OBE) 运行基准地震
operation coils [kɒɪlz] 工作线圈
operation floor 运行层
operational limits and conditions 运行限值和条件
operational records 运行记录
operational states; operating condition 运行工况
operator station 操作员站
optimization of radiation protection [ˌɒptɪmaɪ'zeɪʃən] 辐射防护最优化
organization freedom 组织独立性
organization structure ['strʌktʃə] 组织机构
O-ring O 形环
other BOP building water supply and drainage system 其他 BOP 厂房给排水系统
outgoing breaker 出线断路器
outlet nozzle insulation 出口接管保温层
outlet temperature 出口温度
outline drawing 外形图
outside strap [stræp] 围板
overall responsibility [rɪˌspɒnsɪ'bɪlɪtɪ] 全面责任
overcurrent relay ['əuvəkʌrənt] 过流继电器
overflow 溢出
overhead crane 桥式起重机
overhead position welding 仰焊
overlap 重叠，搭接
overload 过载
overpressure protection 超压保护
overspeed protecting switch 超速保护开关
overspeed safety brake 超速安全制动器
overvoltage relay 过压继电器
oxygen analyzer ['ænəˌlaɪzə] 氧分析器
oxygen free waste gas ['ɒksɪdʒ(ə)n] 无氧废气

P

package unit 柜式机组
page-party system 呼叫通话系统
Pakistan Chashma Nuclear power Project 巴基斯坦恰希玛核电工程
pan-tilt control facility [fəˈsɪlɪtɪ] 摄像机姿态控制器
partial penetration welding [ˈpɑːʃ(ə)l] 部分焊透焊缝
particulate filter [pɑːˈtɪkjʊlət] 微粒过滤器
particulate-iodine sampler [ˈaɪədiːn] 微粒—碘取样器
partition coefficient [ˌkəʊɪˈfɪʃ(ə)nt] 分配系数
partition factor 分配因子
partition plate [pɑːˈtɪʃ(ə)n] 隔板
partition wall 隔墙
pass test 合格试验
passivation [pæsɪˈveɪʃən] 钝化
path selector 路选择器
pcm (pour cent mille) 反应性单位
peak ground acceleration (PGA) 峰值地面加速度
peening 喷丸(处理)
pellet-cladding interaction [ɪntərˈækʃ(ə)n] 芯块与包壳相互作用
penalty factor for rod bow [ˈpen(ə)ltɪ] 燃料棒弯曲亏损因子
pendant pushbutton control box 悬挂式按钮控制项
penetrating radiation [ˈpenɪtreɪtɪŋ] 贯穿辐射
penetration sleeve 穿墙套管
percent by weight 重量百分比
performance curve [pəˈfɔːm(ə)ns] 性能曲线
periodic inspection [ˌpɪərɪˈɒdɪk] [ɪnˈspekʃn] 定期检查
periodic test [ˌpɪərɪˈɒdɪk] 定期试验
peripheral [pəˈrɪfərəl] 外部设备
permanent storage [ˈpɜːm(ə)nənt] 永久储存
permissive circuit [pəˈmɪsɪv] 允许电路
permissive logic module [ˈlɒdʒɪk] [ˈmɒdjuːl] 允许逻辑组件
permit 许可证
perpendicular acceleration [ˌpɜːp(ə)nˈdɪkjʊlə] 垂直加速度
personal dosimeter [dəʊˈsɪmɪtə] 个人剂量计
personnel air lock 人员闸门
pH meter 酸度计
pH value pH 值
phase control and firing circuit [ˈsɜːkɪt] 移相触发电路
phase difference 相位差
phase displacement 相位移
phosphate treatment [ˈfɒsfeɪt] 磷酸盐处理

photomultipliler [ˌfəʊtəʊˈmʌltəˌplaɪə] 光电倍增管
physical separation [sepəˈreɪʃ(ə)n] 实体分离
physics laboratory [ˈfɪzɪks] [ləˈbɒrəˌt(ə)rɪ] 物理测量室
physiography [ˌfɪzɪˈɒgrəfɪ] 自然地理学，地文学
pier foundation 墩式基础
pigtail connection [ˈpɪgˌtel] 软辫线连接
pilot device [ˈpaɪlət] [dɪˈvaɪs] 先导装置
pilot operated pressure relief valve [rɪˈliːf] 导式卸压阀
pilot valve [vælv] 导阀
piping and instrumentation diagram [ˈdaɪəˌgræm] 工艺流程及表计图
piping penetration [penɪˈtreɪʃ(ə)n] 管道贯穿件
piping support 管道支撑件
piping system 管道系统
piston pump [ˈpɪstən] 活塞泵
pitting corrosion [kəˈrəʊʒ(ə)n] 点蚀
plan 平面图
plant alarm processing system 全厂报警处理系统
plant auxiliary electrical system [ɒgˈzɪljərɪ] 厂用电系统（常规岛）
plant computer system 电站计算机系统
plant fire detection and alarm system 电厂火灾探测及报警系统
plant layout drawing [ˈleɪaʊt] 电厂总平面布置图
plant management 电厂运行管理部门
plate tectonic [tekˈtɒnɪk] 板块构造
plenum [ˈpliːnəm] 空腔，腔室
plug valve 旋塞
plug welding 塞焊
plug-in breaker 插入式断路器
plunger pump [ˈplʌn(d)ʒə] 柱塞泵
pneumatic valve [njuːˈmætɪk] 气动阀
pocket dosimeter [dəʊˈsɪmɪtə] 袖珍式剂量计
polar crane [kreɪn] 环形吊车
polarity [pə(ʊ)ˈlærɪtɪ] 极性
policy statement 政策声明
polluted industrial wastewater 污染的工业水
polluted item 污染的工件
population equivalent [ɪˈkwɪv(ə)l(ə)nt] 人口当量
portable air sampler 移动式空气取样器
portable extinguisher [ɪkˈstɪŋgwɪʃə] 手提式灭火器
portable extinguisher system 移动式灭火系统
portable γ exposure rate meter 移动式 γ 照射率仪
portal frame 门式构架，龙门架
portal γ monitor [ˈmɒnɪtə(r)] 门框式 γ 监测仪

position indicating system 位置指示仪

position indicator [ˈɪndɪkeɪtə(r)] 位置指示器

position limit switch 限位开关

position of welding 焊接位置

position switch 行程开关

positioner [pəˈzɪʃənə] 定位器

positioning part 定位部件

positioning system 定位系统

positive pressure 正压

post accident sampling [ˈæksɪdənt] 事故后取样

post weld heat treatment 焊后热处理

post-audit meeting 监查后会议

postheat 后热

post-processing 后处理工艺过程

post-processor 后信息处理器

post-trip review 脱扣追忆记录

postulated accident [ˈpɒstʃəˌleɪtɪd] 假想事故

postulated accident condition 假想事故工况

potential transformer 电压互感器

potentiometer [pəˌtenʃɪˈɒmɪtə(r)] 电位器

power channel comparison facility [ˈtʃænl] 功率通道比较装置

power cord 电源线

power defect 功率亏损

power density 功率密度

power distribution [dɪstrɪˈbjuːʃ(ə)n] 功率分布

power effect 功率效应

power factor 功率因子

power line carrier 电离线载波

power loss 失去电源

power operated relief valve 动力操作卸压阀

power operation 功率运行

power plant operation log 电站运行记录日志

power range 功率量程

power spike factor [spaɪk] 功率峰值因子

power supply 供电

pre-audit meeting 监查前会议

precipitation [prɪˌsɪpɪˈteɪʃ(ə)n] 降雨量

precise-pulse generator 精密脉冲发生器

prefilter [priːˈfɪltə] 预过滤器

preheat 预热

pre-heater 预热器

pre-irradiation [ɪˌreɪdɪˈeɪʃ(ə)n] 辐照前

preliminary safety analysis report [prɪˈlɪmɪn(ə)rɪ] 初步安全分析报告

preliminary treatment water for demineralized water workshop yard pipe network system [dɪˈmɔːrəˌlaɪzd] 化学水源水厂区管路系统

preparation of procedures [ˌprepəˈreɪʃ(ə)n] 规程编制

prescribed limits [prɪˈskraɪbd] 规定限值

pre-service inspection 役前检查

pressure boundary 压力边界
pressure coefficient [ˌkəʊɪˈfɪʃnt] 压力系数
pressure compensation diaphragm [ˈdaɪəˌfræm] 压力平衡膜片
pressure drop 压降
pressure housing assembly [əˈsemblɪ] 耐压壳组件
pressure reduction 减压
pressure relief valve 卸压阀
pressure response 压力响应
pressure tap 测压管
pressure transmitter [trænzˈmɪtə] 压力变送器
pressure vessel [ˈvesl] 压力容器
pressure-retaining boundary [ˈbaʊnd(ə)rɪ] 承压边界
pressurized thermal shock [ˈpreʃəraɪzd] 承压热冲击
pressurizer [ˈpreʃəraɪzə] 稳压器
pressurizer liquid phase [ˈpreʃəraɪzə] [ˈlɪkwɪd] [feɪz] 稳压器液相
pressurizer pressure control system 稳压器压力控制系统
pressurizer relief tank 稳压器卸压箱
pressurizer steam phase 稳压器气相
pressurizer water level control system 稳压器液位控制系统
primary auxiliary building ventilation system [ˌventɪˈleɪʃn] 一回路辅助厂房通风系统
primary neutron source assembly [ˈnjuːtrɒn] 初级中子源组件
primary radiation 初级辐射
primary shielding [ˈʃiːldɪŋ] 一次屏蔽
primary source 初级中子源
primary treatment 初步处理
primary winding 初级绕组
priority call [praɪˈɒrətɪ] 优先传呼
private automatic branch exchange 专用自动小交换机
procedure qualification report [prəˈsiːdʒə] [ˌkwɒlɪfɪˈkeɪʃ(ə)n] 程序鉴定试验报告
procedure qualification test 程序鉴定试验
process control 过程控制
process drain 工艺疏水
process monitoring system 过程监测系统
process radiation monitoring system 工艺辐射监测系统
process specification 工艺规范书
process water 生产用水
procurement specification [prəˈkjʊrmənt] 采购规格书
product analysis 产品分析
programmable logic controller 可编程的控制器
programmed water level 程控水位
project engineering manager 项目工程经理
project management 项目管理
project manager 项目经理
project supervision 项目监督
prompt critical 瞬发临界
prompt gamma radiation 瞬发 γ

辐射

prompt neutron [ˈnuː,trɒn] 瞬发中子

prompt neutron fraction [ˈnuːtrɒn] 瞬发中子份额

proportional boron-lined counter [ˈbɔːrɒn] 正比涂硼计数管

proportional counter 比计数管

proportional heater [prəˈpɔːʃ(ə)n(ə)l] 比例组电加热器

proportional spray valve [spreɪ] 比例喷雾阀

protection relay 保护继电器

protective barrier 防护屏

protective clothing 防护衣具

protective cover [ˈkʌvə(r)] 保护罩

protective glove [ɡlʌv] 防护手套

protective shoe cover 防护鞋套

prototype [ˈprəutətaɪp] 原型机，样机

public address system 有线广播系统

public telephone network 公用电话网

pull-out torque [tɒrk] 失步转矩

pull-out unit 抽芯组件

pulsation [pʌlˈseɪʃən] 脉动

pulse 脉冲

pulse distributor module [ˈmɒdjuːl] 脉冲分配组件

pulse input 脉冲输入

pulse monitor module 脉冲监测组件

pulse signal generator module 脉冲信号发生组件

pulse-code modulation [ˌmɒdjʊˈleɪʃən] 脉冲编码调制

pulse-height analyzer 脉冲幅度分析器

pulser [ˈpʌlsə] 冲发生器

pump characteristic curve [ˌkærəktəˈrɪstɪk] [kɜːv] 泵特性曲线

punching 冲击

purge gas 吹扫气体

purification of coolant [ˌpjʊərɪfɪˈkeɪʃən] 冷却剂净化

purification section 净化段

push-button dial 按钮拨号

Q

quadrant power till ratio [ˈkwɒdr(ə)nt] 象限功率倾斜比

qualification certificate [ˌkwɒlɪfɪˈkeɪʃ(ə)n] [səˈtɪfɪkət] 鉴定合格证

qualification specification [ˌkwɒlɪfɪˈkeɪʃ(ə)n] [ˌspesɪfɪˈkeɪʃ(ə)n] 鉴定规格书

qualification test 鉴定试验

qualified assurance program 质保大纲

quality 含汽量，（蒸汽）干度

quality assurance [ə'ʃʊər(ə)ns] 质量保证
quality assurance record 质保记录
quality audit 质量监查
quality factor 质量因子
quality inspection 质量检查
quality inspector 质量检查员
quality plan 质量计划
quality qualification 质量鉴定
quality release 质量证书
quality trend analysis 质量趋势分析
quenching 骤冷
quoted standard 引用标准

R

raceway 电缆通道
radial and axial temperature profiles 径向和轴向温度分布
radial peaking factor 径向峰值因子
radial peaking hot spot factor 径向峰值热点因子
radial power tilt 径向功率倾斜
radial thermal expansion ['θɜːm(ə)l] [ɪk'spænʃ(ə)n] 径向热膨胀
radial vane 径向翼
radiation accident [reɪdɪ'eɪʃ(ə)n] ['æksɪdənt] 辐射事故
radiation background 辐射本底
radiation chemistry 辐射化学
radiation corrosion [kə'rəʊʒ(ə)n] 辐射腐蚀
radiation detector 辐射探测器
radiation effect 辐射效应
radiation map 辐射分区图
radiation monitor [reɪdɪ'eɪʃ(ə)n] 辐射监测仪
radiation monitoring 辐射监测
radiation monitoring channel 辐射监测通道
radiation monitoring system 辐射监测系统
radiation precaution sign [prɪ'kɔːʃ(ə)n] 辐射危险标志
radiation protection 辐射防护
radiation quantity ['kwɒntɪtɪ] 辐射量
radiation shielding 辐射屏蔽
radiation streaming 辐射漏束
radiation supervisor ['suːpəvaɪzə] 辐射防护监督人
radiation work permit 辐射工作许可证
radiation zone 辐射区
radio paging system 无线移动通信系统
radio transmission [trænz'mɪʃ(ə)n] 无线寻呼系统
radioactive concentration [ˌreɪdɪəʊ'æktɪv] 放射性浓度
radioactive fallout 放射性沉降物
radioactive hydrogenated effluent 放射性含氢排出液
radioactive material 放射性物质
radioactive waste solidification building ventilation system 固化厂房通

风系统
radioactive waste solidification system [ˌsəlɪdɪfɪˈkeɪʃən] 放射性废物固化系统
radioactivity 放射性
radioactivity contamination [kənˌtæmɪˈneɪʃən] 放射性污染
radiochemical analysis [reɪdɪəʊˈkemɪkəl] 放化分析
radio-chemistry [ˈkemɪstrɪ] 放射化学
radiographic examination [ˌreɪdɪəʊˈgræfɪk] 射线检验
radioisotope [ˌreɪdɪəʊˈaɪsətəʊp] 放射性同位素
radio-nuclide [ˈnjuːklaɪd] 放射性核素
rainfall intensity 暴雨强度
ramp insertion of reactivity 反应性线性引入
rapid shutdown 快速停堆
rate of dilution [daɪˈluːʃn] 稀释速率
raw water 原水
raw water pump 原水泵
raw water purification system [ˌpjʊərɪfɪˈkeɪʃən] 原水净化系统
reactivity accident 反应性事故
reactivity change 反应性变化
reactivity coefficient [ˌkəʊɪˈfɪʃ(ə)nt] 反应性系数
reactivity control component 反应性控制组件
reactivity feedback 反应性反馈
reactivity insertion rate 反应性引入组件
reactivity shutdown margin 反应性停堆裕量
reactivity worth of control rod 控制棒反应性价值
reactor building lighting system 反应厂房照明系统
reactor building water supply and drainage system 反应堆厂房给排水系统
reactor cavity 堆腔
reactor cavity cooling system 堆腔冷却系统
reactor containment 安全壳
reactor coolant drain tank 反应堆冷却剂疏水箱
reactor coolant pump 反应堆冷却剂泵
reactor coolant pump seal water system 反应堆冷却剂泵轴封水系统
reactor coolant system 反应堆冷却剂系统
reactor internals 堆内构件
reactor internals lifting device 堆内构件吊具
reactor power control system 反应堆功率控制系统
reactor pressure vessel 压力容器
reactor pressure vessel vent system 反应堆压力容器放气系统
reactor protect system 反应堆保护系统
reactor trip breaker 停堆断路器
reactor vessel head lifting device 反

应堆容器顶盖吊具
reactor water chemistry [ˈkemɪstrɪ] 反应堆水化学
readout and control cabinet 读出和控制柜
real time operating system 实时操作系统
real time spectroscopy detector [spekˈtrɒskəpɪ] 实时能谱探测器
realignment [ˌriːəˈlaɪnmənt] 重新校正（对中）
realistic source terms [rɪəˈlɪstɪk] 现实源项或预期源项
receiving tank 接收槽
reciprocating charging pump [rɪˈsɪprəkeɪtɪŋ] 往复式上充泵
recirculating cooling water 循环冷却水
recirculation phase [riːˌsɜːkju-ˈleɪʃən] 再循环阶段
recirculation system 循环水系统
recommended standard [rekəˈmendɪd] 推荐性标准
recorder 记录仪
recording level 记录水平
recurrence interval [ˈɪntəv(ə)l] 重现期
recycle 再循环
redundancy [rɪˈdʌnd(ə)nsɪ] 冗余度，冗余性，多重性
redundant measurement 多重测量
reentrant program [riːˈentrənt] 重入程序
reference block 基准试块
reference level 参考水平，基准液位
reference value 参考值，基准值
reference voltage 基准电压
refilling stage 再灌水阶段
reflector 反射层
reflector saving 反射层节省
refresh 刷新
refueling 换料
refueling machine 装卸料机
refueling operation 换料操作
refueling pool 换料水池
refueling water tank 换料水箱
regenerated boric acid solution 再生硼酸溶液
regenerative heat exchanger [rɪˈdʒen(ə)rətɪv] 再生热交换器
regular test 定期试验
regulatory guide 管理导则
regulatory requirement [ˈregjulətərɪ] 管理机构要求
reinforcing plate 补强板
relative biological effectiveness [baɪə(ʊ)ˈlɒdʒɪk(ə)l] 相对生物效应
relative humidity [hjʊˈmɪdɪtɪ] 相对湿度
relay actuator module 继电器驱动器组件
release for shipment 运输许可证
reliability 可靠性
relief valve 卸压阀
reload 换料
reload enrichment 换料浓度
remote control 遥控

removal tank 排放箱
repair welding 补焊
representative sample [reprɪ'zentətɪv] 代表性样品
residual heat [rɪ'zɪdjʊəl] 余热
residual heat removal exchanger 余热排出热交换器
residual heat removal pump [rɪ'muːv(ə)l] 余热排出泵
residual heat removal system 余热排放系统
residual sample 残余样品
resin addition tank 树脂添加箱
resin bed 树脂床
resin fine 树脂碎片
resin flush pump 树脂冲排泵
resin flush water 树脂冲排水
resin regeneration 树脂再生
resistance box [rɪ'zɪst(ə)ns] 电阻箱
resistance curve 阻力曲线
resistance temperature detector 电阻温度探测器
resistance to earth 对地电阻
resistance to irradiation damage [ɪ,reɪdɪ'eɪʃ(ə)n] 抗辐照损伤
resistor 电阻器
resolution 分辨，辨别，分解，甄别
resolution time 分辨时间
resonance ['rez(ə)nəns] 共振
resonance absorption [əb'zɔːpʃ(ə)n] 共振吸收
resonance capture 共振俘获
resonance escape probability 逃脱共振几率
resonance integral 共振积分
resonance neutron ['njuːtrɒn] 共振中子
resonance region 共振区
resonance width 共振宽度
resource 资源
respirator ['respɪreɪtə] 呼吸面具
response time 响应时间
responsible organization [rɪ'spɒnsɪb(ə)l] 责任单位
restraint 阻尼器
restricted area boundary [rɪ'strɪktɪd] ['baʊnd(ə)rɪ] 限制区边界
retention [rɪ'tenʃ(ə)n] 滞留，保存
retention area [rɪ'tenʃ(ə)n] 保存区
retention categories 保存分类
retention of records 记录保存
retention period 保存期
retrieval and accessibility [rɪ'triːvl] [əkˌsesə'bɪlətɪ] 检索和查阅
retrieval [rɪ'triːvl] 检索
return chilled water temperature 冷冻水回水温度
return fan 回风机
reusing water 复用水
reusing water pool 复用水池
reusing water pump 复用水泵
reusing water tank 复用水箱
reverse voltage [rɪ'vɜːs] ['vəʊltɪdʒ] 反向电压
rewelding 返修焊
rib（加强）肋
rigid dimple ['rɪdʒɪd] ['dɪmp(ə)l] 刚凸（格架的）

rigid frame 刚架
ring girder [ˈgɜːdə] 环梁
ring support 环形支架
rod cluster control assembly [əˈsemblɪ] 棒束控制组件
rod cluster control assembly and thimble plug assembly [ˈθɪmb(ə)l] [plʌg] 控制棒与阻力塞组件
rod cluster control assembly and thimble plug assembly changing fixture [ˈθɪmb(ə)l] 控制棒与阻力塞抽插机
rod drop 落棒
rod ejection 弹棒
rod ejection accident [ɪˈdʒekʃən] 弹棒事故
rod insertion 插棒
rod position detector [dɪˈtektə] 棒位探测器
rod position digital display 棒位数字显示器
rod position indicator assembly [ˈɪndɪkeɪtə] 棒位指示组件
rod position LED indicator 棒位光点指示器
rod travel housing 棒行程套管
rod withdrawal accident [wɪðˈdrɔː(ə)l] 提棒事故
roof ventilator [ˈventɪleɪtə] 屋顶通风器
root face 钝边
root gap 根部间隙
root mean square value [skweə] 均方根值
root valve 根阀
rotating seal ring 动密封环
routine inspection [ruːˈtiːn] [ɪnˈspekʃn] 例行检查
routine monitoring 常规监测
routine test 例行试验
rubber bumper 橡胶缓冲器
runway track 轨道梁
rupture [ˈrʌptʃə] 断裂，破裂
rupture disk [ˈrʌptʃə] 爆破盘
rupture strength 断裂强度
rupture test 断裂试验

S

saddle support [ˈsæd(ə)l] 鞍式支座
safe end 安全端
safe shutdown earthquake 安全停堆地震
safekeeping [ˌseɪf ˈkiːpɪŋ] 保卫
safety actuator [ˈæktjʊeɪtə] 安全驱动器
safety analysis report 安全分析报告
safety bank 安全组（控制棒）
safety brake 安全制动器
safety class 安全等级
safety class interface 安全等级的接口
safety code [kəʊd] 安全规定
safety factor 安全系数
safety injection [ɪnˈdʒekʃ(ə)n] 安

全注射
safety injection phase [feɪz] 安注阶段
safety injection pump [pʌmp] 安注泵
safety injection signal 安注信号
safety injection system [kəmˈpəunənt] 安注系统
safety injection tank [ɪnˈdʒekʃ(ə)n] 安注箱
safety lighting 安全照明
safety parameter display system [pəˈræmɪtə] 安全参数显示系统
safety rod 安全棒
safety systems 安全系统
safety valve 安全阀
safety-related component [kəmˈpəunənt] 安全相关设备
samarium poisoning [səˈmeərɪəm] [ˈpɒɪzənɪŋ] 钐中毒
sampler [ˈsɑːmplə] 取样器
sampling and detecting assembly 取样和探测装置
sampling cabinet [ˈkæbɪnɪt] 取样箱
sampling cooler 取样冷却器
sampling device 取样装置
sampling of document [ˈdɒkjʊm(ə)nt] 文件抽样
sampling resister [rɪˈzɪstə] 取样电阻
sampling system 取样系统
sand blastin 喷砂
sand filter [ˈfɪltə] 砂滤器
sandstorm 砂暴
sandy silt [sɪlt] 砂质粉土
sanitary sewer treatment system [ˈsuːə] 生活污水处理系统
sanitary sewer yard pipe network system [ˈsænɪt(ə)rɪ] 生活污水厂区管路系统
saturated boiling [ˈsætʃəreɪtɪd] 饱和沸腾
saturated steam/vapor [ˈsætʃəreɪtɪd] [ˈveɪpə] 饱和蒸汽
saturation index [ˈɪndeks] 饱和系数
saturation temperature 饱和温度
scale [skeɪl] 比例尺，结垢
scan rate 扫描速率
scattering [ˈskætərɪŋ] 散射
scattering cross section 散射截面
scattering law 散射定律
scattering matrix [ˈmeɪtrɪks] 散射矩阵
scavenger [ˈskævɪn(d)ʒə] 清洗剂
schedule 进度计划，时间表
scintillator [ˈsɪntɪleɪtə] 闪烁体
seal cover [siːl] 密封盖
seal housing 密封壳
seal water injecting filter 轴封注水过滤器
seal water reflux filter [ˈfɪltə] 轴封回流过滤器
seal water reflux heat exchanger 轴封回流热交换器
sealed plug and socket [ˈsɒkɪt] 密封接插件
sealed source [siːld] 密封源
sealing structure on bottom of the

refueling pool 换料水池底部密封结构

secondary auxiliary building ventilation system [ɔːgˈzɪlɪərɪ] [ˌventɪˈleɪʃ(ə)n] 二回路辅助厂房通风

secondary clock 子钟

secondary neutron source [ˈnjuːtrɒn] 次级中子源

secondary neutron source assembly 次级中子源组件

secondary radiation 次级辐射

secondary shielding [ˈʃiːldɪŋ] 二次屏蔽

secondary steam 二次蒸汽

secondary support assembly 防断组件

secondary support base plate [pleɪt] 防断底板（堆芯）

secondary support plate 防断支撑板

secondary treatment 第二级处理

sectional view [ˈsekʃ(ə)n(ə)l] 剖面图

sedimentation 沉淀

seismic category I [ˈsaɪzmɪk] 抗震I类

seismic classification [ˈsaɪzmɪk] 抗震分级

seismic design 抗震设计

seismic instrumentation system 地震仪表系统

seismic load 地震载荷

seismic risk 地震风险

seismic stabilizer bracket [ˈsteɪbɪlaɪzə] 抗震支架

seismotectonic province [ˌsaɪzmətekˈtɒnɪk] 地震构造区

selection of supplier 供方选择

selector 选择器

selector switch 选择开关

self-contained 成套的，整体式的

self-shielding 自屏蔽

self-shielding factor 自屏蔽因子

semaphore [ˈseməfɔː] 信号灯

semi-automatic welding 半自动焊

semiconductor detector [ˌsemɪkənˈdʌktə] 半导体探测器

semi-graphic display 半图形显示

sensible heat [ˈsensɪb(ə)l] 显热

separate system [ˈsep(ə)rət] 分流制

separator 分离器

septic tank [ˈseptɪk] 化粪池

sequence of events (events sequence) [ˈsiːkw(ə)ns] 事件顺序

service condition 使用工况

service power operation 带厂用电运行

service program 服务程序

service water yard pipe network system 生活给水厂区管系统

servomotor [ˈsɜːvəʊˌməʊtə] 伺服马达

setpoint 整定点

settlement joint 沉降缝

severe accident 严重事故

sewage wastewater engineering [ˈsuːɪdʒ] 排水工程

shaft seal assembly [əˈsemblɪ] 轴封组件

shared files 共享文件
shared resource 共享资源
shear force 剪切力
shear modulus [ʃɪə] [ˈmɒdjʊləs] 剪切弹性膜量
shear test 剪切试验
sheathed cable [ʃiːθt] 铠装热电偶
sheathing and shielding cable [ˈʃiːðɪŋ] [ˈʃiːldɪŋ] 铠装屏蔽电缆
sheave [ʃiːv] 滑轮
shell insulation [ɪnsjʊˈleɪʃ(ə)n] 筒身保温层
shell stream [kənˈsəʊl] 壳程
shell-and-tube heat exchanger 管壳式热交换器
shift (operation personnel) 倒班
shipping cask 运输容器
shipping cask cleaning pool 运输容器清洗池
shipping cask loading pool [ˈʃɪpɪŋ] 运输容器装料池
shock load 冲击载荷
shock resistance [rɪˈzɪst(ə)ns] 抗震能力
shock strength 抗冲击强度
shock test 冲击试验
short life isotope [ˈaɪsətəʊp] 短寿命同位素
short-lived fission product [ˈfɪʃ(ə)n] 短寿命裂变产物
shunt 分流器
shut down 停堆
shutdown bank/group 停堆组
shutdown margin [ˈmɑːdʒɪn] 停堆裕量
shutoff head 关闭扬程
side-by-side 并列
Sievert [ˈsiːvət] 希弗
silencer [ˈsaɪlənsə] 消音器
silty sand [ˈsɪltɪ] 粉砂
silver-indium-cadmium alloy 银-铟-镉合金
silver-plating 镀银
simplified logbook 简化的日志
simulated program [ˈsɪmjʊleɪtɪd] 模拟程序
single failure 单一故障
single failure criteria [kraɪˈtɪərɪə] 单一故障准则
single line diagram [ˈdaɪəgræm] 单线图
single phase flow 单相流
siphon filter [ˈsaɪf(ə)n] 虹吸滤池
sipping device 湿啜吸装置
site acceleration [əkseləˈreɪʃ(ə)n] 当地加速度
site installations [ɪnstəˈleɪʃ(ə)ns] 现场安装
site investigation [ɪnˌvestɪˈgeɪʃ(ə)n] 场地勘察
site personnel [pɜːsəˈnel] 厂区人员
site security system [sɪˈkjʊərɪtɪ] 厂区保安系统
site supervision 现场监督
skeleton [ˈskelɪt(ə)n] 骨架
skip sequence [skɪp] [ˈsiːkw(ə)ns] 跳焊
skirt support 裙式支座，筒式支座
sky shine 天空回散射
slave cycler 从动循环器

slave cycler count and decoder[sleɪv] [ˌdiːˈkəʊdə] 从动计数译码组件
slave cycler input module [ˈmɒdjuːl] 从动输入组件
slide way [slaɪd] 滑道
slightly enriched uranium [ɪnˈrɪtʃt] [jʊˈreɪnɪəm] 低浓缩铀
slip ring brush assembly 滑环电刷组件
sludge [ˈslʌdʒ] 污泥
sludge dewatering [ˈslʌdʒ] [diːˈwɔːtərɪŋ] 污泥脱水
sludge dry 污泥干化
sludge incineration [ɪnˌsɪnəˈreɪʃn] 污泥焚烧
sludge pressure filtration [fɪlˈtreɪʃn] 污泥压滤
sludge thickening [ˈθɪk(ə)nɪŋ] 污泥浓缩
slumping [ˈslʌmpɪŋ] 坍塌
snap switch 快速开关
snubber [snʌb] 阻尼器
sodium vapor lamp [ˈsəʊdɪəm] [ˈveɪpə] [læmp] 钠蒸汽灯
soil corrosion [kəˈrəʊʒ(ə)n] 土壤腐蚀
solenoid [ˈsəʊlənɒɪd] 电磁线圈
solid conductor 单股线
solid waste treatment and storage system [ˈstɔːrɪdʒ] 固体废物处理和存系统
solidification [ˌsəlɪdɪfɪˈkeɪʃən] 固化
sound limiting booth [buːð; buːθ] 隔离电话亭
source file 源文件
source inspection 产地检查
source program 源程序
source range 源量程
source rod [rɒd] 中子源棒
source terms 源项
source verification [ˌverɪfɪˈkeɪʃ(ə)n] 产地验证
source-term evaluation 源项评价
spacer grid 定位格架
span [spæn] 跨距
spatial power osciliation 功率空间振荡器
spatter 飞溅
special computer for radiation monitoring [reɪdɪˈeɪʃ(ə)n] 辐射监测专用计算机
special monitoring 特殊检查
special waste water mixed drainage system [ˈdreɪnɪdʒ] 特种废水混合排放系统
specialized standard 专业标准
specific activity 比活度
specific burnup [ˈbɜːnʌp] 比燃耗
specific gravity [ˈgrævɪtɪ] 比重
specific ion electrode [ˈaɪən] [ɪˈlektrəʊd] 离子选择电极
specific power 比功率
specific radioactivity [ˌreɪdɪəʊækˈtɪvɪtɪ] 比放
specification 规格书，设计说明书
specimen 试样
spectral hardening [ˈspektr(ə)l] 谱硬化
spectrophotometric method [ˌspektrəufəuˈtɒmɪtrɪk] 分光光

度法
speed reducer [rɪ'djuːsə] 减速器
spent fuel assembly handling tool [ə'semblɪ] 乏燃料组件操作工具
spent fuel assembly shipping cask [kɑːsk] 乏燃料组件运输容器
spent fuel cooling pump 乏燃料池冷却泵
spent fuel examination facility 乏燃料检测装置
spent fuel handling tool 乏燃料操作工具
spent fuel pool [fjʊəl] 乏燃料池
spent fuel pool cooling and cleanup system 乏燃料池冷却及净化系统
spent fuel pool cooling water heat exchanger 乏燃冷却器
spent fuel pool purification pump [ˌpjʊərɪfɪ'keɪʃən] 乏燃池净化泵
spent fuel storage pool 乏燃料储存池
spent fuel storage rack 乏燃料储存格架
spent resin collection tank 废树脂收集箱
spherical shell ['sferɪk(ə)l] 球形壳体
spider assembly ['spaɪdə] 连接柄组件
splice box [splaɪs] 分线盒，电缆接头
spray additive tank ['ædɪtɪv] 喷淋添加剂箱
spray ejector [ɪ'dʒektə] 喷淋喷射器
spray header [spreɪ] 喷淋集管
spray heat exchanger 喷淋热交换器
spray line 喷淋管道
spray nozzle 喷嘴
spray pump 喷淋泵
sprayer ['spreə] (稳压器) 喷雾器
spread footing 单独基础
springing line 起拱线
square butt welding ['weldɪŋ] 无坡口对接焊
squirrel-cage motor ['skwɪr(ə)l] 鼠笼电动机
stabilized electrical supply ['steɪbɪlaɪzd] 稳压电源
stack monitor 烟囱监测仪
stage casing 中段泵壳
standard penetration test [penɪ'treɪʃ(ə)n] 标准贯入试验
start up 启动
static balance ['stætɪk] 静平衡
static parameter [pə'ræmɪtə] 静态参数
static pressure 静态压力
station blackout 全厂失电
stationary gripper coil ['grɪpə] [kɒɪl] 传递线圈
stationary pole ['steɪʃ(ə)n(ə)rɪ] 传递磁极
stationary seal ring 静密封环
stator ['steɪtə] 定子
steady state 稳态
steady state operation 稳态运行
steam (bypass) dump system

[dʌmp] 蒸汽旁排系统
steam cushion [ˈkʊʃ(ə)n] 汽垫
steam dryer [ˈdraɪə] 蒸汽干燥器
steam generator 蒸发器
steam generator blowdown system [ˈdʒenəreɪtə] 蒸发器排污系统
steam generator layup system [ˈleɪʌp] 蒸发器保养系统
steam manifold [ˈmænɪfəʊld] 蒸汽总管
steam quality 蒸汽干度
steam restrictor [rɪsˈtrɪktə] 蒸汽限流器
steel support beam 钢支撑梁
steep performance curve [kɜːv] 陡斜性能曲线
stellite clad journal [klæd] [ˈdʒɜːn(ə)l] 钨铬钴合金堆焊的轴颈
stem seal [siːl] 阀杆密封
step insertion of reactivity [ɪnˈsɜːʃ(ə)n] 反应性阶跃引入
step length 步距
step load change 负荷阶跃变化
step load increase [ɪnˈkriːs] 负荷阶跃升高
step load reduction 负荷阶跃降低
stepdown transformer 降压变压器
stepladder [ˈsteplædə] 直梯
stepless speed control 无级调速控制
stepup transformer 升压变压器
stiffening ring [ˈstɪfnɪŋ] 加强环
stochastic effect [stəˈkæstɪk] 随机效应
stone chipping [ˈtʃɪpɪŋ] 小石子
stop 止挡块，停止
storage facility [fəˈsɪlɪtɪ] 储存设施
storage reservoir/tank [ˈrezəvwɑː(r)] 储水池
stored program control 存储程序控制
storm runoff [ˈrʌnɒf] 暴雨水量
strain rate [streɪn] 应变速率
strainer [ˈstreɪnə] 粗滤器
strand [strænd] 钢绞线
stratigraphy [strəˈtɪgrəfɪ] 地层
stress concentration [kɒns(ə)nˈtreɪʃ(ə)n] 应力集中
stress corrosion cracking [kəˈrəʊʒ(ə)n] 应力腐蚀开裂
stress intensity [ɪnˈtensɪtɪ] 应力强度
stress range 应力变化范围
stress relaxation 应力松弛
stress relief [rɪˈliːf] 应力消除
stretching gallery [ˈstretʃɪŋ] [ˈgæl(ə)rɪ] 张拉廊道
strip footing [strɪp] 条形基础
structural shield [ˈstrʌktʃ(ə)r(ə)l] 结构屏蔽
stud tensioner [ˈtenʃənə] 螺栓拉伸机
studhole plug handling tool 螺孔塞装卸工具
sub-channel [ˈtʃæn(ə)l] 子通道
subcompartment 分隔间
sub-contractor [ˈkɒntræktə] 分包商
subcooled blowdown [ˌsʌbˈkuːld]

[ˈbləudaʊn] 过冷喷放
subcritical core [sʌbˈkrɪtɪk(ə)l] [kɔː] 次临界堆芯
sub-item 子项
submerged arc welding [səbˈmɜːdʒd] [ˈweldɪŋ] 埋弧焊
submission of document [səbˈmɪʃ(ə)n] [ˈdɒkjʊm(ə)nt] 文件递交
subsection [ˈsʌbsekʃ(ə)n] 分卷
suction well [ˈsʌkʃ(ə)n] 吸水井
sulfur hexafiuoride breaker [ˈsʌlfə] 六氟化硫断路器
sump pump 污水泵
supervised area 监督区
supplier evaluation [ɪˌvæljuˈeɪʃn] 供方评价
supply chilled water temperature [tʃɪld] 冷冻水供水温度
supply fan 送风机
support 支座
support bracket [ˈbrækɪt] 支撑架
support column [ˈkɒləm] 支承柱
support point 支撑点
supporting plate [pleɪt] 支承板
suppressor [səˈpresə] 抑制器
surface activity 表面活度
surface carburization [ˌkaːbjuraɪˈzeɪʃən] 表面渗碳
surface contamination survey meter [kənˌtæmɪˈneɪʃən] 表面污染测量仪
surface dose rate [dəʊs] 表面剂量率
surface radioactivity contamination [ˌreɪdɪəʊækˈtɪvɪtɪ] [kənˌtæmɪˈneɪʃən] 表面放射性污染
surge nozzle [sɜːdʒ] 波动管接管
surge protection 冲击（电压）保护
surge tank 波动箱
surveillance personnel [səˈveɪl(ə)ns] 监督人员
surveillance plan 监督计划
surveillance program 监督大纲
suspended solid [səˈspendɪd] [ˈsɒlɪd] 悬浮物
swelling [ˈswelɪŋ] 肿胀
swirl vane separator 旋叶式分离器
switch over 切换
switchboard (distribution) 配电盘
switchgear & main electrical equipment fire protect system 电气配电装置及主要电气设备消防系统
switchgear [ˈswɪtʃgɪə] 开关装置，配电装置
switchyard 开关站
symmetrical section [sɪˈmetrɪk(ə)l] [ˈsekʃ(ə)n] 对称截面
synchronized operation [ˈsɪŋkrənaɪzd] 同步操作
system generation [dʒenəˈreɪʃ(ə)n] 系统生成
system load 系统负荷
system operation pressure 系统运行压力
system software 系统软件

T

tack welding [ˈweldɪŋ] 定位焊
tap 分接头，抽头
tap changer 抽头切换开关
tape reader 磁带读出器
tape recorder 磁带录音机
tapered valve [ˈteɪpəd] 针型阀
technical support center 技术支援中心
telescopic tube [telɪˈskɒpɪk] 伸缩套筒
tempering 回火
temporary wiring [ˈtemp(ə)rərɪ] 临时接线
ten path selector 10 路选择器
tendon 钢筋束
tension link 活节拉杆
tension ring 拉紧环
tension structure on top of the CRDMS 驱动机构顶部拉紧装置
terminal [ˈtɜːmɪn(ə)l] 终端，接线端子
terminal box 端子盒
testing program 测试程序，试验大纲
The Institute of Electrical and Electronic Engineering 美国电气与电子工程师协会
thermal barrier [ˈθɜːm(ə)l] [ˈbærɪə] 热屏
thermal capture [ˈkæptʃə] 热中子俘获
thermal expansion [ˈθɜːm(ə)l] [ɪkˈspænʃ(ə)n] 热膨胀
thermal insulation [ɪnsjʊˈleɪʃ(ə)n] 热绝缘
thermal mass flowmeter [ˈfləʊmiːtə] 量热式质量流量计
thermal neutron flux 热中子通量
thermal plume 热烟羽
thermal relay [ˈriːleɪ] 热继电器
thermal shield [ʃiːld] 热屏蔽
thermal shock 热冲击
thermalization [ˌθɜːməlaɪˈzeɪʃən] 热能化
thermocouple [ˈθɜːməʊkʌp(ə)l] 热电偶
thermocouple alloy [ˈθɜːməʊkʌp(ə)l] [ˈælɒɪ] 热电偶合金
thermocouple column [ˈkɒləm] 热电偶导管
thermocouple conduit seal [ˈkɒndjʊɪt；ˈkɒndɪt] 热电偶导管密封
thermoluminescene dosimeter [dəʊˈsɪmɪtə] 热释光剂量计
thermoluminescent detector [ˈθɜːməʊˌljuːmɪˈnesənt] 热释光探测器
thimble plug assembly [ˈθɪmb(ə)l] 阻力塞组件
thimble tube 套管
thin skirt 薄壁（裙）筒
three dimensional [daɪˈmenʃənl] 三维

three valves manifold [ˈmænɪfəʊld] 三阀组
three-pin socket [ˈsɒkɪt] 三芯插座
threshold [ˈθreʃəʊld] 阈值
throttle device [ˈθrɒt(ə)l] [dɪˈvaɪs] 节流装置
throttle valve [ˈθrɒt(ə)l] 节流阀
throttling orifice [ˈɒrɪfɪs] 节流孔板
throughput 吞吐量
thrust bearing 推力轴承
time constant 事件常数
time delay circuit [dɪˈleɪ] [ˈsɜːkɪt] 延时电路
time division multiplex [dɪˈvɪʒ(ə)n] [ˈmʌltɪpleks] 时分多路
time out 超时
time slice [slaɪs] 时间片
toe of weld [weld] 焊趾
top nozzle [ˈnɒz(ə)l] 上管座
torque motor [tɔːk] 力矩电机
total enthalpy rise factor [ˈenθ(ə)lpɪ] 总焓升因子
total head 总扬程
total peaking factor [ˈpiːkɪŋ] 总峰值因子
total power peaking factor 总功率峰值因子
total pressure 全压
trace element [treɪs] 微量元素
tracer ball 跟踪球
track ball 滚球
track/store module 跟踪/储存组件
train 序列
transfer canal/tube [kəˈnæl] 运输通道
transfer of records 记录移交
transfer switch 切换开关
transfer tank 扬液器
transgranular stress corrosion cracking [trænsˈgrænjulə] 穿晶应力腐蚀
transient [ˈtrænzɪənt] 瞬态
transition cone 锥形过渡端
transition ring 过渡段
transition temperature 转变温度
transmittion antenna [ænˈtenə] 发射天线
transport theory 移理论
transverse seam [trænzˈvɜːs] [siːm] 横（焊）缝
transverse stiffness [ˈstɪfnɪs] 横向刚度
trap [træp] 疏水器，捕集器，存水弯管
traveling speed 移动速度
trip 紧急停堆，跳闸
trip (circuit breaker) 跳闸
tritium sampler [ˈtrɪtɪəm] [ˈsɑːmplə] 氚取样器
trolley frame [ˈtrɒlɪ] [freɪm] 小车架
trough duct [trɒf] 电缆沟槽
trunk line 中分线
truss [trʌs] 桁架
tube bundle [ˈbʌnd(ə)l] 管束
tube lane 管廊
tube rolling (or expanding) 胀管
tube sheet 管板
tube spacing (or pitch) 管子间距
tube stream 管程

tube support plate 管子支承板

tubing 装管

turbine building and water treatment building fire protection system 轮机厂房和水处理厂房消防系统

turbine building HVAC system 汽机厂房暖通空调系统

turbine extraction steam system 汽轮机抽汽系统

turbine inlet valve [ˈɪnlet] 汽机进汽阀

turbine power and frequency regulating system [ˈfriːkw(ə)nsɪ] 汽轮机功率和频率调节系统

turbine regulation and control 汽轮机调节与控制

turbine runback module [ˈmɒdjuːl] 汽轮机降功率组件

turbulent flow [ˈtɜːbjʊl(ə)nt] 紊流，湍流

turning gearing (barring gear) [ˈgɪərɪŋ] 盘车装置

turn-key contractor [kənˈtræktə] 交钥匙承包商

two group theory 两群理论

two phase flow 两相流

typical cell 典型栅元

U

ultimate heat sink [ˈʌltɪmət] 最终热阱

ultimate heat sink cooling tower and water storage tank system 最终热阱冷却塔储水池系统

ultrasonic decontaminating tank 超声波去污槽

ultrasonic decontamination [ʌltrəˈsɒnɪk] [ˈdiːkənˌtæmɪˈneɪʃən] 超声去污

ultrasonic examination [ʌltrəˈsɒnɪk] 超声检验

ultrasonic liquid level meter 超声波液位计

uncontrolled release 失控释放

uncontrolled rod withdrawal accident [wɪðˈdrɔː(ə)l] 控制棒失控提升事故

underclad cracking 堆焊层下裂纹

undercut 咬边

underground wiring 地下配线

underload 低载，轻载

underwater lighting device 水下照明装置

underwater lights 水下照明灯具

underwater transfer tube 水下运输通道

uniform corrosion 均匀腐蚀

uninterruptible power supply 不间断电源

unit weight 重度

unsealed source 开放源

upending device 倾翻机

upper core plate 堆芯上板

upper core support structure 堆芯上部构件

upper head 上封头
upper internals storage stand 压紧部件存放架
upper shell 上筒体
upper support column 上支撑筒
upsetting 锻粗
upstream 上游
utility program 应用程序
U-tube bundle [ˈbʌnd(ə)l] U形管束

V

vacuum break valve [ˈvækjʊəm] 真空破坏阀
vacuum cleaner 真空吸尘器
vacuum degasifier [dɪˈgæsɪfaɪə] 真空脱气器
vacuum pump 真空泵
valve bonnet 阀盖
valve disc 阀盘
valve position meter 阀位指示器
valve seat 阀座
valve stem 阀杆
valve trim 阀芯
vane axial fan 轴流风机
vapor extractor [ɪkˈstræktə] 排风机
vapor phase 汽相
variable area flow converter 转子流量转换器
variable area flowmeter 转子流量计
variable voltage & variable frequency speed regulation 变压变频调速
velocity pressure (or head) 速度压头
vent cabinet (or ventilation cabinet) [ˈkæbɪnɪt] [ˌventɪˈleɪʃ(ə)n] 通风柜
vent condenser 排器冷凝器
vent pipe 通风管
vent stack 排气烟囱
vent valve 放气阀
ventilated protective suit [ˈventɪleɪtɪd] 通风防护服
ventilated suit 防护服，气衣
ventilation control room system [ˌventɪˈleɪʃ(ə)n] 通风控制室系统
verification [ˌverɪfɪˈkeɪʃ(ə)n] 验证
vertical pipe 立管
very soft silty clay 淤泥质粉质黏土
vessel support bracket insulation [ˈves(ə)l] [ɪnsjʊˈleɪʃ(ə)n] 容器支架保温层
vibration frequencies 振动频率
viewing window 观察窗
vinyl paint [ˈvaɪnɪl] 乙烯基漆
visual display 荧屏显示器
visual inspection report 目视检查报告
visual inspection (or examination) [ɪnˈspekʃn] 目视检查
vital instrumentation power supply [ˌɪnstrʊmenˈteɪʃ(ə)n] 重要仪表

电源
void ratio 空泡比
void volume 空腔体积
voltage ratio 变压比
voltage tolerance [ˈtɒl(ə)r(ə)ns] 允许电压偏差
volume average effective stress 体积平均有效应力
volume change 容积变化
volume control tank 容控箱
volume defect 体积缺陷
volumetric examination [ˌvɒljʊˈmetrɪk] [ɪgzæmɪˈneɪʃ(ə)n] 体积检查
voting logic module 表决逻辑组件

W

warning sign 警告标志
wastage 耗蚀
waste collecting tank (waste collector tank) 废物收集箱
waste decay tank 废物衰变箱
waste gas after-cooler 废气后置冷却器
waste gas compressor [kəmˈpresə] 废气压缩机
waste gas pre-cooler 废气前置冷却器
waste gas sampling box 废气取样箱
waste hold-up tank 废物暂存箱
waste sampling box 废物取样箱
wastewater flow [ˈweɪstˌwɔːtə] 污水量
water chemistry [ˈkemɪstrɪ] 水化学
water consumption [kənˈsʌm(p)ʃ(ə)n] 用水量
water content 含水量
water inventory [ˈɪnv(ə)nt(ə)rɪ] 水装量
water level 水位
water level transmitter 液位变送器
water line corrosion [kəˈrəʊʒ(ə)n] 水线腐蚀
water proofing material 防水材料
water quality 水质
water seal 水封
water sprinkler [ˈsprɪŋklə] 水喷淋器
water supply system 给水系统
water treatment 水处理
water treatment building HVAC system 水处理厂房暖通空调系统
water-steam equilibrium [ˌiːkwɪˈlɪbrɪəm] 水-汽平衡
watt-hour meter 电度表
wattles power meter 无功功率表
wattmeter [ˈwɒtmiːtə] 功率表
wear allowance 磨损裕量
weight percent 重量百分比
weighting factor 权重因子
weld metal 焊缝金属
weld seam 焊缝
weldability 可焊性
welder qualification [ˌkwɒlɪfɪˈkeɪʃ(ə)n] 焊工资格

评定
welding by both sides 双面焊
welding by one side 单面焊
welding condition 焊接条件
welding parameter [pə'ræmɪtə] 焊接工艺参数
welding procedure qualification [ˌkwɒlɪfɪ'keɪʃ(ə)n] 焊接程序鉴定
welding process 焊接过程
welding sequence ['siːkw(ə)ns] 焊接顺序
welding symbols ['sɪmbəls] 焊缝代号
welding technique 焊接技术
welding technology [tek'nɒlədʒɪ] 焊接工艺
welding with flux backing 焊剂垫焊
wet bulb temperature 湿球温度
wet deposition [ˌdepə'zɪʃ(ə)n] 湿沉降
wet layup 湿保养（蒸发器）
whirlpool pump ['wɜːlpuːl] 旋涡泵
whole body contamination monitor [kənˌtæmɪ'neɪʃən] 全身污染监测仪
whole-body counter 全身计数器
winch [wɪn(t)ʃ] 卷扬机
Winchester disk 温（磁）盘
winding 绕组
wire fence 铁丝网
wire-wireless switching controller 有线、无线转换器
withdrawable [wɪð'drɔːəbl] 可拔出的
withdrawal speed(rate) 提棒速度(率)
withstand voltage [wɪð'stænd] 耐压
witness point 见证点
work order 工作指令
work schedule ['ʃedjuːl] 工作进度表
working authorization [ɔːθəraɪ'zeɪʃ(ə)n] 工作许可(证)
working instruction [ɪn'strʌkʃ(ə)n] 操作规程，工作细则
working medium 工作介质
workload 工作负荷
worm gear reducer 蜗轮减速器
wrapper 围筒

X

xenon buildup ['zenɒn] 氙积累
xenon equilibrium [ˌiːkwɪ'lɪbrɪəm] 平衡氙
xenon poisoning 氙中毒

Z

zero power 零功率
zircalloy cladding tube ['zɜːkəlɒɪ]

锆合金包壳管

zircalloy-4 锆-4 合金

zirconium-water reaction [zɜːˈkəʊnɪəm] 锆—水反应

zone-selected controller 区域选择控制器

zoning 分区

水能发电

"C" clamp C形夹

"O" -ring "O" 形环

"T" connection; junction (tee); 3-way 三通

230kV side overcurrent protection 高压侧过流保护

230kV side zero-sequence overcurrent protection 高压侧零序过流保护

3 way ball valve (lockable) 三相球阀（可锁定的）

33kV side overcurrent protection 中压侧过流保护

33kV side zero-sequence overcurrent protection 中压侧零序过流保护

3-phase wind asynchronous motor [ə'sɪŋkrənəs] 三相绕线式电动机

A

aberrance [æ'berəns, æ'berənsɪ] 畸变

abrasion of water turbine [ə'breɪʒ(ə)n] 水轮机磨损

abrasion performance 耐磨性

absolute pressure 绝对压力

acceptance test 验收试验

access of pit 机坑进人门

accessories 附件

accident of turbine 水轮机事故

accumulator [ə'kjuːmjʊleɪtə] 蓄能器

active value 动作值

adjustable spanner; monkey wrench 活动扳手

aging resistance [rɪ'zɪst(ə)ns] 抗老化性

air admission drainage pipe 大轴中心补气排水管

air admission system 补气系统

air admission test 补气试验

air compressor 空压机

air cooler 空冷器

air cooler ring pipe 空冷器环管

air driving spanner 风动扳手

air filter 空气过滤器

air passage 进风槽

air receiver 储气罐

air shovel 风铲

air shroud 挡风板

air supply pipe 供气管

air tank for braking system 制动气罐

air tight test 气密性试验

air ventilation [ˌventɪ'leɪʃ(ə)n] 通风

allowable variation 允许偏差

ambient humidity ['æmbɪənt]

环境湿度

ambient temperature 环境温度

anchor bolt [ˈæŋkə] 基础螺栓，地脚螺栓

anti-corona section of the bar [kəˈrəʊnə] 线棒防晕段

anti-corrosion 抗蚀

anticorrosion [ˌæntɪkəˈrəʊʒən] 防腐

anti-fire cloth 防火布

anti-rust paint 防锈漆

apply resin for the bar lining paper 线棒衬纸刷胶

assembly block cutting 组装把合块切割

assembly pedestal for the hub [ˈpedɪst(ə)l] 中心体安装支墩

assembly test 装配试验

asynchronous [əˈsɪŋkrənəs] 异步

atmpspheric pressure 环境压力（或大气压）

automatic control of hydrogenerator set 水轮发电机组自动控制系统

automatic element 自动化元件

automatic instrumentation 自动化仪器

automatic operation 自动操作

automatic switch 自动开关

automatic synchronizer 自动同期

automatic turbine run up system 水轮机自动起转系统

automatic-restoration [restəˈreɪʃ(ə)n] 自恢复

auxiliary contact [ɔːgˈzɪlɪərɪ] 辅助触点

auxiliary equipment 辅助设备

auxiliary equipment of hydraulic turbine 水轮机辅助设备

average current 均流

average voltage 均压

axial-flow adjustable blade propeller turbine 轴流转桨式水轮机

axial storage pump 轴流式蓄能泵

axial turbine 轴流式水轮机

axiality [ˌæksɪˈælətɪ] 同轴度

axis displacement 轴向移位

B

Babbit metal 巴氏合金

back gouging 背面清根

back strip 背板

backfilling；backfill 回填

backing plate 背缝垫板

baffle plate 导流板

balance pipe 均压管

balancing allowance 静平衡余量

ball valve 球阀

ballast 配重

band 下环

bar high voltage test；HIPOT 线棒耐压试验

bar insulation 线棒绝缘

barb bolt 地脚螺栓

base 底座

base plate 底板

base ring 瓦架

basic parameters of hydrogenerator 水轮发电机基本参数
battery charger 蓄电池充电器
bearing box 轴承箱
bearing bush 轴瓦，轴套
bearing carrier 轴承座
bearing collar 轴承环，推力头
bearing control 轴承控制
bearing cover 轴承盖
bearing housing 轴承座
bearing journal 轴颈
bearing neck 轴承挡油环
bearing segment; guide bearing shoe 轴瓦
bend radius 弯曲半径
bending strength 抗弯强度
best center 最佳中心
bifurcation [baɪfəˈkeɪʃ(ə)n] 喷嘴支管
biplane butterfly valve 平板蝶阀
blade 叶片
blade angle 叶片安放角
blade force character 叶片力特性
blade opening 叶片开口
blade rotating angle 叶片转角
blade tilt angle 叶片倾角
bottom outlet 底孔
bottom outlet diversion 底孔导流
bottom ring/cover 底环
box for guide bearing oil pump 水导轴承油泵坑衬
box for head cover drainage pump 顶盖排水泵坑衬
box for servomotor 接力器坑衬
box opening for the bar 线棒开箱
brake 风闸，制动器
brake control board 风闸控制板
brake nozzle 制动喷嘴
brake valve 制动阀
braking system of hydrogenerator 水轮发电机制动系统
branch pipe 叉管
break down voltage 击穿电压
breakdown maintenance [ˈmeɪnt(ə)nəns] 故障维修
breaker 断路器
breaker fail protection 失灵保护
bridge 电桥
brush 电刷
bucket 水斗
buffer; buffer device 缓冲装置
bulb 灯泡体
bulb support 灯泡体支柱
bulb turbine 灯泡式水轮机
busbar 母线
busbar coulper breaker charge protection 母联充电保护
busbar coupler breaker failure 母线失灵保护
busbar coupler breaker overcurrent protection 母联过流保护
busbar coupler breaker protection panel 母联继电器保护柜
busbar coupler incomplete phase protection 母联非全相保
busbar dead-zone protection 死区保护
busbar differential [ˌdɪfəˈrenʃ(ə)l] 母线差动
busbar protection panel 母线保

护屏

bushing 轴衬；轴套

butterfly valve 蝴蝶阀

by-pass valve 旁通阀

C

cable box 电缆盒

cable casing 电缆套

cable clamp/clip 电缆夹

cable cutter 电缆切割机

cable duct 电缆沟

cable gallery 电缆廊道

cable pipe 电缆管

cable rack/stand 电缆架

cable terminal end 电缆终端

cable tray 电缆盘

calibration block [kælɪˈbreɪʃ(ə)n] 校准试块

capacity of storage battery 蓄电池容量

cavitation [ˌkævɪˈteɪʃ(ə)n] 空化；空蚀

cavitation factor/coefficient of storage pump 蓄能泵空化系数

cavitation factor/ coefficient of turbine 水轮机空化系数

cavitation inception factor 初生空化系数

cavitation margin 空化裕量

cavitation of hydraulic turbine [ˌkævɪˈteɪʃ(ə)n] 水轮机空蚀

cavitation performance of hydraulic turbine 水轮机气蚀特性

cavitation test 空化试验

center line 中心线

centrifugal storage pump; mixed-flow storage pump [ˌsentrɪˈfjuːg(ə)l] 混流式（离心式）蓄能泵

characteristic of hydraulic turbine [kærəktəˈrɪstɪk] 水轮机特性

characteristic test 特性试验

check valve 止回阀

chock 垫块

circuit breaker failure 断路器失灵保护

circuit ring 汇流铜环

circuit ring neutral point [ˈnjuːtr(ə)l] 汇流铜环中性点

circular "H" steel; circular channel steel 挡风板环形工字钢

circular inlet-water pipe 进水环管

circular return-water pipe 回水环管

circumferential and longitudinal seam [səˌkʌmfəˈrenʃəl] [ˌlɒn(d)ʒɪˈtjuːdɪn(ə)l] 环纵焊缝

circumferential rib 环筋

clad weld 被覆熔接

clamp 卡头，夹板

clamping ring 卡环

clearance in diameter 总间隙

clearing of fault 故障排除

clockwise 顺时针

close order 合闸命令

closing edge 导叶关闭接触线

CO_2 shielding welding CO_2 保护焊

column-type insulator 支柱绝缘
combination device
[kɒmbɪ'neɪʃ(ə)n] 协联装置
(turbine) combined characteristic curve (水轮机) 综合特性曲线
combined condition 协联工况
commissioning [kə'mɪʃənɪŋ] 调试
commissioning test 调试试验
communication 通信，通讯
compensator 补偿器 (节)
complete characteristics of pump-turbine 水泵水轮机全特性
compressed air 压缩空气
compressed air piping for blow down 调相压气管路
compressing plate 压板
concentricity [kɒnsen'trɪsɪtɪ] 同心度
concentricity deviation [diːvɪ'eɪʃ(ə)n] 同心度偏差
conduct grease 导电膏
conductivity 导电率
conical draft tube 锥形尾水管
connect; connection 连接
connection box for head cover pressure relief 顶盖卸压管连接座
connection shaft/rod 连接杆
conservator ['kɒnsəˌveɪtə, kən'sɜːvətə] 油枕
constant close contact 常闭触点
constant open contact 常开触点
contact agent 接触剂
contact; contactor 接触器
container 集装箱
continuance current [kən'tɪnjʊəns] 持续电流
control and hydraulic unit for braking and lifting 制动及顶起系统控制柜
control cable 控制电缆
control circuit 控制电路
control method 控制方式
conventional test [kən'venʃ(ə)n(ə)l] 常规试验
cooling liquid ['lɪkwɪd] 冷却液
cooling water equipment room 冷却水设备室
cooling water piping for heat exchanger 空水冷却器水管
core 铁芯
core lamination [ˌlæmɪ'neɪʃən] 铁芯片
core lost 铁损
corona [kə'rəʊnə] 电晕
corona loss 电晕损失
corrosion protection [kə'rəʊʒ(ə)n] 防腐保护
coupling flange 连接法兰
cover for test ring 环筒盖
crank; lever 拐臂
critical cavitation factor ['krɪtɪk(ə)l] 临界空化系数
cross-flow turbine 双击式水轮
crosshead 操作架
crown 上冠
CT abnormal protection CT 异常保护
current breaker (CB) 电流断路器
current divider 分流器
current intensity [ɪn'tensɪtɪ] 电流

强度

current transformer（CT） 电流互感器

current-voltage characteristics [ˌkærəktəˈrɪstɪks] 电流电压特性

cushioning closing [ˈkʊʃənɪŋ] 分段关闭

cushioning pipe 分段关闭管路

cylinder；oil cylinder 油缸

cylindrical valve 圆筒阀

D

damper connection 阻尼连接

damper；absorber；cushion 阻力器，减震器

datum mark 基准点

DC resistance 直流电阻

DC resistance checking 直流电阻试验

DC resistance tester 直流电阻测试仪

dead band/ zone 死区

decompression plate 转轮减压板

de-excitation 失磁

delay 延时

delayed sweep 延时扫描

delivery valve 输水阀

depth of elbow draft tube 肘形尾水管深度

Deriaz turbine 斜流转桨式水轮机

design head 设计水头

design of hydraulic turbine 水轮机设计

dewatering header pipe 尾水层排水总管

diagonal storage pump 斜流式蓄能泵

diagonal turbine 斜流式水轮机

diesel generator [ˈdiːz(ə)l] 柴油发电机

difference 误差

differential pressure transducer 压差传感器

diffuser 扩散管

direct current 直流电（DC）

direct current electorde positive (DCEP) 直流电极接正

direct current electrode negative (DCEN) 直流电极接负

direction of rotation 旋转方向

discharge specific speed of storage pump 蓄能泵（水泵水轮机的泵工况）的流量比转速

discharge voltage 放电电压

disconnected switch 隔离开关

displacement sensor 位移传感器

distance from end of bar to stator core 线棒电接头离铁芯的距离

distance protection 距离保护

distinguishing rate [dɪˈstɪŋgwɪʃɪŋ] 分辨率

distributing valve 配水阀

distributor centerline 导水机构中心线

distributor system [dɪˈstrɪbjʊtə] 导水机构

dovetail bar [ˈdʌvteɪl] 定位筋
dovetail bar strap 定位筋托板
dowel pin 定位销
draft tube 尾水管
draft tube blow down level monitoring 尾水管调相压水液位监测
draft tube cone 尾水管锥管
draft tube cone pressure pulsation measuring point 尾水锥管压力脉动测点
draft tube cooling water drainage pipe 尾水管冷却水排水管
draft tube cooling water intake pipe 尾水管冷却水取水管
draft tube diffuser outlet pressure measuring point 尾水管扩散段出口压力测点
draft tube diffuser side wall 尾水管扩散段侧墙
draft tube drainage pipe 尾水排水管
draft tube elbow 尾水管肘管
draft tube elbow 尾水肘管
draft tube elbow pressure pulsation measuring point [pʌlˈseɪʃən] 尾水肘管压力脉动测量点
draft tube gate 尾水门
draft tube inlet pressure 尾水管进口压力
draft tube inlet pressure measuring panel 尾水管进口压力测量仪表盘
draft tube inlet pressure measuring point 尾水管进口压力测量点
draft tube liner 尾水管里衬
draft tube mandoor 尾水管进人门
draft tube outlet part 尾水管扩散段
draft tube outlet pressure measuring panel 尾水管出口压力测量仪表盘
draft tube pier 尾水管支墩
draft tube pressure pulsation [pʌlˈseɪʃən] 尾水管压力脉动
drainage canal for emptying of coolers 发电机冷凝排水
drainage pipe 排水管
drilling machine with magnetic base; magnetic base drilling machine 磁力钻
dry test 无水试验
duct lamination 通风槽片
dust protection 防尘
dynamic mode 动态控制
dynamic test 动态试验
dynamoelectric spanner; motor spanner 电动扳手

E

earth electrode [ɪˈlektrəʊd] 地电极
earth potential [pə(ʊ)ˈtenʃ(ə)l] 地电势
earthing switch 接地开关
earthing transformer 接地变
eccentric cam 偏心凸轮

eccentricity 偏心，偏心率，偏心度

echo 回波

effective head 有效水头

efficiency hill diagram （水轮机）综合特性曲线

efficiency test 效率试验

elastic bearing 弹性轴承

elastic deformation [ˌdiːfɔːˈmeɪʃ(ə)n] 弹性变形

elastic layer 弹性垫层

elastic limit 弹性极限

elastic material 弹性材料

elbow draft tube 肘形尾水管

electric discharge；discharge 放电

electric leakage 漏电

electric lighting 电气照明

electric loss 电损耗

electric quantity transducer 电量变送器

electrical brake 电制动器

electrical braking 电制动

electricity generation operating condition 发电工况

electromagnet [ɪˌlektrə(ʊ)ˈmægnɪt] 电磁铁

electromagnetic interference [ɪˌlektrə(ʊ)mægˈnetɪk] [ɪntəˈfɪər(ə)ns] 电磁干扰

electromechanical installation [ɪˌlektrəumɪˈkænɪk(ə)l] 机电安装

elevation 高程

elevation difference 高差

ellipse head [ɪˈlɪps] 封头曲面

ellipsed of inclined jet （when reaching the wheel vane） 射流椭圆

embedded component 埋入部件

embedded part 预埋件

embedded plate 预埋板

emergency lighting 事故照明

emergency shut down 紧急事故停机

emergency valve 事故阀

energizing coil 励磁线圈

epoxy resin [ɪˈpɒksɪ] 环氧树脂

equalizing air tank only for blaking 空气压力罐

equalizing air tank only for blow down 空气稳压罐

erection [ɪˈrekʃ(ə)n] 架立

erection of water turbine [ɪˈrekʃ(ə)n] 水轮机安装

erosion resistant coating [ɪˈrəʊʒ(ə)n] 抗磨涂层

excitation lead；slip ring leads 励磁引线

excitation off [ˌeksɪˈteɪʃ(ə)n] 灭磁

excitation switch 灭磁开关

excitation system fault 励磁系统故障

excitation transformer instaneous overcurrent 励磁变速断

excitation transformer overcurrent protection 励磁变过流

excitation transformer overload protection 励磁变过负荷

expansion joint [ɪkˈspænʃ(ə)n] 伸缩节

extinguisher; fire extinguisher 灭火器

eye bolt 吊耳螺栓

F

facing plates 抗磨板
factory assembly 工厂装配
fault 故障
female end 内止口
female screw thread 内螺纹
fictive synchronize [ˈsɪŋkrənaɪz] 假同期
filter 过滤器
final pressing 最终压紧
fineness 过滤精度
finger inclination 压指上翘量
fire fighting control cubicle 消防控制柜
fire protection 消防
fire resistant cloth [rɪˈzɪstənt] 防火布
firing angle 导通角
fish eye; flake 白点
fitting allowance 凑合余量
fitting allowance 装配公差
fixed blade of Deriaz turbine 斜流定桨式水轮机
flange 法兰，法兰面
flange checking 检查法兰面
flange of hydrocyclone [ˌhaɪdrəuˈsaɪkləun] 旋流器法兰
flange sealing surface 法兰密封面
flange seat 法兰座
flash test 高压闪络试验
flexible connection 软接头
flowing water 动水
(turbine) flume 引水室
flywheel effect 飞轮效应
force character 力特性
force characteristic test 力特性试验
foundation 基础
foundation of spiral case 蜗壳基础
foundation plate; sole plate 基础板
foundation ring; discharge ring 基础环
Francis turbine 混流式水轮机
frequency limiter 频率限制器
friction device 摩擦装置
function 功能

G

gas 瓦斯
gas insulated switchyard 气体绝缘开关站
gas leak detector 漏气检验器
gas relay 瓦斯继电器
gate lever 导叶拐臂
gauge panel 仪表盘
gauge pressure [geɪdʒ] 表计压力

（简称为压力）

gear speed detector 齿盘测速

generator 发电机

generator active power 发电机有功功率

generator air cooler condensating drainage 发电机空冷器冷凝水管

generator armature [ˈɑːmətʃə] 发电机电枢

generator braking equipment 发电机制动设备

generator breaker; generator cut-out 发电机断流器

generator bus 发电机母线

generator complete differential [ˌdɪfəˈrenʃ(ə)l] 发电机完全差

generator cover/shield [ʃiːld] 发电机机罩

generator CT fault 发电机CT断线

generator efficiency 发电机效率

generator incomplete differential [ɪnkəmˈpliːt] 发电机不完全差

generator loss 发电机损失

generator low-voltage 发电机低电压

generator neutral grounding equipment 发电机中性接地设备

generator over-voltage 发电机过电压

generator pit 发电机机坑

generator protection panel 发电机保护屏

generator PT fault 发电机PT断线

generator reactive power 发电机无功功率

generator room 发电机室

generator rotor 发电机转子

generator shaft 发电机大轴

generator stator 发电机定子

generator stator overload protection 发电机定子过负

generator step-up transformer 发电机升压变压器

generator support 发电机机墩

generator unit 发电机机组

globe valve 球阀

governor 调速器

governor cubicl/cabinet 调速器柜

gross head 毛水头

grouding resistance 接地电阻

grouding; earthing 接地

ground connector 接地体

ground strip 槽底垫条

ground wire 接地线

guide bearing 导轴承

guide bearing collar 轴领

guide bearing housing 轴承体

guide bearing oil pipe from cooler 水导轴承冷油管

guide bearing oil pipe to cooler 水导轴承热油管

guide blade loss 导叶损失

guide block 导向块

guide vane 固定导叶

guide vane arm 导叶拐臂

guide vane bearing 导叶轴承

guide vane control valve 导叶控制阀

guide vane end seal 导叶端面密封

guide vane force character 导叶力

特性
guide vane height 导叶高度
guide vane lever 导叶臂杆
guide vane lever; wicket gate lever 导叶臂
guide vane link; wicket gate link 导叶连杆
guide vane of hydraulic 水轮机导水机构
guide vane opening; gate-opening 导叶开度
guide vane overload protection device 导叶过载保护装置
guide vane packing 导叶填料
guide vane profile 活动导叶型线
guide vane ring 导叶环
guide vane seal 导叶立面密封
guide vane servomotor 导叶接力器
guide vane stem 导叶叶柄
guide vane stem seal 导叶轴密封
guide vane stop block 导叶限位块
guide vane thrust bearing 导叶止推轴承
guide vane; wicket gate 导叶
gulp valve 补气阀

H

hanging and installing 挂装
hardener 固化剂
head cover forced drainage 顶盖强迫排水
head cover forced drainage pipe 顶盖强迫排水管
head cover gravity drainage pipe 顶盖自流排水管
head cover pressure measuring 顶盖测压
head cover pressure measuring point 顶盖压力测点
head cover pressure relief pipe 顶盖卸压管支管
head cover 顶盖
head measurement 水头测量
heat exchanger 热交换器
heat running test 温升试验
hexagon spanner [ˈheksəg(ə)n] 梅花扳手
high frequency 高频
high oil level 高油位
high pressure measuring section 高压测量断面
high pressure oil injecting system 高压油顶起系统
high pressure oil pipeline 高压油管路
high pressure system for lower combined bearing 推力轴承高压油系统
high pressure water pipe 高压冲水管
high pressure water pump 高压水泵
high speed earthing switch 快速接地开关
high voltage tes 耐压试验

high-fit operation 高空作业
high-voltage equipment 高压电气设备
hinged shaft 铰轴
hinged support 铰座
hollow-cone valve；howell-Bunger valve 盘形阀
hook bolt 钩头螺
hose connection 软管连接
housing 机壳
hub transportation [trænspɔː] [trænspɔːˈteɪʃ(ə)n] 中心体转运
humidity 湿度
hydraulic control valve 水控阀
hydraulic efficiency of storage pump 蓄能泵水力效率
hydraulic efficiency of turbine 水轮机水力效率
hydraulic generator operation 水轮发电机组运行
hydraulic generator protection 水轮发电机组保护
hydraulic generator；hydrogenerator 水轮发电机
hydraulic jack 液压千斤顶
hydraulic machinery 水力机械
hydraulic measuring 水力测量
hydraulic system 液压系统
hydraulic thrust 水推力
hydraulic turbine 水轮机
hydraylic level indicator 液位计
hydrocyclone [ˌhaɪdrəʊˈsaɪkləʊn] 水力旋流器
hydrocyclone drainage pipe 水力旋流器排水管
hydrodynamic head [ˌhaɪdrəʊdaɪˈnæmɪk] 动水头
hydrogenerator shaft 水轮发电机轴
hydrogenerator unit 水轮发电机组

I

impact 冲击
impeller 叶轮
impeller back shroud 叶轮后盖
impeller blade；impeller vane 轮叶
impeller front shroud 叶轮前盖
impulse turbine 冲击式水轮机
impulse turbine 冲击式水轮机
incipient cavitation factor [ɪnˈsɪpɪənt] [ˌkævɪˈteɪʃ(ə)n] 初生空化系数
inclined jet turbine 斜击式水轮机
incompatibility 不相容性
individual guide vane servomotor [ɪndɪˈvɪdjʊ(ə)l] 单导叶接力器
inductance character [ɪnˈdʌkt(ə)ns] 感性
induction coil 感应线圈
inflexion；inflexibility 拐点
inherent stress [ɪnˈhɪər(ə)nt] 固有应力
initial cavitation factor [ˌkævɪˈteɪʃ(ə)n] 初生空化系数
initial pressure 初始压力

initial speed 初始转速

inlet measuring section of storage pump 蓄能泵进口测量断面

inlet measuring section of turbine 水轮机进口测量断面

inlet valve 进水阀

inner guide ring 内导水环

inner head cover; inner top cover 内顶盖（支持盖）

inner support 内支撑

input power of impeller 叶轮输入功率

input power of runner 转轮输入功率

inspection trolley [ɪn'spekʃn] 检修小车

installation fundamental circle [fʌndə'ment(ə)l] 安装基准圆

instant fault 瞬间故障

instrument panel of turbine floor 水轮机层仪表盘

insulating resistance 绝缘电阻

insulation coating 绝缘涂层

insulation grade 绝缘等级

insulation paint 绝缘漆

insulation plate 绝缘板

insulation shoes 绝缘鞋

insulation test 绝缘试验

insulator 绝缘子

intake valve 进口阀

interbus 旁路母线

interface 分界面，接触面，接口

interlocker; interlocking 连锁装置

intermediate pressing 中间压紧

intermediate strip [ˌɪntə'miːdɪət] 层间垫条

inverter 电流换向器

isolating switch 隔离开关

isolating transformer 隔离变压器

isolation 隔离

J

jack screw 顶起螺钉

jet deflector 折向器

jet diameter 射流直径

jet inclined angle [ɪn'klaɪnd] 射流入射角

jet ratio 射流直径比

K

Kaplan turbine 轴流转桨式水轮机

key 键

key slot 键槽

kilowatt-hour 千瓦小时，度（电）

knife switch 刀闸开关

knob-and-tube wiring 穿墙布电线

L

labyrith ring 迷宫环
lateral force 径向力
leading (trailing) edge 叶片进(出)水边
leak test 密封性试验
leakage inspecting; leakage checking 检漏
leakage monitor 泄漏监控
leakage protection 漏电保护
level; levelness 水平度
lift into position 起吊到位
lifting beam 起吊梁
lifting hole 起吊孔
lifting link 拉杆
lifting lug 吊耳
light resistor 光敏电阻
lighting distribution box (board) 照明配电箱
lighting equipment 照明设备
limit switch 限位开关
line-to-line short circuit 线间短路
link 连杆
load capacity 负载能力
load distribution 载荷分配
load generation balance 电力电量平衡
load rejection 甩负荷；卸负荷
load rejection test 甩负荷试验
load sensor 荷载传感器
load test 负载试验
load-ratio voltage protection 有载调压保护
local control 现地控制
local control cubicle 现地控制柜
local control unit 现地控制单元
local distortion/deformation 局部变形
local stress 局部应力
locate 定位
lockable 3-way ball valve 可锁定的三通球阀
locking glue 锁定胶剂
loss of excitation [ˌeksɪˈteɪʃ(ə)n] 发电机失磁
low alloy steel 低合金钢
low carbon steel 低碳钢
low head 低水头
low oil level 低油位
low pressure measuring section 低压测量断面
low resistance damping winding [rɪˈzɪst(ə)ns] 低电阻阻尼绕组
low voltage cable end box 低压电缆箱
low voltage overcurrent 低压启动过流保护
low voltage terminal 低压接线端子
low water level operation 低水位运行
lower axial thrust bearing 下轴径向推力轴承
lower bar 下层线棒
lower bracket assembly 下机架组装

lower bracket hub 下机架中心体
lower bracket oil painting 下机架油漆
lower bracket oil pot 下机架油槽
lower bracket sole plate 下机架基础板
lower bracket spider 下机架支臂
lower bracket welding 下机架焊接
lower guide bearing clearance 下导瓦间隙
lower guide bearing jacking bolt 下导瓦调整丝顶
lower guide bearing journal 下导轴领
lower guide bearing metal temperature 下导瓦温
lower guide bearing segment 下导瓦
lower oil pressure 低油压
lower pressure finger 下压指
lower ring plate 下环板
lower wearing ring cooling water pipe 下止漏环冷却水供水管
low-voltage electric appliance 低压电器
lubrication circuit [ˌluːbrɪˈkeɪʃən] 润滑管路
lug plate 吊板

M

magnet rim/yoke 磁轭
magnetic coil 电磁线圈
magnetic field density 磁场强度
magnetic field winding 磁场绕组
magnetic flux 磁通量
magnetic particle inspection；magnetic particle testing 磁粉探伤
magnetic pole；pole 磁极
magnetic switch 磁开关
magnetization [ˌmægnɪtaɪˈzeɪʃən] 磁化
magnetizing current 磁化电流
magnetizing test 磁化试验
main distribution valve 主配压阀
main girder 主梁
main hoisting cabinet 主起升柜
main hoisting resistor 主起升电阻器
main lifting system 主起升机构
main shaft 主轴
main shaft seal 主轴密封装置
main transformer 主变压器
main transformer differential protection 主变差动
main transformer overcurrent with voltage blocking 主变复合电压过流保护
main transformer overload protection 主变过负荷保护
main transformer zero-sequence overcurrent protection 主变零序过流保护
main valve 主阀
main/spare switch over 主/备用切换
maintenance [ˈmeɪnt(ə)nəns]

检修

maintenance air supply 检修密封供气

maintenance seal 检修密封

malfunction 误动作

mandoor; manhole 进人门

manifold 分流管

manifold [ˈmænɪfəʊld] 汇流管

man-machine communicate [kəˈmjuːnɪkeɪt] 人机交换

man-machine dialogue [ˈdaɪəlɒg] 人机对话

master switch 主令开关

maximal head 最大水头

maximum momentary overspeed of turbine 水轮机最大瞬态转速

maximum momentary pressure of turbine 水轮机最大瞬态压力

maximum (minimum) head 最大(最小)水头

maximum (minimum) head of storage pump 蓄能泵最大(最小)扬程

maximum (minmum) storage pump discharge 蓄能泵最大(最小)流量

mechanical breakdown 机械故障

mechanical efficiency of storage pump 蓄能泵机械效率

mechanical efficiency of turbine 水轮机机械效率

mechanism of runner blade 转叶机构

medium loss 介质损耗

medium position 中间位置

medium value 中间值

megohm [ˈmegəʊm] 兆欧

mica plate 云母板

mica tape 云母带

minimal head 最小水头

minimum impedance [ɪmˈpiːd(ə)ns] 低阻抗保护

minimum input power of storage pump 蓄能泵最小输入功率

misalignment; dislocation 错边

mis-operation 误操作

mixed-flow turbine 混流式水轮机

model 模型(机)

model test 模型试验

momentary counterrotation speed of storage pump (pump-condition of pump-turbine) 蓄能泵(水泵水轮机的泵工况)的最大瞬态反向转速

momentary pressure variation ratio 瞬态压力变化率

momentary speed variation ratio 瞬态转速变化率

monitor terminal [ˈtɜːmɪn(ə)l] 监视终端

multi-stage pump-turbine 多级水泵水轮机

multi-stage storage pump 多级式蓄能泵

multi-voltmeter 毫伏

mushroom valve 菌形阀

mushroom valve 盘形阀

N

name plate 铭牌
needle 喷针
needle servomotor 喷针接力器
needle valve 针阀
negative phase sequence current ['si:kw(ə)ns] 发电机负序过流
negative sequence ['negətɪv] ['si:kw(ə)ns] 负序
net head 净水头
net positive suction head of storage pump 蓄能泵空化余量(NPSH)(蓄能泵净吸上扬程)
net turbine power 水轮机净出力
neutral point 中性点
neutral terminal 中性点引线
no-discharge head of storage pump 蓄能泵零流量扬程
no-discharge input power of storage pump 蓄能泵零流量功率
noise 噪声
no-load characteristic 空载特性
no-load discharge of turbine 水轮机空载流量
no-load operating condition 空载工况
no-load status 空载状态
no-load test 空载试验
nominal diameter ['nɒmɪn(ə)l] 公称直径
nominal dimension 标称尺寸
nondestructive inspection 无损探伤
nondestructive test 无损检验
non-electric quantity transducer 非电量变送器
non-full phase protection 非全相保护
non-magnetic material 非磁性材料
non-regulated hydraulic machinery 不可调式水力机械
nose angle 蜗壳包角
nose vane 鼻端导叶
nozzle 喷嘴

O

O ring protection O形密封圈
observation hole [ɒbzə'veɪʃ(ə)n] 观察孔
oil baffle ring 挡油圈
oil cooler 油冷却器
oil drain hole 排油孔
oil filling 充油
oil filter; oil filtering machine 油过滤器
oil head 受油器
oil in; oil inlet 进油
oil inlet valve 进油阀
oil leakage 漏油
oil level at standstill 停机油位
oil level extreme low [ɪk'stri:m] 油位极低

oil level in operation 运行油位
oil level indicator 油位计
oil level switch 油位信号指示器
oil mist absorber [əb'sɔːbə] 油雾吸收器
oil or water pipeline 油或水管路
oil pan 油盆
oil pipe 油管
oil piping for lower combined bearing 下导轴承油管
oil pot cover 油槽盖板
oil pressure 油压
oil pressure tank/vessel 压力油罐
oil pump 油泵
oil sample 油样
oil steam exhaust piping for lower combined bearing 下导轴承排油雾
oil sump tank 回油罐
oil tank 油箱
oil well tube 下机架挡油管
open circuit 开环
open spanner 开口扳，开口扳手
open system 开放式系统
opening 开度
opening control/limit 开度控制
operating condition 运行情况
operating gap 运作间隙
operating oil pipe 操作油管
operating pump 工作泵
operating ring 控制环
operating ring machining 控制环加工
operating system 操作系统
operation and maintenance manual 操作及维修手册
operation condition 运行工况
operation flowchart 操作流程图
operation instruction 操作须知
operation lever 操作拐臂
operation mechanism 操作机构
operation sequence control ['siːkw(ə)ns] 操作规程
optic isolator 光频隔离器
optimum operating condition 最优工况
optimum specific speed 最优比转速
O-ring O形环
outdoor operation/ work 户外作业
outer guide ring 外导水环
outer micrometer; outside micrometer 外径千分尺
outlet 引出线，电源插座
outlet box 引出盒（箱），接线盒
outlet measuring section of storage pump 蓄能泵出口测量断面
outlet measuring section of turbine 水轮机出口测量断面
output power of impeller 叶轮输出功率
output power of runner 转轮输出功率
ovality [əu'vælɪtɪ] 椭圆度
over current protection 电源过载保护
over current protection 过流保护
over current relay 过流继电器
over excitation 过励
over load 过负荷

over speed 过速
over speed pendulum [ˈpendjʊləm] 过速飞摆
over stroke 压紧行程
over-excitation protection 过激磁保护
overflow ring covering cap 溢流环盖板
overfrequency/underfrequency protection 发电机频率异常保护
overhang distance between bars 线棒间斜边距离
over-limit 越限
overtravel-limit switch 行程开关

P

pack 衬垫
parallel operation 并行操作
parameter modify 修改参数
parameter of hydraulic 水轮机参数
partial discharge 局部放电
partial discharge device 局部装置
peak value 峰值
pedestal adjustment 支墩调整
pedestal installation [ˈpedɪst(ə)l] 支墩安装
pedestal leveling check 支墩水平度测量
peening 锤击
Pelton turbine 水斗式水轮机
(turbine) performance curve (水轮机)运转特性曲线
performance diagram of hydraulic turbine 水轮机特性曲线
performance test [pəˈfɔːm(ə)ns] 性能试验
periodic inspection [ˌpɪərɪˈɒdɪk] [ɪnˈspekʃn] 定期检查
peripheral equipment [pəˈrɪf(ə)r(ə)l] 外围设备
permanent speed droop [ˈpɜːm(ə)nənt] 永态转差系数(Bp)
phase angle 相位角
phase modulation [ˌmɒdjʊˈleɪʃən] 调相
phase sequence [ˈsiːkw(ə)ns] 相序，相位
phase terminal 相引线
piling platform 叠片平台
pin 销钉
pipe clamp 管夹
pipe crossing without connection 管交叉没有连接点
pipe support 管架
pipe thickness 管壁厚度
pit liner 机坑里衬
pit turbine 竖井贯流式水轮机
pitch diameter 节圆直径
planimetric average efficiency [ˌpleɪnəˈmetrɪk] 积分平均效率
plant cavitation factor 电站空化系数
plant dewatering header pipe 电站排水总管
plant drainage gallery [ˈdreɪnɪdʒ] 厂房排水廊道

plant electrical consumption 厂用电
plant service 厂用
polarity 极性
polarization index [ˌpəʊləraɪˈzeɪʃən] 极化指数
pole connection 磁极连接
pole installation 磁极挂装
pole ring 磁极环
pole strength 磁极强度
pole winding 磁极绕组
pop safety valve 紧急安全阀
porcelain insulator [ˈpɔːs(ə)lɪn] 瓷绝缘子
position accurately 精确定位
position sensor 高度传感器
positive 正极
potential energy 位置比能
potential head 位置水头
potential transformer (PT) 电压互感器
power angle 功角
power cable 电力电缆
power capacitor [kəˈpæsɪtə] 电力电容器
power circuit 电源电路
power control 功率控制
power discharge; power flow 发电机流量
power factor 功率因数
power feeder 电力馈线
power frequency high-voltage 工频耐压
power generation 发电
power resistance [rɪˈzɪst(ə)ns] 电源电阻
power storage 发电机库容
power switch 电源开关
power switching 电源切换
power system 电力系统
power transformer 电源变压器
pre-assembly 预装
pre-heat; pre-heating 预热
premier painting; prime painting; primer paint 底漆
pre-set 预留，预置
pressing bolt; punching bolt 压紧螺杆
pressure balance water pipe 冲水平压管
pressure control device 压力控制装置
pressure difference 压差
pressure energy 压力比能
(turbine) pressure fluctuation (水轮机) 压力脉动
pressure fluctuation test 压力脉动试验
pressure gauge 压力计
pressure head 压力水头
pressure pulsation [pʌlˈseɪʃən] 压力脉动测盒
pressure pulsation measuring 压力脉动测量
pressure pulsation measuring seat 压力脉动测量座
pressure release 压力释放
pressure relief /reducing valve 减压安全阀，降压阀
pressure sensor/transducer 压力传

感器

pressure switch 压力开关

pressure test 耐压（水压）试验；压力试验

pressure test for air cooler 空冷器打压

pressure transducer 压力变送器

pressure（suction）side of blade 叶片正（背）面

pretreatment 预处理

primary circuit 一次电路

productive head 发电水头

propeller storage pump 轴流式蓄能泵

propeller turbine 轴流定桨式水轮机

protective measure 防护措施

proto type 原型（机）

prototype turbine 原型水轮机

Pt resistance（platinum） 铂电阻

pump-turbine 水泵水轮机

push and pull rod 推拉杆

R

radial force 径向力

radial thrust bearing 径向推力轴承

radius alignment［ə'laɪnm(ə)nt］半径调整

radius of blade-stem 叶片轴颈圆角

random check for the bar 线棒抽样检查

ratchet spanner 棘轮扳手

rated current 额定电流

rated discharge 额定流量

rated head 额定水头

rated input power of storage pump 蓄能泵最大输入功率

rated output power of turbine 水轮机额定输出功率

rated speed 额定转速

rated voltage 额定电压

reaction turbine 反击式水轮机

reactive power 无功

reactor 电抗器

real-time control 实时控制

real-time multi-tasking system 实时对多任务系统

record for temperature and humidity 温度湿度记录表

rectify 整流

reduce conductivity 降低导电率

refuse to active 拒动

regulated hydraulic machinery 可调式水力机械

regulating device 回复机构

regulating guarantee 调节保证

regulating mode 调节模式

regulating ring 调速环

regulating ring 控制环

regulating shaft 调速轴

relative allowable remaining unbalance 允许相对残余不平衡力矩（转轮）

relative efficiency 相对效率

relief valve 减压安全阀，降压阀

remaining rest unbalance 允许不平

衡重量
remote control 远方控制
remote diagnostics 远方诊断
repair welding 补焊
reserved head cover air admission pipe 顶盖预留（强迫）补气管
reservoir for power generation [ˈrezəvwɑː(r)] 发电水库
resin 环氧胶
response time; reactive time 反应时间
retaining plate 支撑板
retraction 回缩
return value 返回值
reverse/reversing power 逆功率
reverse regulation 反调
reverse runaway speed of storage pump 蓄能泵（水泵水轮机的泵工况）反向飞逸转速
reverse slide block 反向滑块
reversible turbine 水泵水轮机
reversing time limit 反时限
rim alignment [əˈlaɪnm(ə)nt] 磁轭调整
rim compensating 磁轭补偿
rim-generator unit 全贯流式水轮机
rim lip 磁轭挂钩
rim piling 磁轭堆片
rim plate 磁轭片
rim shrinking 磁轭收缩
rim spacer 磁轭隔板
rim stud 磁轭螺杆
rim stud nut 磁轭螺母
rim stud torque [tɔːk] 磁轭螺杆扭紧
ring gate 圆筒阀
rocker arm 转臂
root bend test 背弯试验
rotating component 转动部件
rotating ring 镜板
rotor 转子
rotor earth fault 转子一点接地保护
rotor hub 转子中心体
rotor on point earthing 转子一点接地
rotor roundness measuring device 转子测圆架
rotor roundness measuring device seat 转子测圆架底座
rotor spider 转子支臂
rotor temperature monitoring 转子温度保护
round bar for gauge panel 支撑柱
roundness adjustment 圆度调整
roundness control device 测圆架
roundness measuring device pedestal [ˈpedɪst(ə)l] 测圆架中心柱
routine maintenance 常规维修
routine test 常规试验
routine work 常规作业
rubber shim 橡皮垫
run out 摆度
runaway 飞逸
runaway protection 防飞逸保护
runaway speed 机组过速
runaway speed curve 飞逸特性曲线
runaway speed of turbine 水轮机飞

逸转速
runaway speed operating condition 飞逸工况
runaway speed test 飞逸试验
runner 转轮
runner balancing 转轮平衡
runner blade servomotor 转轮叶片接力器
runner blade trunnion [ˈtrʌnjən] 叶片枢轴
runner blade/bucket 转轮叶片
runner chamber 转轮室
runner cone 转轮泄水锥
runner hub 转轮体
runner lower band 转轮下环
runner seal 转轮密封装置
runner upper crown 转轮上冠
runner wearing ring 转轮止漏环
runner (impeller) diameter 转轮(叶轮)公称直径

S

scaffold [ˈskæfəuld] 脚手架
scratch; stria; striation 擦痕
seal; rubber seal 水封
secondary circuit [ˈsek(ə)nd(ə)rɪ] 二次电路
secondary coil 二次绕组
secondary winding 二次线圈
segments bracket 瓦块托架
self-diagnostic [daɪəgˈnɒstɪk] 自诊断程序
self-start 自启动
semi-automatic welding 半自动焊缝
sensitive 反应灵敏的
sensor for resistance thermometer [rɪˈzɪst(ə)ns] [θəˈmɒmɪtə] 电阻温度计传感器
sensor for temperature contact instrument 温度接触设备传感器
separate adjustment 分别调整
sequence [ˈsiːkw(ə)ns] 顺序
servomotor [ˈsɜːvəʊˌməʊtə] 接力器
servomotor foundation plate 接力器基础板
servomotor oil leakage pipe 接力器漏油总管
set pin 定位销
set up 架设
setting elevation [ˌelɪˈveɪʃ(ə)n] 安装高程
shaft air admission pipe 大轴补气器
shaft coupling bolt 联轴螺栓
shaft coupling flange 联轴法兰
shaft current 轴电流
shaft current protection 轴电流保护
shaft seal 主轴密封
shaft seal booster pump 主轴密封加压泵
shaft seal cooling water header pipe 主轴密封冷却水供水总管
shaft seal cooling water supply pipe 主轴密封冷却供水管

shaft seal drainage pipe 主轴密封排水管

shaft seal lubrication water 主轴密封润滑水

shaft seal maintenance seal air pipe 主轴密封检修密封供气管

shaft seal spare water supply pipe 主轴密封供水备用水管

shaft seal water supply 主轴密封供水

shaft seal water supply pipe 主轴密封供水管

shear pin 剪断销

shift phase 移相

shim 薄垫片

shim plate 垫片

short back plate 短压条

short-circuit 短路

short-circuit characteristic [kærəktə'rɪstɪk] 短路特性

short-ciruit breaker 短路开关

shortdown relay 短路继电器

shrinking and shimming 磁轭加垫

shut-off valve 截止阀

similar operating condition 相似工况

single phase earthing 单相接地

single stage pump-turbine 单级水泵水轮机

site instruction 现场指令

slip ring 集电环

slip ring housing 集电环室

slot liner 槽衬纸

sole plate for brake cylinder 制动器基础

sole plate for top cover 上盖板基础

soleplate for the pedestal 支墩基础板

soleplate pin ['səʊlpleɪt] 基础板定位销

spacer 衬套

spacing 间距

spare automatic operation 备用自投

specific energy [spə'sɪfɪk] 比能

specific speed of turbine 水轮机比转速

speed sensor 测速装置

spherical valve ['sferɪk(ə)l] 球阀

spinning status 空转状态

spiral case 蜗壳

spiral case drainage pipe 蜗壳排水管

spiral case inlet pressure measuring point 蜗壳进口压力测点

spiral case inlet pressure pulsation measuring point 蜗壳进口压力脉动测点

spiral case mandoor 蜗壳进人门

spiral case nose 蜗壳鼻端

spiral case segment 蜗壳瓦片

spiral case Winter-Kennedy 蜗壳压差测量

spiral housing 蜗室

split flange 分瓣法兰

split key 分半键

split line 分瓣线

split pin 开口销

spot check for the bar 线棒抽样

检查
spring washer 弹性垫圈
square butt welding I 形坡口对边焊
square washer 方形垫圈
stacking platform 叠片平台
stainless steel welding 不锈钢焊接
stainless steel welding electrode 不锈钢焊条
standard pressure testing table 标准压力试验台
standby pump 备用泵
star connection 星形
start cooling fans protection 通风保护
start cooling fans protection 主变通风保护
starter 起动器
static characteristic 静特性
static loading test 静载试验
static pressure measuring 静压压力测量
static suction of storage pump 吸入高度
static suction of turbine 吸出高度
static test 静态试验
station service 厂用
station service power 厂用电
stationary wearing ring ['steɪʃ(ə)n(ə)rɪ] 固定止漏环
stator 定子
stator adjustment 定子调整
stator assembly 定子组装
stator assembly welding 定子组焊
stator case 定子罩
stator coil 定子线圈
stator core 定子铁芯
stator current 定子电流
stator earth fault 定子接地保护
stator earth fault protection of startup condition 起停机定子接地保
stator frame 定子机座
stator frame segment ['segm(ə)nt] 机座分瓣
stator frame sole plate 定子机架基础
stator frame sole plate 定子机座基础板
stator intermediate ring plate [ˌɪntə'miːdɪət] 定子中环板
stator interturn fault 横差保护
stator lifting and installing 定子吊装
stator lower ring plate 定子下环板
stator one point earthing 定子一点接地
stator soleplate ['səʊlpleɪt] 定子基础板
stator upper ring plate 定子上环板
stator winding 定子绕组
stator winding pitch 定子绕组节距
stay ring 座环
stay ring lower barrel 座环下筒体
stay ring lower flange 座环下法兰面
stay ring machining 座环加工
stay ring upper deck ring 座环上环板
stay vane 固定导叶
steel liner 钢衬

steel scaleboard [ˈskeɪlbɔːd] 钢衬板
step transformer 分降压变压器
step voltage 跨步电压
still water 静水
stop lock 止动块
storage battery 蓄电池（组）
storage pump 蓄能泵
storage pump discharge/flow rate 蓄能泵流量
storage pump head 蓄能泵扬程
storage pump input power 蓄能泵的输入功率
storage pump output power 蓄能泵的输出功率
straight flow turbine 全贯流式水轮机
straightness accuracy 直线度
stress relaxation 应力松弛
stress relived 消除应力处理
strip 挡风围带
stroke 行程
submersible pump [səbˈmɜːsɪb(ə)l] 潜水泵
suction head loss of storage pump 蓄能泵吸入扬程损失
suction tube 吸水管
supervisory circuit [ˈsjuːpəˌvaɪzərɪ] 监控电路
support 支撑件
support frame 机架
support plate 支撑板
support steel plate for adjusting balance 调平垫板
surface crack 表面裂纹
suspended insulator 悬式绝缘子
switchboard 配电盘
switching-in 合闸，接通（入）
synchronize 同期
synchronize device 同期装置
synchronize; synchronous 同步

T

tap 攻丝
taper key 楔键
taper pin 锥销
temperature measuring instrument 温度测量仪器
temperature transmitter 温度变送器
temporary speed droop 暂态转差系数（Bt）
terminal block 接线板，接线盒
terminal box [ˈtɜːmɪn(ə)l] 端子箱，集线箱，接线盒
test ring 环筒
thermal relay 热继电器
thermal stability 热稳定性
thermal stress 热应力
thickness of lamination [ˌlæmɪˈneɪʃən] 定子铁片厚度
Thoma turbine 轴流调桨式水轮机
three phase circuit 三相电路
three way valve 三通阀，三向阀
three-phase four-wire system 三相

四线制
throttle plate 节流片
throttle plate 截流板
throttle; throttle valve 节流阀
throttling hole 节流孔
through flow butterfly valve 平板蝶阀
thrust bearing 推力轴承
thrust bearing pads 推力瓦
thrust bearing spring 推力瓦弹簧
thrust block 止推块
thrust ring 推力头
thrust skirt ring 挡油裙环
tight welded 工地封焊
tightness test 气密性试验
time constant 时间常数
time constant of accelerating 加速时间常数（T_n）
time constant of damping 缓冲时间常数（T_d）
time limit 定时限
times of moving 摆动次数
tolerance difference 允许偏差
top cover 顶盖
top cover sole plate 上盖板基础
total allowable remaining unbalance 总体允许不平衡量（转轮）
transformer fire fighting 主变火警
transformer oil over temperature 主变油温过高
transformer winding over temperature 主变绕组温度过高
transient current [ˈtrænzıənt] 瞬时电流
transient voltage 瞬时电压
transition plate [trænˈzıʃ(ə)n] 过流板
transition process 过渡过程
transition process of water turbine 水轮机过渡过程
transversal key 切向键
travelling mechanism 运行机构
trip 脱扣
trip order 跳闸命令
tubular turbine（S-type turbine）轴伸贯流式水轮机（S形水轮机）
tubular turbine; through flow turbine 贯流式水轮机
turbine air vent pipe 水轮机通气管
turbine auxiliary components 水轮机辅助部件
turbine blade 水轮机叶片
turbine bypass 水轮机旁通管
turbine cap 水轮机顶盖
turbine case 水轮机罩
turbine characteristics [kærəktəˈrıstık] 水轮机特性
turbine control panel 水轮机控制盘
turbine design data 水轮机设计参数
turbine discharge/flow rate 水轮机流量
turbine draught tube 水轮机尾水管
turbine efficiency 水轮机效率
turbine gallery 水轮机廊道
turbine gate 水轮机闸门
turbine generator set 水轮发电

机组

turbine guard valve 水轮机保护阀

turbine guide bearing 水导轴承

turbine guide bearing clearance 水导轴承间隙

turbine guide bearing oil pot 水导油槽

turbine guide bearing oil supply pipe 水导轴承供油管路

turbine guide bearing pad 水导轴瓦

turbine idling 水轮机空转

turbine inlet bend 水轮机进水弯管

turbine inlet valve 水轮机进水阀

turbine input power 水轮机输入功率

turbine instrument cubicle [ˈɪnstrʊm(ə)nt] 水机仪表柜

turbine major axis 水轮机主轴

turbine oil 透平油

turbine output power 水轮机输出功率

turbine output test 水轮机功率试验

turbine parameter 水轮机参数

turbine pit 水轮机机坑

turbine pit contour 水轮机机墩外形

turbine pit liner 水轮机机坑里衬

turbine runner 水轮机转轮

turbine shaft 水轮机大轴

turbine shaft 水轮机轴

turbine shaft gland 水轮机轴密封套

turbine sitting 水轮机安装高程

turbine stay ring 水轮机座环

turbine terminal cubicle 水轮机端子控制箱

turbine type selection 水轮机选型

turbine venting; turbine air supply 水轮机补气

turbo-generator set installation 水轮发电机组安装

turnbuckle 拉紧器

two way valve 双向阀

T-wrench 丁字柄套筒扳手

U

ultrasonic testing/examination/inspection 超声波探伤

unbalance 不平衡

under excitation 欠励

under water head condition 水压作用条件下

unit 机组

unit capacity 单机容量

unit discharge 单位流量

unit disconnect 解列

unit emergency shut down flow 事故停机流程

unit hydraulic thrust 单位水推力

unit hydraulic torque 单位水力矩

unit power 单位功率

unit runaway speed 单位飞逸转速

unit speed 单位转速

unit starting flow 开机流程

upper axial thrust bearing 上轴向推力轴承
upper bar 上层线棒
upper bracket assembly 上机架拼装
upper bracket cover 上机架盖
upper bracket hub 上机架中心体
upper bracket pedestal 上机架支柱
upper bracket soleplate 上机架基础板
upper bracket spider 上机架支臂
upper bracket welding 上机架焊接
upper guide bearing 上导轴承
upper guide bearing clearance 上导轴承间隙
upper guide bearing metal temperature 上导瓦温
upper guide bearing oil cooler 上导油冷器
upper guide bearing oil pot 上导油槽
upper guide bearing oil temperature 上导油温
upper guide bearing segment 上导瓦
upper pressure finger 上压指
upper ring plate 上环板
upper shaft 上端轴
upper wearing ring cooling water pipe 上止漏环冷却水供水管
upper wearing ring cooling water supply 上止漏环冷却供水
upstream drainage trench 上游排水沟

V

vacuum break valve 直空破坏阀
valve for filling and drainage of the water 进水排水阀门
vane apparatus [ˌæpəˈreɪtəs] 导叶装置
vapour pressure [ˈveɪpə] 汽化压力
velocity energy 速度比能
velocity head 速度水头
vent hole 通气孔
vertical reinforcement [riːɪnˈfɔːsm(ə)nt] 立筋
vertical，horizontal and inclined unit 立式、卧式和倾斜式机组
vibration and run out measuring system 测振测摆系统
vibration control [vaɪˈbreɪʃ(ə)n] 机组振摆
vibration of hydraulic turbine 水轮机振动
virtual value 有效值
voltage amplifier [ˈæmplɪfaɪə] 电压放大器
voltage blocking 电压闭锁
voltage limiter 电压限制器
voltage ratio 变压比
voltage regulator 调压器
voltage transducer [trænzˈdjuːsə] 电压变送器
voltage-current characterisitic 伏安

特性

vortex 涡流

W

washer; gasket 垫圈
water chiller 水冷却器
water connection brazing 水接头铜焊
water connection fitting 水接头配置
water connection installation 水接头安装
water connection measurement 水接头测量
water connection pressure test 水接头打压试验
water connection PT test 水接头PT探伤试验
water connection seal ring 水接头密封圈
water connection support for bars 线棒水接头支撑
water connection support for brazing 线棒水接头支撑铜焊
water connection; hydraulic connection 水接头
water filling 充水
water flow meter 水流量计
water hammer 水锤
water hammer 水击
water in oil 油混水
water in oil detector 油混水探测器
water in/inlet 进水
water leakage 漏水
water passage 过流面
water passing parts of turbine 水轮机过流部件
water pressure 水压
water pressure test; hydraulic test 水压试验
water pump 水泵
water pump motor 水泵电机
water turbine regualtion system 水轮机调节系统
wattmeter [ˈwɒtmiːtə] 功率表
wave 纹波
wave record for faults 故障录波
wearing block 抗磨块
wearing ring 抗磨环，止漏环
web plate 腹板
wedge plate 楔子板
weighted average head 加权平均水头
weighted (arithmetic) average efficiency 加权（算术）平均效率
welding of turbine 水轮机焊接
wicket gate 活动导叶
wicket gate actuating rod 导叶操纵杆
wicket gate adjust servomotor [ˈsɜːvəʊˌməʊtə] 导叶调节接力器
wicket gate closing gap 导叶立面间隙
wicket gate operating ring 导叶操纵环
wicket gate side gap 导叶端面间隙

wicket gate stem 导叶转动轴，导叶叶柄

winding condition 下线环境

winding documentation 下线资料

winding material 下线材料

winding tools and devices 下线工装

wing nut 蝶型螺母

winter kennedy measuring point 蜗壳流量测量点

wiring drawing；wiring scheme 接线图

wiring terminal 接线端子

working grounding 工作接地

workshop 车间

workshop assembly 车间装配

汉 英 词 汇

风 能 发 电

50 年一遇的阵风 50 year return gust
A 计权滤波器 A-weighted filter
A 计权声压级 A weighted sound pressure level
A 计权噪声测量 A-weighted noise measurements
CO_2 排放 CO_2 emission [ɪˈmɪʃən]
IEC 风区 IEC wind class
MW 级风电机组 Megawatt turbine
NACA 翼型 NACA airfoil
N 年一遇风速 N-year wind speed
PID 控制器 PID controller
PI 控制器 PI controller
S—N 曲线 S—N curve [kɜːv]
S 形—达里厄组合装置 composite Savonieus-Darrieus systems
S 形风轮 Savonieus rotor
T 形头螺栓 t-head bolt
T 形线夹 T-connector
U 形挂钩；U 形螺栓 U-bolt
V 形螺纹 v-thread
X 轴；横坐标轴 x-axis
Y 形达里厄式风轮 Y-Darrieus rotor

A

安全标志 safety marking
安全带 safety belt
安全等级 safety class（IEC WTGS Class）
安全阀 safety valve
安全方案 safety concept [ˈkɒnsept]
安全风速 survival wind speed [səˈvɑɪvəl]
安全隔离变压器 safety isolating transformer [ˈɑɪsəleɪtɪŋ]
安全环境温度 survival ambient temperature
安全距离 safety distance
安全开关 safety switch
安全联轴器 security coupling
安全链路 safety loop
安全帽 safety helmet [ˈhelmɪt]
安全色 safety color
安全寿命 safe life
安全停止位置 safety stop position
安全系数 safety factors
安全系统 safety system
安全阻抗 safety impedance[ɪmˈpiːdəns]
安装 installation/erection [ɪˈrekʃən]
安装地点 site location
安装费 installation costs[ɪnstəˈleɪʃən]
安装高度 installation height
安装角 pitch setting angle
(叶片)安装角 setting angle [ˈæŋg(ə)l]
安装与运行 installation and operation

B

摆动角 angle of oscillation[ˌɒsɪˈleɪʃən]
摆振运动 lagging motion
板式基础 slab foundation
半模 half-moulds
薄板（片） laminate [ˈlæmɪneɪt]
薄壁钢管塔筒 thin-walled steel tubular towers [ˈtjuːbjʊlə]
薄壳结构 shell structure
薄木片 wood veneer [vəˈnɪə]
饱和特性 saturation characteristic [ˌkærəktəˈrɪstɪk]
暴风（十一级风） violent storm
杯式风力机 cupped wind machine
贝塞尔函数 Bessel functions
贝兹 Betz
贝兹定律 Betz' law
贝兹极限 Betz limit
贝兹理论 Betz' theory
备用电池 battery backup
备用电源 backup power
备用容量 reserve capacity
备用系统 back-up system
背景响应 background response
被动变桨控制 passive pitch control
被动调向风轮，被动对风风轮 yaw-passive rotor [ˈpæsɪv]
被动偏航 passive yawing
被动失速 passive stall
被动失速控制 passive stall control
本征模 eigenmode
本征值 eigenvalue
比恩法 method of Bins [ˈmeθəd] [bɪnz]
毕奥一萨伐尔定律 Biot-Savart law
避雷带 lightning belt [ˈlɑɪtnɪŋ]
避雷器 arrester [əˈrestə]
避雷器的残压 residual voltage of lightning arrester [rɪˈzɪdjʊəl]
避雷网 lightning-protection net
避雷线 shield wire
避雷针 lightning rod，lightning conductor [kənˈdʌktə]
避雷针基础 lightning rod base
避雷针支架 lightning rod support
避雷装置 lightning protector
变桨距调节机构 regulating mechanism by adjusting the pitch of blade [ˈmekənɪzəm]
边界层 boundary layer
边界层（速度）廓线 boundary layer profile
边界层；附面层 boundary layer
边界层堆积；边界层增厚 boundary layer accumulation [əkjuːmjʊˈleɪʃ(ə)n]
边界层分离 boundary layer separation [sepəˈreɪʃ(ə)n]
边界层回流 boundary layer reversal
边界层结构 boundary layer structure
边界层涡量 boundary layer vorticity [vɔːˈtɪsətɪ]
边界层增长 growth of boundary layer
边界条件 boundary conditions

边界效应；（风洞）洞壁效应 boundary effect
边界元法 BEM method ['meθəd]
边界元模型 BEM model
边界约束；（风洞）洞壁约束 boundary constraint
边缘 edgewise
边缘弯曲 edgewise bending
边缘效应 edge effect
边缘整流 edge fairing
变薄；稀薄化；变细；衰减 attenuation [əˌtenjʊ'eɪʃən]
变桨 variation of blade pitch
变桨风力机 pitch regulated wind turbines
变桨距控制 active pitch control
变桨距控制方案 variable-pitch control scheme [ski:m]
变桨距系统 pitch systems
变桨力矩；仰俯力矩 pitching moment
变桨力矩系数 pitching moment coefficient
变桨驱动器 pitch actuator
变桨驱动系统 pitch actuation systems
变桨式风力机线性变参数模型 LPV model of variable-pitch WECS ['veərɪəbl]
变桨式机组 pitch-regulated machines
变桨速度 pitch speed
变桨速率限制 pitch rate limits
变桨轴承 pitch bearing
变截面叶片 variable chord blade ['veərɪəbl]
变频顺桨 converter feathering
变速/恒速运行 variables/constant operation
变速变桨距 variable-speed variable-pitch (VS-VP)
变速定桨距 variable-speed fixed-pitch (VS-FP)
变速风力机 variable-speed wind turbine
变速风轮 variable-speed rotor
变速杆 gear lever
变速恒频风电机组 VSCF (variable speed constant frequency) WTGS
变速控制方案 variable-speed control scheme
变速驱动系统 variable speed drive system
变速箱 gearbox ['gɪəbɒks]
变速运行 variable speed operation
变位齿轮 gears with addendum modification [ə'dendəm]
变弦长叶片 variable chord blade
变形齿 reduction gear
标准大气 standard atmosphere ['stændəd] ['ætməsfɪə]
标准大气压 standard atmosphere pressure
标准大气状态 standard atmospheric state [ætməs'ferɪk]
标准风速 standardized wind speed ['stændəˌdaɪzd]
标准空气 normal air
标准气象条件 meteorological standard condition

标准状态 normal state
表面粗糙度 surface roughness[ˈrʌfnɪs]
表面粗糙度比 surface roughness ratio
表面厚度 skin thickness
表面击穿 surface breakdown
表面摩擦 skin friction
表面温度 surface temperature
玻璃钢 glass fibre reinforced plastic [ˈfɑɪbə]
玻璃钢叶片 fiberglass blades [ˈfɑɪbəglɑːs]
玻璃绝缘子 glass insulator[ˈɪnsjʊleɪtə]
玻璃纤维 E-glass; fiberglass
玻璃纤维增强塑料 glass fibre reinforced plastic
玻璃纤维增强塑料叶片 GFRP blade [bleɪd]
伯努利常数 Bernoulli constant
伯努利定律 Bernoulli's law
伯努利二项分布 Bernoulli binomial distribution [bɑɪˈnəʊmɪəl]
伯努利方程 Bernoulli's equation [bɜːˈnuːlɪ]
伯努利面 Bernoulli surface
伯努利矢量 Bernoulli vector [ˈvektə]
伯努利尾迹 Bernoulli trail
不均匀场 non-uniform field
不均匀地形 nonuniform topography [təˈpɒgrəfɪ]
不均匀介质 nonuniform medium [nɒnˈjuːnɪfɔːm]
不均匀流 non-uniform flow
不可压缩流 incompressible flow [ˌɪnkəmˈpresəbl]
不可再生能源 non-renewable energy source [rɪˈnuəbl]
不稳定分层 unstable stratification
不匀称 unsymmetry [ˈʌnˈsɪmɪtrɪ]
部分安全系数 partial safety factors
部分安全系数法 method of partial safety factors
部分跨度间距控制 partial-span pitch control

C

材料的安全因素 material safety factors
材料属性 material properties
参考风速 reference wind speed
侧风 sidewind
侧风轮 auxiliary wind wheel [ɔːgˈzɪljərɪ]
侧滑流 yawing flow
侧滑速度，滑移速度 slip speed
侧力 lateral/side force
侧逆风 side/cross headwind
侧顺风 side leading wind
侧向挥舞 lateral flapping
侧向挠度 lateral deflection [dɪˈflekʃ(ə)n]
侧向倾斜（度） side rake
侧向屈曲 lateral buckling
侧向位移 lateral displacement [dɪsˈpleɪsm(ə)nt]
侧压孔 tapping hole

侧翼 lateral plane [ˈlætərəl]
侧阵风 side gust
测得数据比较 measured data comparisons
测风 wind survey
测风杆 measurement mast
测风塔 anemometer tower
测风塔 wind mast/tower
测量参数 measurement parameters
测量读数；测量值 metered reading
测量功率曲线 measurement power curve [kɜːv]
测量扇区 measurement sector [ˈsektə]
测量位置 measurement seat
测量误差 uncertainty in measurement [ʌnˈsɜːtəntɪ]
测量周期 measurement period [ˈpɪərɪəd]
测试设备 measuring equipment
测试仪器 measuring instrument
层流 laminar flow [ˈlæmɪnə]
层流边界层 laminar boundary layer [ˈlæmɪnə] [ˈbaʊnd(ə)rɪ]
层流底层 laminar sub layer
层流对流 laminar convection
层流分离 laminar separation [sepəˈreɪʃ(ə)n]
层流环型对流 laminar cellular convection [ˈseljʊlə] [kənˈvekʃ(ə)n]
层流尾流 laminar wake
层流涡街 laminar vortex street
层流翼型 laminar aerofoil；laminar flow airfoil [ˈeəfɒɪl]
层流黏性 laminar viscosity [vɪˈskɒsɪtɪ]
层流阻力 laminar drag
颤振 flutter [ˈflʌtə]
长度尺度 length scales
长期风资料 long-term wind data
长期观测系统 long-term observation system [ɒbzəˈveɪʃ(ə)n]
长期平均风速 long-term mean wind speed
长期效应 long-term effect
长期预报 long-range prediction；long-term prediction [prɪˈdɪkʃ(ə)n]
长时间闪变值 long term flicker severity [sɪˈverɪtɪ]
常规试验 routine test [ruːˈtiːn]
常温常压 normal temperature and pressure（NTP）
常现风速 frequent wind speed
厂址分析 site analysis
场地评估 site assessment；site evaluate [ɪˈvæljʊeɪt]
场梯度 field gradient [ˈgreɪdɪənt]
场址勘测 site survey
场址选择 site selection
超前—滞后时刻 lead-lag moment
超声波 ultrasonic [ˌʌltrəˈsɒnɪk]
超速保护 overspeed protection
超速保护扰流板 overspeed spoiler [ˈspɒɪlə]
超速保护装置 overspeed device
超速控制 overspeed control
超载；过载；过负荷 overload
沉降风 fallout wind
承载结构 load-bearing structure
承载能力 load-carrying capability [ˌkeɪpəˈbɪlətɪ]
承重 load-bearing

[能] 持续功率（最坏风况下）continuous power

持续性预报 persistence forecast [pə'sɪstəns] ['fɔːkɑːst]

持续运行 continuous operation

持续运行的闪变系数 flicker coefficient for continuous operation ['flɪkə] [ˌkəʊɪ'fɪʃənt]

赤道地区 equatorial region ['riːdʒən]

冲击动载荷试验 impulse load tests

冲击荷载 impulsive/shock loads [ɪm'pʌlsɪv]

冲击闪络 impulse flashover ['flæʃˌəʊvə]

冲角，攻角 angle of attack

出口速度，排出速度 outlet velocity [vɪ'lɒsɪtɪ]

出现频率 frequency of occurrence [ə'kʌr(ə)ns]

初始风速 initial wind speed

初始扰动 initial disturbance [dɪ'stɜːb(ə)ns]

初始条件 initial condition

触动力矩 fluctuating moment

传递比 transfer ratio

传递函数 transfer function

传动比 transmission ratio

传动精度 transmission accuracy [trænz'mɪʃən]

传动误差 transmission error

传输线 transmission line

吹流；风积 wind drift [drɪft]

垂直风廓线 vertical wind profile ['prəʊfaɪl]

垂直风切变 vertical wind shear

垂直轴风力发电机 vertical axis wind turbine ['æksɪs]

磁感风杯风速计 cup generator anemometer

粗糙长度 roughness length ['rʌfnɪs]

粗糙度 roughness ['rʌfnɪs]

粗糙度测量仪 roughness measuring machine

D

达里厄风轮 Darrieus rotor

达里厄竖轴风轮机 vertical axis Darrieus turbine ['tɜːbaɪn]

大尺度环流 large-scale circulation [sɜːkjʊ'leɪʃ(ə)n]

大尺度湍流 Macro turbulence ['tɜːbjʊl(ə)ns]

大尺度涡旋 large-scale eddy

大范围对流 Macro scopic convection

大风速 high wind speed

大件与超重货物运输 Odc & Owc cargo service (overdimensional cargo and overweight cargo)

大陆风 continental wind

大陆性气候 continental climate [ˌkɒntɪ'nentəl] ['klaɪmɪt]

大气边界层 atmospheric boundary layers ['baʊnd(ə)rɪ]

大气层；空气圈 aerosphere

[ˈeərəʊsfɪə]

大气对流 atmospheric convection [kənˈvekʃ(ə)n]

大气环流 general atmospheric circulation [ˌætməsˈferɪk]

大气扩散方程 atmospheric diffusion equation

大气密度 atmospheric density[ˈdensɪtɪ]

大气平流 atmospheric advection [ætməsˈferɪk] [ədˈvekʃ(ə)n]

大气上界 aeropause [ˈeərəpɔːz]

大气湿度；大气水分 atmosphere moisture [ˈætməsfɪə] [ˈmɒɪstʃə]

大气湍流；大气湍流度 air turbulence [ˈtɜːbjʊl(ə)ns]

大气稳定性 atmosphere stability [ˈætməˌsfɪə]

大气压力；大气压强 atmospheric pressure

大气阻尼 atmospheric damping

大缩尺比模型 large-scale model

大型风力机 large-sized wind machine

大型风能转换系统 LWECS (large scale wind energy conversion system)

当地方位角 local angle of latitude

当地风力资源 local wind resource

当地马赫数 local Math number

当地速度比 local speed ratio

当地湍流（度） local turbulence [ˈtɜːbjʊl(ə)ns]

当地压力 local pressure

当地叶片弦(长) local blade chord

当地迎角 local angle-of-attack

当量浓度 normality [nɔːˈmælətɪ]

挡风玻璃；风挡 windscreen

导风板；导流板 air/airstream deflector [dɪˈflektə]

导风器式风力机 deflector wind machine

导流（叶）片 guide blade/vane

导流罩；导流板 air flow guide

等变压风 isallobaric wind [ɑɪˌsæləˈbærɪk]

等风速线 isanemone；isotach [ˈɑɪsəʊtæk]

等角投影；正等侧投影 isometric projection [ˌɑɪsəʊˈmetrɪk]

等截面叶片 constant chord blade [kɔːd]

等倾线 isoclines [ˈɑɪsə(ʊ)klaɪn]

等容过程 isochoric process [ˌɑɪsəʊˈkɔːrɪk]

等熵过程 isentropic process

等熵流 isentropic flow

等熵线 isentropic [ˌɑɪsenˈtrɒpɪk]

等速线 isovel(isovelocity) [ˈɑɪsəvel]

等梯度线 isogradient [ˌɑɪsəʊˈgreɪdɪənt]

等温过程 isothermal process

等弦长截面叶片 constant chord blade

等效负荷 equivalent loading

等效连续声级（A 计权声级） equivalent continuous A-weighted sound pressure level

等效满负荷小时数 full-load equivalent hours [ɪˈkwɪvələnt]

等压线 isobar；isobaric line

[ˌɑɪsəʊˈbærɪk]
低层风 low-level wind
低电压穿越 LVRT（low voltage ride through）
低风速 low wind speed
低速风洞 low spend wind tunnel [ˈtʌnl]
低速风轮 slow-speed rotor
低速特性 low-speed characteristics [ˌkærəktəˈrɪstɪk]
低速翼型 low-speed aerofoil [ˈeərəfbɪl]
低阻翼型 low-drag airfoil
笛卡儿坐标系 Cartesian coordinate system
底座（风电塔架底座）pedestal [ˈpedɪstəl]
底座（机舱）bedplate base；mainframe base (nacelle)
底座支撑 pallet
地方电网 local power network
地理变异 geographical variation [dʒɪəˈgræfɪkəl] [ˌveərɪˈeɪʃən]
地貌，地形 landform [ˈlæn(d)fɔːm]
地貌特征 morphological characteristics [ˌmɔːfəˈlɒdʒɪkəl] [ˌkærəktəˈrɪstɪks]
地面边界层 ground boundary [ˈbɑʊndərɪ]
地面粗糙度 ground roughness [ˈrʌfnɪs]
地球自转偏向力 deflection force of earth rotation
地形；地势 topography [təˈpɒgrəfɪ]
地形改造 topographic modification [ˌmɒdɪfɪˈkeɪʃ(ə)n]
地形干扰 topographic interference [ɪntəˈfɪər(ə)ns]
地形开敞度 topographic exposure factor [ɪkˈspəʊʒə]
地形起伏 topographic relief
地形气候 topoclimate
地形气候学 topoclimatology
地形特征 topographic feature
地形条件 orographic condition [ˌɔːrəʊˈgræfɪk]
地形狭道风 topography channel wind [ˈtʃæn(ə)l]
地形效应 topographic effect [ˌtɒpəˈgræfɪk]
地形因素 topographic element
地形影响 orographic effect；effects of terrain [teˈreɪn]
地转风 geo-strophic winds [ˈstrɒfɪk]
第一层方向本征 first edgewise eigenmodes
第一副翼方向本征 first flapwise eigenmodes
点腐蚀 spot corrosion/pitting [kəˈrəʊʒən]
点蚀 pitting
电磁制动系 electromagnetic braking system [ɪˌlektrəʊmægˈnetɪk]
电动风速计 anemocinemograph
电缆解绕 untwist the cables [ˌʌnˈtwɪst]
调节特性 regulating characteristics [ˌkærəktəˈrɪstɪks]
调频 frequency modulation (FM)

[ˌmɒdjʊˈleɪʃən]
调试界面 debugging interface [diːˈbʌɡɪŋ]
调速机构 regulating mechanism [ˈmekənɪzəm]
调向，对风 orientation control; veering; windseeking; yaw orientation [ˌɔːrɪənˈteɪʃ(ə)n]
调向机构，对风机构 yaw mechanism [ˈmek(ə)nɪz(ə)m]
调向灵敏度（顺风） sensitivity of following wind
调向死区 yaw drive dead band
调向尾舵，对风尾舵 yaw vane
调向稳定性 stability of following wind
调向装置，对风装置 yawing device
调整板 adjusting plate
顶风 upwind
顶环（风机） top ring
定桨距 fixed-blade pitch
定桨式风力机线性变参数模型 LPV model of fixed-pitch WECS
定向风轮 free-yaw rotor; yaw-fixed rotor
定址 siting
定址准则 siting criteria [kraɪˈtɪərɪə]
东北信风带 northeast trades [ˈtreɪdz]
动(势)能 kinetic(potential)energy
动触头 moving contact [ˈkɒntækt]
动量边界层 momentum boundary layer
动量方程 momentum equation [ɪˈkweɪʒ(ə)n]
动量方法 momentum method
动量厚度 momentum thickness [ˈθɪknɪs]
动量矩 moment of momentum
动量理论 Momentum Theory
动量守恒 momentum conservation [kɒnsəˈveɪʃ(ə)n]
动态分析；动力特性分析 dynamic analysis [daɪˈnæmɪk] [əˈnæləsɪs]
动态风载 wind dynamic load [daɪˈnæmɪk]
动态结构模型 dynamic structural models [ˈstrʌktʃərəl]
动态控制 dynamic control
动态模型 dynamic models
动态刹车 dynamic brakes
动态失速 dynamic stall
动态尾涡模型 dynamic wake models
动态响应；动力特性 dynamic response
动载模拟 dynamic load simulation
动载系数 dynamic factor
冻雨 freezing rain
独立变桨 independent electric drive
独立式塔架 free stand tower
独立运行系统 stand-along system
堵塞油路控制 blocking control
度电成本 cost per kilowatt hour of the electricity generated by WTGS [ˈkɪləuwɒt]
短期风资料 short-term wind data [ˈdeɪtə]
短期荷载 short duration loading [djʊˈreɪʃ(ə)n]
短期阵风 short-duration gust

[djʊ'reɪʃ(ə)n]
短时间闪变值 short term severity of flicker
短时切出风速 short-term cut-out wind speed
断面系数；剖面模数 section modulus
对称层板 symmetric laminate [sɪ'metrɪk] ['læmɪneɪt]
对风惯性 yaw inertia [ɪ'nɜːʃə]
对风控制机构 yaw control mechanism ['mek(ə)nɪz(ə)m]
对风制动 yaw brake
对数分布 logarithmic distribution [ˌlɒgə'rɪðmɪk]
对数分布律 logarithmic distribution law
对数风（速）廓线 logarithmic wind profile
对数风切变律 logarithmic wind shear law [ˌlɒgə'rɪðmɪk]
对数频谱 logarithmic frequency spectrum ['friːkw(ə)nsɪ] ['spektrəm]
对数剖面 logarithmic profile ['prəʊfaɪl]
对数衰减 log dec；logarithmic decrement [ˌlɒgə'rɪðmɪk]
对数速度廓线 logarithmic velocity profile [vɪ'lɒsɪtɪ]
对数线性分布 log linear distribution
对数线性风廓线 log linear wind profile
对数正态分布 log-normal distribution
钝体空气动力学 bluff body aerodynamics [ˌeərə(ʊ)daɪ'næmɪks]
多变风 variable-wind
多点测压系统 multipressure measuring system ['meʒrɪŋ]
多风性 windiness
多级轴流通风机 multistage axial-flow fan ['mʌltɪsteɪdʒ] ['æksɪəl]
多模态 multimode ['mʌltɪˌməʊd]
多向风；不定向风 multi-directional wind
多项式分布 multinomial distribution
多叶片风车 multibladed windmill ['wɪn(d)mɪl]
多叶片风轮 multi-bladed rotor
多自由度 multi-degree of freedom
舵轮；方向盘 steering wheel
舵盘 steering box ['stɪərɪŋ]
惰轮；空转轮；导轮 idler pulley ['pʊlɪ]

E

额定风速 rated wind speed
额定工况 rated condition
额定力矩系数 rated torque coefficient
额定输出功率 rated power output
额定叶尖速度比 rated tip-speed ratio

额定载荷 rated load

额定值 rated value

额定转矩 rated load torque [tɔːk]

额定转速 rated speed

恶劣气候运行 operation in severe climates

二维流动 bidimensional flow [ˌbaɪdɪˈmenʃənl]

二项分布 binomial distribution [baɪˈnəumɪəl]

二项展开式 binomial expansion [ɪkˈspænʃ(ə)n]

二自由度系统 2-DOF system

F

发动；开动；启动 start up

法向力 normal force

法向湍流模型 normal turbulence model [ˈtəːbjuləns]

法向位移 normal displacement [dɪsˈpleɪsm(ə)nt]

法向压力梯度 normal pressure gradient [ˈgreɪdɪənt]

法向诱导速度 normal induced velocity [vɪˈlɒsətɪ]

反弯度翼型 airfoil with reverse camber [rɪˈvəːs]

方位；朝向 orientation [ˌɔːrɪənˈteɪʃ(ə)n]

方位角 azimuth angle [ˈæzɪməθ]

方位角影响 effects of azimuth angle

方向探头 yaw probe [prəub]

方向性（风力发电机组） directivity (for WTGS)

防尘 dust-protected

防尘罩 dust cap

防滴 protected against dropping water

防毒面具 gas mask [mɑːsk]

防风林；风障 wind break

防溅 protected against splashing

防浸水 protected against the effects of immersion [ɪˈməːʃən]

防雷区 lightning protection zone

防雷系统 lightning protection system

防冷凝加热器 anti condensation heater

防水篷布罩 tarpaulin covering (for wind turbines)

防雨罩 rain deflector

防坠落防护 fall protection system

防坠落轨道（爬梯） (ladder with) preassembled fall arrest rail

放电 sparkover

飞溅润滑 splash lubrication

飞逸 run-away

非承重部件 non-load bearing element

非定常边界层 unsteady boundary layer

非定常边界元法 unsteady BEM model

非定常伯努利方程 unsteady Bernoulli's equation [bəːˈnuːlɪ]

非定常流 unsteady flow

非定常气动力 unsteady aerodynamics [ˌeərəʊdaɪˈnæmɪks]
非定常翼型理论 unsteady airfoil theory
非对称力 asymmetric force
非对称模态 asymmetric mode
非对称阻尼 asymmetric damping [ˌesɪˈmetrɪk]
非均匀流动 inhomogeneous flow [ˌɪnhɒmə(ʊ)ˈdʒiːnɪəs]
非均匀湍流 inhomogeneous turbulence; non-isotropic turbulence [ˈtɜːbjʊl(ə)ns]
非流线型的 unstreamlined
非破坏性试验 non-destructive testing [dɪˈstrʌktɪv]
非设计工况 off-design condition
非收缩性 non-shrinking [ʃrɪŋkɪŋ]
非线性 non-linearity
非线性空气动力学 nonlinear aerodynamics [ˌeərə(ʊ)daɪˈnæmɪks]
非线性增益 non-linear gains
非再生能源 non-renewable fuels [rɪˈnjuːəbl] [fjʊəlz]
非黏性流 inviscid flow [ɪnˈvɪsɪd]
分布函数 distribution function
分类排列 assorted files
分流；汽流分离 slow separation
分组方法 method of bins
风（力）级 wind class/scale
风杯风速计 cup anemometer [ˈænɪˈmɒmɪtə]
风变性 wind variability [ˌveərɪəˈbɪlətɪ]
风标；导风板 air vane
风标；对风尾舵 wind vane
风不连续性 wind discontinuity
风侧滑角，风横偏角 wind yaw angle [ˈæŋg(ə)l]
风场 wind power plant; wind farm; wind site
风场布局 wind farm layout [ˈleɪaʊt]
风场电器设备 site electrical facilities
风场评价 wind assessment
风车 windmill [ˈwɪndmɪl]
风车式风速计 windmill anemometer [ˈwɪn(d)mɪl] [ænɪˈmɒmɪtə]
风车制动状态 windmill brake state [ˈwɪndmɪl]
风程 wind run
风挡刮水器；风挡雨雪刷 windscreen wiper
风道；烟道 air flue
风的测定 wind measurement
风电场 wind power station
风电场服务器 farm server
风电场功率曲线 farm power curve
风电场过滤器 wind farm filter
风电机组停机 parked wind turbine
风动涡旋 wind spun vortex [spʌn] [ˈvɔːteks]
风洞 wind tunnel [ˈtʌnəl]
风洞（试验段）观察窗 wind tunnel window
风洞；风道 wind channel [ˈtʃæn(ə)l]
风洞吹程 wind tunnel fetch
风洞吹风时间；风洞占用时间 wind tunnel time

风洞导流片 wind tunnel turning vane
风洞顶壁 wind tunnel roof
风洞洞壁干扰 wind tunnel wall interference [ɪntə'fɪər(ə)ns]
风洞洞壁效应 wind tunnel wall effect
风洞风速 wind tunnel speed
风洞回流道 wind tunnel circuit ['sɜːkɪt]
风洞截面几何形状 wind tunnel geometry [dʒɪ'ɒmɪtrɪ]
风洞扩散段 wind tunnel diffuser [dɪ'fjuːzə]
风洞雷诺数 wind tunnel Reynolds number ['renəldz]
风洞模拟技术 wind tunnel modelling technique [tek'niːk]
风洞模型 wind tunnel model
风洞能量比 wind tunnel energy ratio ['reɪʃɪəʊ]
风洞实验 wind tunnel tests
风洞实验室 wind tunnel laboratory ['læb(ə)rət(ə)rɪ]
风洞实验数据 wind tunnel data
风洞试验段 wind tunnel test section
风洞收缩段；风洞收缩比 wind tunnel contraction [kən'trækʃ(ə)n]
风洞湍流（度） wind tunnel turbulence ['tɜːbjʊl(ə)ns]
风洞效率 wind tunnel efficiency
风洞壅塞 wind tunnel choking
风洞噪声 wind tunnel noise
风洞轴系 wind tunnel axes system ['æksiːz]
风洞轴线 wind tunnel centerline ['sentəlɑɪn]
风方位角 wind azimuth ['æzɪməθ]
风感受器 wind sensor ['sensə]
风工程 wind engineering
风工程师 wind engineer
风功率密度 wind power density
风荷载 wind load
风环流；气流循环 wind circulation [sɜːkjʊ'leɪʃ(ə)n]
风机 wind turbine
风机工况 turbine state
风机性能曲线 $C_p-\lambda$ curve
风机用螺母 wind nut
风镜；护目镜 goggles ['gɒglz]
风口；风隙 wind gap
风况 wind regime
风况类别（IEC） IEC wind class
风廓线 wind profile ['prəʊfaɪl]
风切变律 wind shear law
风廊 wind porch [pɔːtʃ]
风力 wind force/strength/stress
风力泵 wind pump
风力发电厂 wind power plants
风力发电机 wind turbine generator; aerogenerator [ˌeərəʊ'dʒenəreɪtə]
风力发电机端口 wind turbine terminals ['tɜːmɪnlz]
风力发电机停机 parked wind turbine
风力发电机组 wind turbine generator system (WTGS)
风力发电机组输出特性 output characteristic of WTGS [ˌkærəktə'rɪstɪk]

风力发电机组效率 efficiency of WTGS [ɪ'fɪʃənsɪ]

风力发电机最大功率 maximum power of wind turbine ['mæksɪməm]

风力发电系统 WECS

风力发电装置 wind-power installation [ˌɪnstə'leɪʃən]

风力功率 power of the wind

风力机 wind turbine

风力机端口 wind turbine terminal

风力机械 wind powered machinery [mə'ʃiːn(ə)rɪ]

风力机噪音 wind-turbine noise

风力机最大功率 maximum power of wind turbine

风力级 wind force scale

风力计；风速仪 anemometer [ˌænɪ'mɒmɪtə]

风力记录仪 anemograph [ə'neməgrɑːf]

风力加热器 wind furnace ['fɜːnɪs]

风力热泵 wind-powered heat pump

风力提灌系统 wind powered irrigation system [ˌɪrə'geʃən]

风力微区划 wind microzonation

风力影响；船身挡风面 windage

风力助航系统 wind propulsion system [prə'pʌlʃ(ə)n]

风力资源 wind resource

风力资源评价 wind resource assessment

风力自记曲线 anemogram [ə'neməgræm]

风流，气流 wind current ['kʌr(ə)nt]

风流能密度 wind stream power density

风轮 wind rotor

风轮额定转速 rated turning speed of rotor

风轮机 aeroturbine

风轮机停机 shutdown for wind turbine

风轮机支撑结构 support structure for wind turbine

风轮空气动力特性 aerodynamic characteristics of rotor[ˌeərəudaɪ'næmɪk]

风轮偏侧式调速机构 regulating mechanism of turning wind rotor out of the wind sideward ['mekənɪzəm] ['saɪdwəd]

风轮偏航角 yawing angle of rotor shaft

风轮扫风面积 rotor swept area

风轮实度 rotor solidity [sə'lɪdɪtɪ]

风轮推力 rotor-thrust

风轮尾流 rotor wake

风轮仰角 tilt angle of rotor shaft [tɪlt]

风轮叶片扭转 rotor blade twist

风轮直径 rotor diameter [daɪ'æmɪtə]

风轮质量不平衡 rotor mass imbalance

风轮轴偏航角 yawing angle of rotor shaft

风轮转速 rotor speed

风轮最高转速 maximum turning speed of rotor

风玫瑰图 wind rose

风门；阻尼器；防震器；防振锤 damper

风弥散（作用） wind dispersion [dɪˈspɜːʃ(ə)n]

风描述 wind description [dɪˈskrɪpʃ(ə)n]

风模拟 wind simulation

风磨蚀 wind abrasion [əˈbreɪʒ(ə)n]

风能 wind energy/power

风能持续时间曲线 wind power duration curve [djʊˈreɪʃ(ə)n]

风能分布曲线 energy distribution curve [ˌdɪstrɪˈbjuːʃən]

风能工程 wind energy engineering [endʒɪˈnɪərɪŋ]

风能获取极限值 wind energy extraction limit [ɪkˈstrækʃ(ə)n]

风能廓线 wind power profile [ˈprəʊfaɪl]

风能利用系数 rotor power coefficient

风能利用系数跟踪 C_p tracking

风能利用系数曲线 C_p performance curve [kɜːv]

风能玫瑰（图） wind energy rose

风能密度 wind energy density; wind power density

风能密度 wind energy density [ˈdensətɪ]

风能频率分布直方图 frequency distribution (histogram) diagram

风能谱 wind energy spectrum [ˈspektrəm]

风能潜力 wind energy potential; wind power potential [pə(ʊ)ˈtenʃ(ə)l]

风能区别 wind energy region [ˈriːdʒ(ə)n]

风能收集系统，风力发电系统 Wind Energy Collection Systems (WECS)

风能收集装置 wind energy collector

风能田，风车田 wind energy farm

风能通量 wind energy flux

风能溢出（限速法） wind spilling [spɪlɪŋ]

风能溢出式限速装置 wind spilling type of governor [ˈgʌv(ə)nə]

风能蕴藏量 wind energy content

风能转换 wind energy conversion

风能转换系统 Wind Energy Conversion System (WECS)

风能转换效率 wind conversion efficiency

风能转换装置 wind energy converter

风能资源 wind energy resource [rɪˈsɔːs; rɪˈzɔːs]

风能资源估计 wind resource estimation [ˌestɪˈmeɪʃən]

风能资源图谱 wind resources atlases

风频率分布 wind frequency distribution [ˈfrɪkwənsɪ]

风频曲线 wind frequency curve

风谱 wind spectrum [ˈspektrəm]

风气候 wind climate

风强度 wind intensity [ɪnˈtensɪtɪ]

风切变 wind shear

风切变对数幂次法则 power law for wind shear
风切变律 wind shear law
风切变幂律 power law for wind shear
风切变影响 influence by the wind shear
风切变指数 wind shear exponent [ɪkˈspəunənt]
风区 wind sector
风沙气候 sand storm climate
风扇轮毂 fan hub
风扇叶片 fan blade
风蚀 wind corrosion [kəˈrəuʒ(ə)n]
风矢 wind arrow
风矢杆；风轮转轴 wind shaft
风矢量 wind velocity
风数据，风资料 wind data
风速 wind speed
风速测量 wind speed measurement
风速持续时间曲线 wind speed duration curve [djʊˈreɪʃ(ə)n]
风速分布 wind speed distribution
风速概率分布 probability distribution of wind speeds
风速级 wind speed scale
风速计 anemometer；wind gauge；wind meter [ˌænɪˈmɒmɪtə][geɪdʒ]
风速计系数 anemometer factor
风速较差直方图 amplitude histogram of wind-speed [ˈæmplɪtjuːd]
风速廓线 wind (speed/velocity) profile
风速雷诺数 wind speed Reynolds number [ˈrenəldz]
(埃克曼）风速螺线 wind velocity spiral [ˈspɑɪr(ə)l]
风速轮廓线 wind profile
风速频率 wind speed frequency
风速频数曲线 wind speed frequency curve
风速谱 wind speed spectrum [ˈspektrəm]
风速一时间曲线 speed duration curve [djʊəˈreɪʃən] [kɜːv]
风速梯度 wind velocity gradient [vɪˈlɒsɪtɪ] [ˈgreɪdɪənt]
风特性 wind characteristic [ˌkærəktəˈrɪstɪk]
风梯度 wind gradient [ˈgreɪdɪənt]
风统计(学) wind statistics [stəˈtɪstɪks]
风图 wind chart/map
风图谱 wind-atlas [ˈætləs]
风湍流（度） wind turbulence [ˈtɜːbjʊl(ə)ns]
风涡 wind eddy [ˈedɪ]
风吸力：风抽吸 wind suction
风向 wind direction
风向标 wind vane
风向波动 wind direction fluctuation [ˌflʌktjʊˈeɪʃən]
风向测量 wind direction measurement
风向传感器 wind transducer [trænzˈdjuːsə]
风向袋 wind cone
风向风速仪 aerovane [ˈeərəˌveɪn]
风向感受器 wind direction sensor [ˈsensə]
风向跟踪，对风 wind following

风向界线 wind divide [dɪˈvaɪd]
风向玫瑰图 rose diagram
风向偏转 wind deflection [dɪˈflekʃ(ə)n]
风向频率 wind direction frequency
风向仪 anemoscope [əˈneməskəʊp]
风向转变 wind shift/veering [ˈvɪərɪŋ]
风效应 wind effect
风压 wind pressure
风压测量孔 wind pressure tap
风迎角，风角 wind angle
风影；背风区 wind shadow
风载标准 wind loading standard [ˈstændəd]
风载谱 wind load spectrum [ˈspektrəm]
风载设计准则 wind loading design criterion [kraɪˈtɪərɪən]
风噪声 wind noise
风障 wind break
风致荷载 wind induced load [ɪnˈdjuːst]
风致压力 wind induced pressure
风致振动 wind induced vibration [vaɪˈbreɪʃ(ə)n]
风资料 wind information
风阻 windage resistance; wind drag [rɪˈzɪst(ə)ns]
风阻荷载 wind drag load
风阻力 drag
风阻损失 windage loss
峰荷 peak-load
峰值 peak value
冯卡曼常数 von Karman constant
浮力 buoyancy [ˈbɒɪənsɪ]
浮力；浮升 buoyancy lift [ˈbɒɪənsɪ]
浮力修正 correction for buoyancy [ˈbɒɪənsɪ]
浮力中心 center of buoyancy
浮式风电机组 floating wind turbine
浮质；气溶胶；气雾剂；烟雾剂 aerosol [ˈeərəsɒl]
辐射 radiation [ˌreɪdɪˈeɪʃən]
辐射通量 radiant flux [ˈreɪdɪənt]
俯角 angle of depression [dɪˈpreʃ(ə)n]
俯仰力矩 pitching moment
辅助失速 assisted stall
辅助装置 auxiliary device [ɔːgˈzɪljərɪ]
腐蚀 corrosion
腐蚀性气候 corrosive climate
负变桨控制 negative pitch control
负浮力 negative buoyancy [ˈbɒɪənsɪ]
负荷调整 load leveling
负气动刚度 negative aerodynamic stiffness [ˌeərəʊdaɪˈnæmɪks] [ˈstɪfnɪs]
负气动阻尼 negative aerodynamic damping [ˌeərəʊdaɪˈnæmɪks]
负相关系数 negative correlation coefficient [ˌkɒrəˈleɪʃ(ə)n] [ˌkəʊɪˈfɪʃ(ə)nt]
负压 negative pressure
负压系数；吸力系数 suction coefficient [ˌkəʊɪˈfɪʃ(ə)nt]
负迎角 negative angle of attack

[ˈæŋg(ə)l]
负载比 duty ratio [ˈreɪʃɪəʊ]
负载特性 load characteristic [ˌkærəktəˈrɪstɪk]
负阻尼 negative damping
附加荷载 extraneous loading [ɪkˈstreɪnɪəs]
附加质量 added mass
附着环量 bound circulation
附着涡；约束涡 bound vortex[ˈvɔːteks]
附着涡量 bound vorticity
附着涡系 bound vortex system
附着系数 attachment coefficient [əˈtætʃmənt] [ˌkəʊɪˈfɪʃənt]
复合材料 composite material
复合材料性质 properties of composites [kəmˈpəʊzɪt]
复合材料叶片 composite blade
复合式轮毂（联轴器用） composite spacer
复合填料 jointing compound
复杂地形带 complex terrain[teˈreɪn]
副翼 aileron [ˈeɪlərɒn]
副翼方向弯曲力矩 flapwise bending moment
副翼控制 aileron control
傅立叶变换 Fourier transform
傅立叶分量 Fourier component [kəmˈpəʊnənt]
傅立叶分析 Fourier analysis [ˈfʊrɪeɪ]
傅立叶级数；傅立叶序列 Fourier series [ˈfʊrɪeɪ]
傅立叶系数 Fourier coefficient
傅立叶相位谱 Fourier phase spectrum [ˈspektrəm]
傅立叶变换 Fourier transformation [trænsfəˈmeɪʃ(ə)n]
覆盖；敷层；有效区域 coverage [ˈkʌvərɪdʒ]

G

伽马函数 gamma function [ˈgæmə]
概率密度分布函数 probability density functions [ˌprɒbəˈbɪlətɪ]
干扰区域 interference regions [ˌɪntəˈfɪərəns]
干涉；干预；插入；介入 intervene
刚性塔 stiff towers
钢管塔筒 steel tubular tower [ˈtjuːbjʊlə]
钢筋混凝土 reinforced concrete
高度表 altimeter [ˈæltɪmiːtə]
高度影响；高度效应 effect of altitude
高固成分润滑(风电机组润滑系统用语) lubricants with high solid contents
高空风 wind aloft [əˈlɒft]
高空观测 aerological ascent [ˌeərəʊˈlɒdʒɪkəl]
高空观测站 aerological station
高空气象学 aerology [eəˈrɒlədʒɪ]
高轮廓线 high profile [ˈprəʊfaɪl]
高斯分布 Gauss distribution [gaʊs]

高速闸 high speed brake
高速轴 high speed shaft
高速转子 high-speed rotor
高黏度液体 high viscosity fluid
格里戈—普特南级数（植物风力指标） Griggs-Putnam index
格子形钢塔架 lattice steel towers [ˈlætɪs]
各向同性点源 isotropic point source
各向同性弥散 isotropic dispersion [ˌɑɪsə(ʊ)ˈtrɒpɪk] [dɪˈspɜːʃ(ə)n]
各向同性湍流 isotropic turbulence [ˈtɜːbjʊl(ə)ns]
各向异性湍流 anisotropic turbulence [ˌænɑɪsə(ʊ)ˈtrɒpɪk] [ˈtɜːbjʊl(ə)ns]
根涡 root vortex [ˈvɔːteks]
工况 operating condition [ˈɒpəreɪtɪŋ]
工频续流 power-flow current
工作齿面 working flank
工作点 operating point
工作环境 operational environment [ɪnˈvɑɪərənmənt]
工作时间 operating time
工作状态 operating states
功率谱 power spectrum [ˈspektrəm]
功率谱密度 power spectral density (PSD)
功率曲线 power curve [kɜːv]
功率失常激增 power excursions
功率—时间曲线 power duration curve [djʊəˈreɪʃən]
功率输出 power output
功率特性 power performance
功率特性测试数据组 data set for power performance measurement [ˈmeʒəmənt]
功率系数 power coefficient/factor [ˌkəʊɪˈfɪʃənt]
功率摇摆 power swings
功率与风速曲线 power versus wind speed curve [ˈvɜːsəs]
攻角 angle of attack
共振 resonance [ˈrezənəns]
共振尖响应 resonant tip response
共振弯矩 resonant bending moment
共振响应 resonant response
共振引起的载荷 resonance-induced loads
孤立地区 isolated area
孤立翼型 isolated aerofoil [ˈeərəfɒɪl]
孤立运行 isolated operation
古德曼曲线 Goodman diagram [ˈgʊdmæn] [ˈdɑɪəgræm]
谷底 valley floor
谷风 valley breeze/wind [briːz]
谷风环流 valley wind circulation
谷坡风 valley slope wind
固定轮毂 fixed hub
固定叶片 stationary blade
固定坐标系统 fixed coordinate systems [kəʊˈɔːdɪnɪt]
固有频率 natural frequency[ˈfrɪkwənsɪ]
固有振荡 natural oscillation [ˌɒsɪˈleɪʃən]
固有振动 natural vibration [vɑɪˈbreɪʃ(ə)n]
固有周期 eigenperiod

故障 fault [fɔːlt]
故障安全组件 fail-safe component
故障查找 fault finding
故障穿越 fault ride through (FRT)
故障电流 fault current [ˈkʌrənt]
故障风险 fail-safe
故障接地 fault earthing
故障失效，故障自动保护 fail-safe
故障条件 fault conditions
刮风小时数 hours of wind
关机 shutdown for wind turbine
观测气球 observation balloon
观测误差 observation error
管状塔 tubular towers [ˈtjuːbjʊlə]
惯性滑行 freewheeling
惯性矩，转动惯量 moment of inertia [ɪˈnɜːʃɪə]
惯性开关 inertia switch
惯性力 inertia forces [ɪˈnɜːʃɪə]
惯性载荷 inertia loads
光电器件 photoelectric device
光伏发电 photovoltaic power [ˌfəʊtəʊvɒlˈteɪɪk]
光滑涂料 gloss paint
光谱 spectrum [ˈspektrəm]
广义力向量 generalized force vector [ˈvektə]
广义载荷 generalized load
广义质量 generalized mass
广义坐标 generalized coordinates [kəʊˈɔːdɪneɪts]
归共同谱 normalized co-spectrum
归化条件 normalization condition
归一化；正态化 normalization [ˌnɔːməlaɪˈzeɪʃən]
归一化常数 normalization constant
归一化功率谱函数 normalized power spectrum function [ˈspektrəm]
归一化互谱 normalized cross spectrum
归一化谱 normalized spectrum
归一化系数 normalized coefficient [ˌkəʊɪˈfɪʃ(ə)nt]
规定；保证 stipulate [ˈstɪpjʊleɪt]
规定的最初起动转矩 specifies breakaway torque [spesɪfaɪs]
硅胶加热垫 silicone heating pad [ˈsɪlɪkəʊn]
硅胶加热器 silicone heater
硅橡胶 silicon rubber
硅油密封 silicon oil sealed
国际标准化组织 ISO (international standardization organization)
国际电工（技术）委员会 IEC (international Electrotechnical Commission)
国际风工程学会 IAWE (International Association for Wind Engineering)
国际能源署 International Energy Agency (IEA)
国家电网 national grid [grɪd]
国家风力资源 national wind resource
过速 overspeed
过速控制 over speed control
过压 overpressure
过载度 ratio of over load
过载功率 over power

H

海岸风 shore wind
海岸区 on-shore area
海拔 altitude [ˈæltɪtjuːd]
海拔高度的影响 effect of altitude
海底光纤电缆 submarine fibre optic cable system
海风 landward wind
海陆风 land-sea breezes
海上风力发电 offshore wind turbines
海上风能 offshore wind energy [ˈɒfˈʃɒ]
海上平台 off-shore oil rig
海洋大气 marine atmosphere
海洋风 ocean wind
海洋风况 marine wind regime [reɪˈʒiːm]
海洋环境 marine environment [ɪnˈvɑɪrənm(ə)nt]
海洋气团 maritime air mass
海洋气象学 marine meteorology [ˌmiːtɪəˈrɒlədʒɪ]
海洋性环境 marine environment
海洋性气候 ocean climate [ˈəʊʃən]
亥姆霍兹定理 Helmholtz theorem [ˈhelmhəʊlts]
含有周期分量的功率谱 power spectra containing periodic components [ˌpɪərɪˈɒdɪk]
航空风洞 aeronautical type wind tunnel [ˌeərəˈnɔːtɪkl]
航空学 aeronautics [eərəˈnɔːtɪks]
核电站 nuclear power station
核反应堆 nuclear reactor
核能 nuclear power
赫兹 Hz (Hertz)
恒速 constant rotational speed [rəʊˈteɪʃənəl]
恒速变桨距 fixed-speed variable-pitch (FS-VP)
恒速定桨距 fixed-speed fixed-pitch (FS-FP)
恒速发电机 fixed speed generator
恒速风力机 fixed-speed turbines
恒速运行 fixed speed operation
恒星齿轮 sun gear
桁架式塔架 truss tower [trʌs]
横摆惯性矩；偏航惯性矩 yawing moment of inertia [ɪˈnɜːʃə]
横档（爬梯用） rung
横风 cross wind; wind abeam
横风抖振 cross-wind buffeting
横风扩散 cross-wind diffusion
横风向的 cross-wind direction
横风轴风力机 cross-wind-axis wind machine
横剖 transversal cut [trænzˈvɜːsəl]
横切的；横轴 transverse
横纹（防滑） crosswise ribbing
横向尺度 lateral scale
横向荷载 lateral load
横向剪切（力） lateral shear
横向扩散 lateral diffusion [dɪˈfjuːʒ(ə)n]

横向流动，横流向的 cross flow
横向弯曲 chordwise bending [ˈkɔːdwaɪz]
横向相关函数 lateral correlation function [ˌkɒrəˈleɪʃ(ə)n]
横向阵风 lateral gust
横向振动 lateral oscillation [ˌɒsɪˈleɪʃən]
横向阻力 lateral resistance [rɪˈzɪst(ə)ns]
横轴风车；水平轴风力机 horizontal axis windmill [ˈhɒrɪˈzɒntəl]
宏观尺度 macroscale [ˈmækrəskeɪl]
宏观气象的 macro-meteorological [ˈmækrəʊ] [miːtɪərəˈlɒdʒɪkəl]
宏观黏性 macro viscosity [vɪˈskɒsɪtɪ]
虹吸管 siphon [ˈsɑɪfən]
后缘 trailing edge
后缘失速 trailing edge stall
后缘旋涡 trailing vortex
厚度函数 thickness function of airfoil [ˈθɪknɪs] [ˈeəfɒɪl]
厚弦比 thickness/chord ratio
弧线 camber line
胡克定律 Hooke's law
互连(风力发电机组) interconnection (for WTGS) [ˌɪntəkəˈnekʃən]
护栏 guard rail
滑环 slip rings
化学腐蚀 chemical corrosion
环境湿度 ambient humidity [ˈæmbɪənt] [hjʊˈmɪdɪtɪ]
环境条件 environment condition [ɪnˈvɑɪərənmənt]
环境温度 ambient temperature [ˈæmbɪənt] [ˈtempərɪtʃə]
环境污染负荷 ambient pollution burden
环境噪声 ambient noise
环境振动 ambient vibration
环流 circulatory flow
环面 annulus [ˈænjʊləs]
环网系统 loop system
环涡 ring vortices [ˈvɔːtɪˌsiːz]
环形扳子 ring spanner
环形接地体 ring earth external [ɪkˈstɜːnəl]
环氧基树脂 epoxy [epˈɒksɪ]
环氧树脂 epoxy resin [ˈrezɪn]
缓冲装置 shock absorber
换流器；变换器；变流器 converter
换向片 commutator segment [ˈkɒmjʊteɪtə] [ˈsegmənt]
换向器 commutator
挥舞—摆振 flap-lag
辉光放电 glow discharge
回复力 restoring force
回油（口） return flow
回转支承轴承 slewing ring bearings
混合系统 Hybrid systems [ˈhɑɪbrɪd]
混凝土塔筒 concrete tower [kənˈkriːt]

J

击穿 breakdown
机舱 nacelle [nəˈsel]
机舱底座 nacelle bedplate
机舱加速度 nacelle acceleration [əkˌseləˈreɪʃən]
机舱罩 nacelle cover
机舱装载 nacelle loading
机侧额定容量 rotor side rated capacity
机器生产率曲线 machine productivity curve
机械零件；机械元件 machine elements
机械寿命 mechanical endurance [mɪˈkænɪkəl] [ɪnˈdjʊərəns]
机械图 mechanical drawing
机械载荷 mechanical loads
机械噪声 mechanical noise
机械黏合剂 machinery adhesives
机械制动 mechanical brake [mɪˈkænɪkəl]
机械制动系 mechanical braking system
机翼 aerofoil [ˈeəfɒɪl]
机组变压器 unit transformers
机组效率 efficiency of WTGS
积分时间和长度尺度 integral time and length scales [ˈɪntɪgrəl]
基础；根本；建立；创立；地基 foundation [faʊnˈdeɪʃən]
基础环 foundation ring
基础接地体 foundation earth electrode [ɪˈlektrəʊd]
基架；底座；基础 pedestal [ˈpedɪstəl]
基准粗糙长度 reference roughness length [ˈrʌfnɪs]
基准风速 acoustic reference wind speed [əˈkuːstɪk]
基准高度 reference height [hɑɪt]
基准距离 reference distance
基准误差 basic error
畸变 distortion
畸变率（电流）(current) distortion rate
激光对中仪 laser shaft alignment instrument
激光跟踪仪 laser tracker
激紊线 trip wire
极端大气现象 extreme atmospheric events [ætməsˈferɪk]
极端风 extreme wind
极端风速 extreme wind speed [ɪkˈstriːm]
极端梯度风 extreme climate wind
极端载荷 extreme load
极端阵风 extreme gust [ɪkˈstriːm]
极端最高 extreme maximum [ˈmæksɪməm]
极限环境条件 extreme environmental condition
极限限制状态 ultimate limit state [ˈʌltɪmət]
极限载荷 ultimate loads

极限值 limiting value [ˈvæljuː]
极限状态 limit state
极限状态设计 limit-state design
极性效应 polarity effect
[pəʊˈlærətɪ]
极值 extreme values
疾风(七级风) moderate gale [ɡeɪl]
集成变速箱 integrated gearbox
集成设计 bearing and generator are integrated
集风器 wind concentrator
集中荷载 concentrated load
集中控制 centralized control
集中润滑系统 central lubrication/centralized lubricating system
集中式风能转换系统 centralized wind energy systems
[ˈsentrəlɑɪzd]
几何攻角 attack of blade
几何桨距 geometric pitch
[dʒɪəʊˈmetrɪk]
几何弦长 geometric chord of airfoil
[dʒɪəʊˈmetrɪk] [ˈeəfɒɪl]
几何相似风力机 similar geometric wind machine [dʒɪəʊˈmetrɪk]
挤压铝叶片 extruded aluminium blade [æl (j) ʊˈmɪnɪəm]
计数风杯风速计 cup-counter anemometer
计算空气动力学 computational aerodynamics [ˌkɒmpjʊˈteɪʃənl]
计算流体力学 computational fluid dynamics (CFD)
技术问题 technical issues
[ˈɪʃjuːz]
风电机组技术许可证 licensing of wind turbine technology
加强器 intensive
加热；过度；使发展过快；使过热 overheat [ˌəʊvəˈhiːt]
加速 accelerating [ækˈseləreɪtɪŋ]
夹板；托板 splint [splɪnt]
夹层构造 sandwich construction
夹线器 conductor holder
夹心材料 sandwich material
尖速比 tip speed ratio; tip speed-to-wind speed ratio
尖削结构 tapered structure
尖削叶片 tapered blade [ˈteɪpəd]
间隔棒 spacer
减风效率 wind reduction efficiency
减压阀 reducing valve
减载 load relief [rɪˈliːf]
减震器 vibration isolator
[vɑɪˈbreɪʃən] [ˈɑɪsəleɪtə]
减震器；震动吸收器 shock absorber [əbˈsɔːbə]
剪切变形；切变 shear deformation
剪切层；切变层 shear layer
剪切负荷 shear loads
剪切角 angle of shear [ʃɪə]
剪切流 shear flow
剪切模量 shear modulus [ˈmɒdjʊləs]
剪切速度 shear velocity [vɪˈlɒsɪtɪ]
剪切湍流 shear turbulence
剪切中心 shear center [ʃɪə]
剪应力 shear stress
检测阻抗 detection impedance
[ɪmˈpiːdəns]
检修接地 inspection earthing

建设管理费 construction administration expense
建设用地费 expenses for occupying construction land
渐近法 asymptotic method
渐近分布 asymptotic distribution [ˌæsɪmpˈtɒtɪk] [dɪstrɪˈbjuːʃ(ə)n]
渐近稳定性 asymptotic stability
桨距，叶片节距 blade pitch
桨距调节 pitch regulation
桨距调节气缸 pitch cylinder [ˈsɪlɪndə]
桨距角 pitch angle
桨距角气动力矩增益 pitch angle to aerodynamic torque gain [ˌeərəʊdaɪˈnæmɪk]
桨距控制 blade pitch control
桨叶根套 bladed cuff
桨叶空气动力（学） blade aerodynamics [ˌeərə(ʊ)daɪˈnæmɪks]
桨叶锥角 blade coning angle
降水 precipitation [prɪˌsɪpɪˈteɪʃən]
交变荷载 alternating load
交变压力 alternating pressures
交变应力 alternating stress
交叉逆风 cross headwind
交叉相关函数 cross correlation function
交互作用（与电网） interaction (with grid)
交流电动机的最初启动电流 break-away starting current of Ac motor
交替分布涡 alternately spaced vortex [ˈvɔːteks]
交替旋涡脱落 alternating vortex shedding
胶衣 gel coated
角动量；动量矩 angular momentum [məˈmentəm]
角度位置编码器 angle position encoder
角分布 angular distribution [dɪstrɪˈbjuːʃ(ə)n]
角盘 angle plate
角偏转 angular deflection [dɪˈflekʃ(ə)n]
角速度 angular velocity [vɪˈlɒsɪtɪ]
角位移 angular displacement [dɪsˈpleɪsm(ə)nt]
角相关 angular correlation [ˈæŋgjʊlə]
角形绕组 Delta winding
绞车；绞盘 winch [wɪntʃ]
脚手架 scaffolding [ˈskæfəldɪŋ]
铰接 articulation [ɑːtɪkjʊˈleɪʃ(ə)n]
铰接叶片 articulated blade [ɑːˈtɪkjʊletɪd]; hinged blades [hɪndʒd]
轿厢（升降梯用） cabin
阶跃响应 step response
接地基准点 earthing reference points [ˈrefərəns]
接近（距离）感测器 proximity sensor
接近开关 proximity switch
接口 interface [ˈɪntəfeɪs]
接闪器 air-termination system [ˌtɜːmɪˈneɪʃən]
接续管 splicing sleeve [ˈsplaɪsɪŋ]
节点 pitch point
节流板；挡板 cut-off plate

节流阀 throttle valve
结构性质 structural properties
结构载荷 structural loads
结构阻尼 structural damping [ˈstrʌktʃərəl]
结合涡；附着涡 bound vorticity
截止频率 cut-off frequency
解缆 untwist [ˌʌnˈtwɪst]
介质常数 dielectric constant [ˌdɑɪɪˈlektrɪk]
介质试验 dielectric test
介质损耗 dielectric loss
襟翼；副翼 flap
紧固件 fasteners
紧急出口（机舱） emergency escape
紧急关机 emergency shutdown for wind turbine [ɪˈmɜːdʒənsɪ]
紧急顺桨 emergency feathering
紧急停车按钮 emergency stop push-button
紧急停机（风力机） emergency shutdown (for wind turbines)
紧急制动系统 emergency braking system [ɪˈmɜːdʒənsɪ]
近地风 near-surface wind
近地面层 near-surface layer
近海风电场 offshore wind farm
近海工程 off shore engineering [endʒɪˈnɪərɪŋ]
近海环境 off shore environment
近海区 off shore area
近似相似型 approximate similarity [əˈprɒksɪmət] [sɪməˈlærətɪ]
近尾流 near wake
近尾流区 near wake region [ˈriːdʒən]
进谷风 up-valley wind
进气口 air inlet/intake
进人孔（机舱） hatch
浸入式加热器 immersion heater
经纬仪 theodolite [θɪˈɒdəlaɪt]
晶闸管软启动装置 thyristor soft-start unit
景观评估 landscape assessment [ˈlændskeɪp]
景观影响 visual impact
净辐射 net radiation
净荷载 net load
静压测量孔 orifice static tap [ˈstætɪk]
静载荷 static loads
静止 standstill
静止状态；停顿；静止；停机 standstill
局部地形 local topography
局部放电 partial discharge
局部分离 local separation [sepəˈreɪʃ(ə)n]
局部风 local wind
局部风况 local wind regime [reɪˈʒiːm]
局部风廓线 local wind profile
局部环流 local circulation [sɜːkjʊˈleɪʃ(ə)n]
局部气候 local climate
局部气候条件 local climate condition
局部气流，当地流动 local flow
局部阵风 local gust

局域网 LAN (local area network)
距离常数 distance constant
飓风（十二级以上） hurricane [ˈhʌrɪk(ə)n]
飓风边界层 hurricane boundary layer [ˈbaʊnd(ə)rɪ]
飓风核心 hurricane core
飓风眼 hurricane eye
绝对湿度 absolute humidity [hjuːˈmɪdətɪ]
绝缘比 insulation ratio
绝缘表 insulation tester
绝缘材料 insulant [ˈɪnsjʊlənt]
绝缘层材料 lagging
绝缘配合 insulation coordination [kəʊˌɔːdɪˈneɪʃən]
绝缘手套 insulating glove [ɡlʌv]
绝缘套管 insulating bushing
绝缘靴 insulating boots
均方根值 rms (root mean square)
均压环 grading ring
均匀场 uniform field [ˈjuːnɪfɔːm]
均匀定常风 uniform steady wind
均匀风 uniform wind

K

卡尔曼功率谱 Kalman power spectrum
卡尔曼滤波器 Kalman filter
开敞状态 open condition
开放式桁架塔；桁架式塔架 open-truss tower
开环的 open-loop
开路式风洞 open-ended wind tunnel
坎贝尔图 Campbell diagram
抗点蚀 pitting resistance
抗风；风阻（力） wind resistance [rɪˈzɪst(ə)ns]
抗风桁架 wind truss [trʌs]
抗风设计 wind resistant design
抗风特性 wind-resistant feature
抗剪腹板 shear webs
抗龙卷风设计 tornado-resistant design [rɪˈzɪstənt]
抗蚀剂；保护层 resist
抗台风型风机（沿海） (coastal) anti-typhoon type
抗弯刚度 flexural rigidity
抗压强度与重量比 compressive strength-to-weight ratio [kəmˈpresɪv]
抗氧化剂 water-antioxidant [ˌæntɪˈɒksɪdənt]
科氏参数 Coriolis parameter [pəˈræmɪtə]
科氏力（地球自转偏向力） Coriolis forces
可动叶片 moving blade
可靠性测定试验 reliability determination test [rɪˌlaɪəˈbɪlətɪ]
可利用率（风力发电机组） availability (for WTGS) [əˌveɪləˈbɪlətɪ]
可锁可解装置（爬梯） unlockable barrier

可用风能特性系数 usable energy pattern factor [ˈjuːzəbl] [ˈpætən]
可用功率；有效功率 available power
可用系数 availability factor
空间电荷 space charge
空气边界层 air boundary layer
空气动力 aerodynamically [ˌeərədaɪˈnæmɪkəlɪ]
空气动力表面 aerodynamic surface
空气动力刹车；气动刹车 aerodynamic brake [ˌeərəudaɪˈnæmɪk]
空气动力学 aerodynamics [ˌeərəudaɪˈnæmɪks]
空气动力学半径 aerodynamic radius [ˈreɪdɪəs]
空气动力学设计 aerodynamic design
空气动力制动器 aerodynamic brake; airbrake
空气静力学 aerostatics [ˌeərəuˈstætɪks]
空气扩散 air diffusion [dɪˈfjuːʒ(ə)n]
空气流量计 air flow meter
空气流泄；通风管 air drain
空气密度 air density
空气摩擦阻力 air friction [ˈfrɪkʃ(ə)n]
空气湿度 air humidity [hjuːˈmɪdətɪ]
空气温度测量 air temperature measurement
空气压力 air pressure
空气压力测量 air pressure measurement
空气制动系统 air braking system
空气轴承 air bearing
空气阻力；气动阻力 air resistance/drag [rɪˈzɪst(ə)ns]
空气阻尼 air damping/dam
空心轴 hollow shaft [ˈhɒləu]
空载最大加速度 maximum bare table acceleration [əkˌseləˈreɪʃən]
空转 idling [ɑɪdlɪŋ]
控制策略 control strategy [ˈstrætɪdʒɪ]
控制齿轮 control-gear
控制机构 control mechanism [ˈmekənɪzəm]
控制系统（风电机） control system (for wind turbines)
控制翼面 control surfaces
控制装置 control device
库塔—儒科学夫斯基定理 Kutta-Joukowski theorem
库塔—儒科学夫斯基方程 Kutta-Joukowski equation
库塔条件 Kutta condition
跨度中点；桥跨中点 center of span
快速傅立叶变换 FFT (fast Fourier transform) [ˈfʊrɪeɪ]
快速重合闸 fast reclosure
馈电线 feeder
馈送 feeding
扩风器 wind diffuser [dɪˈfjuːzə]
扩散加力式风力机 diffuser augmenter [dɪˈfjuːzə] [ɔːgˈmentə]
扩散器 diffuser [dɪˈfjuːzə]

扩散增强风力涡轮机 diffuser augmented wind turbine [ɔːgˈmentɪd]

L

拉格朗日方程 Lagrangian equation [ɪˈkweɪʒ(ə)n]

拉格朗日积分间尺度 Lagrangian integral time scale [ˈɪntɪgr(ə)l]

拉格朗日谱函数 Lagrangian spectral function [ˈspektr(ə)l]

拉格朗日相关 Lagrangian correlation

拉格朗日相似理论 Lagrangian similarity theory [sɪməˈlærətɪ] [ˈθɪərɪ]

拉格朗日协方差 Lagrangian covariance [kəʊˈveərɪəns]

拉格朗日应变 Lagrangian strain

拉格朗日自协方差 Lagrangian autocovariance

拉力系数 thrust coefficient

拉普拉斯 Laplace

拉普拉斯变换 Laplace transform

拉普拉斯方程 Laplace equation [lɑːˈplɑːs]

拉伸负荷 tension loading

拉绳 guy-ropes

拉丝；拨丝；使延长 wiredrawing

拉索 guy wires

拉索（支撑）结构 guyed structure

拉索式塔架 guyed tower

拉索塔架 guyed tower

拉索支撑悬臂 guyed cantilever [ˈkæntɪliːvə]

来流，迎面流 oncoming flow

来流风迎角；风接近角 wind-approach angle

来流湍流（度） oncoming turbulence [ˈtɜːbjʊl(ə)ns]

雷暴 thunderstorm [ˈθʌndəstɔːm]

雷电过电压 lightning overvoltage [ˌəʊvəˈvəʊltɪdʒ]

雷电计数器 flash counter

雷电流 lightning current

雷击 lightning strike

雷击电涌 lightning surges

雷诺数 Reynolds number（Re） [ˈrenəldz]

雷诺数修正 correction for Reynolds number [ˈrenəldz]

雷诺应力 Reynolds stress

累计值 integrated value

离岸风 offshore wind [ɒfˈʃɔː]

离合器 clutches

离散傅立叶变换 discrete Fourier transformation (DFT)

离散傅立叶逆变换 inverse discrete Fourier transformation [ˈfʊrɪeɪ]

离散化 discretization [dɪsˈkriːtaɪˈzeɪʃən]

离散机械系统 discretized mechanical systems [mɪˈkænɪkəl]

离散控制器 discrete controller

离散涡 discrete vortices [ˈvɔːtɪsiːz]

离散阵风模型 discrete gust models

离心调节器 fly-ball governor [ˈgʌvənə]

离心力 centrifugal forces [senˈtrɪfjʊgəl]

离心式通风机 centrifugal fan [senˈtrɪfjugəl]

离心式限速器 centrifugal governor [ˈgʌv(ə)nə]

离心释放装置 centrifugal release units

离心应力 centrifugal stresses

离心载荷 centrifugal loads

离心重量 fly-ball weight

李雅普诺夫函数 Lyapunov function

理查森数 Richardson number [ˈrɪtʃədsən]

理论空气动力效率 theoretical aerodynamic efficiency [ˌeərəudaɪˈnæmɪk]

理论最大功率系数 maximum theoretical power coefficient [ˌkəʊɪˈfɪʃənt]

理想风力机 ideal wind turbines

理想功率曲线 ideal power curve [kɜːv]

理想转子 ideal rotors

立式达里厄风机 vertical-axis Darrieus wind turbine

立式风洞 vertical wind tunnel

立式风力机；垂直轴风力发电机 vertical axis wind turbine

连接件（爬梯竖杆间用）connector

连续运行的闪变系数 flicker coefficient for continuous operation

联板 yoke plate

联接点（电网）（风力发电机组）network connection point (for WTGS)

联锁装置 interlocker [ˌɪntə(ː)ˈlɒkə]

联轴器 coupling

烈风（九级风） strong gale [geɪl]

临界发散风速 critical divergence wind speed [daɪˈvɜːdʒ(ə)ns]

临界功率 activation power [ˌæktɪˈveɪʃən]

临界击穿电压 critical breakdown voltage

临界雷诺数 critical Reynolds number [ˈrenəldz]

临界扭曲应力 critical buckling stress [ˈbʌklɪŋ]

临界频率 critical frequency

临界倾覆风速 critical overturning wind speed

临界设计情况 critical design case

临界条件；临界状态 critical condition [ˈkrɪtɪk(ə)l]

临界压力梯度 critical pressure gradient [ˈgreɪdɪənt]

临界应力 critical stresses

临界转速 activation rotational speed [rəʊˈteɪʃənəl]

临界状态 critical regime [reɪˈʒiːm]

临界阻尼 critical damping

灵敏（度）；灵敏性 sensitivity [ˌsensɪˈtɪvɪtɪ]

零风面位移 zero plane displacement

零上穿频率 zero-upcrossing frequency [ˈfrɪkwənsɪ]

零升力角 zero lift angle

零升力矩 zero-lift moment
零升力线 zero lift line
零升迎角 zero-lift angle of attack
零升阻力 zero lift drag
零速力矩 zero-torque speed [tɔːk]
零序电流 zero sequence current [ˈsiːkwəns]
零压穿越 ZVRT
龙卷风；陆龙卷 tornado [tɔːˈneɪdəʊ]
龙卷风 F 等级 tornado F-scale
龙卷风参数 tornado parameter [pəˈræmɪtə]
龙卷风带 tornado belt
龙卷风核 tornado core
龙卷风路径 tornado path
龙卷风模拟模型 tornado simulation model [ˌsɪmjʊˈleɪʃən]
龙卷风气旋 tornado cyclone
龙卷风式旋涡 tornado-like vortex [ˈvɔːteks]
龙卷风通道 tornado alley [ˈælɪ]
龙卷风效应 tornado effect
龙卷风影响区 tornado-affected area [tɔːˈneɪdəʊ]
龙卷风中心 tornado center
龙卷风轴 tornado axis [ˈæksɪs]
露 dew [djuː]
露天；户外 open air
露天气候 open air climate
鲁棒控制 robust control [rəʊˈbʌst]
鲁棒稳定性 robust stability
鲁棒性能 robust performance
陆地风 land wind
陆地蒲福风级 land Beaufort scale [ˈbəʊfət]
陆风 land breeze
陆上风机 onshore WTGS
铝棒 aluminum bars [əˈljuːmɪnəm]
铝爬梯 aluminium vertical ladder
绿色能源 green energy
掠面速度 area velocity
掠射角 grazing angle [ˈgreɪzɪŋ]
轮毂 hub
轮毂刚度 hub rigidity [rɪˈdʒɪdətɪ]
轮毂高度 hub height [haɪt]
轮毂控制器 hub controller
轮毂罩 hubcap
轮换抽样 rotational sampling [rəʊˈteɪʃənəl]
轮廓，外形；等值线 contour [ˈkɒntʊə]
轮缘 flange
洛克数 Lock number

M

马达启动 motor start
马尔可夫矩阵 Markov matrix [ˈmeɪtrɪks]
马格纳斯效应 Magnus effect [ˈmægnəs]
马格努斯力 Magnus force
马格努斯效应 Magnus effect [ˈmægnəs]

马格努斯效应转子 Magnus effect rotor

马赫数 Mach number

马赫数效应 Mach number effect

马克斯韦尔分布 Maxwell distribution

马蹄涡 horseshoe vortex

马蹄涡系 horseshoe vortex system

脉动荷载 fluctuating load

脉动气动力 fluctuating aerodynamic force

脉动压力 fluctuating pressure

每小时风速 hourly wind speed

每小时平均风速 hourly mean wind speed

幂律廓线 power law profile [ˈprəufaɪl]

面积（一次）矩 area moment

面积惯性矩；面积二次矩 area moment of inertia [ɪˈnɜːʃə]

面积平均压力 area-mean pressure

面内 in-plane

面内疲劳载荷 in-plane fatigue loads [fəˈtiːg]

模拟风 simulated wind

模态分析 modal analysis [əˈnæləsɪs]

模态阻力系数 modal damping coefficient [ˌkəuɪˈfɪʃənt]

摩擦边界层 frictional boundary layer [ˈbaund(ə)rɪ]

摩擦层 friction layer

摩擦力矩 frictional moment [ˈfrɪkʃ(ə)n]

摩擦扭矩 frictional torque [tɔːk]

摩擦速度 friction velocity

摩擦系数 friction coefficient [ˌkəuɪˈfɪʃənt]

摩擦应力 frictional stress

摩擦阻力 drag friction；frictional resistance；friction drag [ˈfrɪkʃ(ə)n] [rɪˈzɪst(ə)ns]

N

内部防雷系统 internal lightning protection system

内部放电 internal discharge

内衬；衬里 lining

内在转子阻尼 intrinsic rotor damping [ɪnˈtrɪnsɪk] [ˈdæmpɪŋ]

纳维尔—斯托克斯方程 Navier-Stokes equation

奈斯勒数 Nusselt number

耐风设计 wind-proof design [pruːf]

耐风性能 wind proof performance [pəˈfɔːm(ə)ns]

耐腐试验 corrosion resistance tests [rɪˈzɪstəns]

耐久性 durability

耐久性试验 endurance test

耐受电压 withstand voltage

耐压试验 withstand test

耐张线夹 strain clamp

挠度；偏移 deflection

挠曲应力 flexural stress

挠性风轮 bearingless rotor

能量玫瑰图 energy rose
能量平衡 energy balance
能量需求 energy demand
逆变器 inverter
逆风 upwind
逆风，迎面风 contrary wind; opposing wind [ˈkɒntrərɪ] [əˈpɒzɪŋ]
逆四象限运行 bidirectional and reversible 4 quadrant operation
黏度指教 viscosity index [ˈɪndeks]
黏性 viscosity [vɪˈskɒsətɪ]
黏性流动 viscous flow
黏性流体 viscosity fluid
黏性摩擦 viscosity friction [ˈfrɪkʃ(ə)n]
黏性系数 viscosity coefficient [vɪˈskɒsɪtɪ]
黏性效应 viscosity effect
黏滞阻力 viscosity resistance; viscous drag [rɪˈzɪst(ə)ns]
年变化 annual variation [ˈænjʊəl]
年度能源计算 annual energy calculation
年发电量 annual energy production
年风况 annual wind regime
年风向图 annual wind direction diagram
年极端风速 extreme annual wind speed [ɪkˈstriːm]
年际变化 interannual variation
年平均 annual average
年平均风能密度 annual average wind-power density; mean annual wind power density
年平均风速 annual average wind speed; yearly wind speed
年有效风能 annual available wind energy [ˈænjʊəl] [əˈveɪləb(ə)l]
年增长速度 annual growth rate
年最高 annual maximum
年最高日平均温度 annual extreme daily mean of temperature
牛顿第二定律 Newton's second law
扭（转力）矩 moment of torsion; torque [ˈtɔːʃ(ə)n] [tɔːk]
扭矩耐力曲线 torque-endurance curves
扭缆 twisting of the power cable
扭曲叶片 (rotor blade) twist
扭转刚度 torsional rigidity
扭转刚度系数 coefficient of torsional [ˈtɔːʃənəl]
扭转角 torsion angle; angle of twist [ˈtɔːʃən]
扭转频率 torsion frequency
扭转柔度；扭转挠度 torsional flexibility [ˌfleksɪˈbɪlɪtɪ]
扭转振动 torsional vibration [vaɪˈbreɪʃən]

P

爬梯固定件 ladder bracket
攀爬设备（用于风电塔） ladder equipment
配重重锤 counter weight

疲劳负荷 fatigue load [fə'tiːg]
疲劳光谱组合 fatigue spectra combination
疲劳极限 fatigue limit; endurance limit [ɪn'djʊərəns]
疲劳临界 fatigue criticality [fə'tiːg] [ˌkrɪtɪ'kælətɪ]
疲劳螺栓 fag bolt
疲劳判据 fatigue criterion [fə'tiːg] [krɑɪ'tɪərɪən]
疲劳评定 fatigue evaluation [ɪˌvælju'eɪʃən]
疲劳破坏 fatigue failure ['feɪljə]
疲劳强度 fatigue strength [streŋθ]
疲劳设计 fatigue design
疲劳寿命 fatigue life
疲劳损伤 fatigue damage
疲劳系数 factor of fatigue [fə'tiːg]
疲劳性能 fatigue properties
疲劳循环计数 fatigue cycle counting [fə'tiːg] ['sɑɪkl]
疲劳应力 fatigue stresses [stres]
疲劳应力范围 fatigue stress ranges
疲劳载荷 fatigue loading
偏差；偏移 deviation [ˌdiːvɪ'eɪʃən]
偏航 yaw [jɔː]
偏航齿轮 yaw gear
偏航机构 yaw mechanism
偏航基座；偏航盘 yaw base
偏航加速度；侧滑加速度；横摆加速度 yaw acceleration [əkselə'reɪʃ(ə)n]
偏航角 yaw angle
偏航控制 yaw control
偏航力矩 yaw moments
偏航力矩系数 yawing moment coefficient
偏航模式 yaw model
偏航偏移 yaw offset [ˌɒf'set]
偏航驱动 yawing driven
偏航误差 yaw error
偏航系统 yawing drive
偏航制动 yaw brake
偏航轴承 yaw bearings
偏离偏心；主距 offset ['ɒfset]
偏离轴线 off-axis ['æksɪs]
偏球风速计 yaw sphere anemometer [sfɪə] [ˌænɪ'mɒmɪtə]
偏斜角 angle of declination [ˌdeklɪ'neɪʃ(ə)n]
偏心载荷 eccentric loading
偏移角；挠度 angle of deflection [dɪ'flekʃ(ə)n]
偏置叶片风力机 bent-bladed wind machine
偏转风速计 deflection anemometer [ˌænɪ'mɒmɪtə]
偏转计；挠度计 deflection gauge [geɪdʒ]
偏转角 deflection angle
拼接板 splice plates
频率 frequency ['frɪkwənsɪ]
频率变化率 frequency variation rate ['frɪkwənsɪ]
频率分布 frequency distribution
频率响应 frequency response
频率效应 frequency effect
频谱 frequency spectrum ['spektrəm]
频谱分析 spectral analysis

[ˈspektrəl]
频谱密度 frequency-spectral-density
频闪效应 stroboscopic effect [ˌstrəʊbəʊˈskɒpɪk]
频移 frequency drift
频域 frequency domain [dəʊˈmeɪn]
品质因数 quality factor
平板颤振 flat plate flutter [ˈflʌtə]
平板绕流 flat-plate flow
平板阻力 flat-plate drag
平衡尾涡 equilibrium wake [ˌiːkwɪˈlɪbrɪəm]
平均分子速度 mean molecular velocity [məʊˈlekjʊlə]
平均风速 average/mean wind speed
平均风速廓线 mean wind profile
平均风速模型 mean wind speed model
平均海平面 mean sea level
平均环境风 mean environmental wind [ɪnˌvaɪrənˈmentl]
平均几何弦（长） mean geometric chord [ˌdʒɪəˈmetrɪk]
平均内部风荷载 mean internal wind loading
平均年降水量 mean annual precipitation [ˈænjʊəl] [prɪˌsɪpɪˈteɪʃ(ə)n]
平均气动弦（长） mean aerodynamic chord [ˌerəʊdaɪˈnæmɪks] [kɔːd]
平均气动中心 mean aerodynamic center
平均寿命 mean life
平均温度年较差 mean annual range of temperature
平均小时风速 mean hourly wind speed
平均应变 mean strain
平均应力 mean stress
平均噪声 average noise level [ˈævərɪdʒ]
平均自由行程 mean free path
平面键 parallel key [ˈpærəlel]
平面弯矩 out-of-plane bending moment
平台 platform
平坦地形 flat terrain
平行四边形 parallelogram [ˌpærəˈleləgræm]
平行轴布置 parallel shaft arrangement
平行轴齿轮副 gear pair with parallel axes [ˈpærəlel] [ˈæksiːz]
屏蔽；罩；防护；庇护 shield [ʃiːld]
屏蔽环 shielding ring
坡度角 slope angle [ˈæŋg(ə)l]
破坏 failure load
破坏荷载 failing load
破坏判据 failure criterion [kraɪˈtɪərɪən]
破坏应力 failure stress
蒲福（级）数 Beaufort number
蒲福风级 Beaufort scale
蒲福风力 Beaufort force
蒲田风级表 Beaufort scale [ˈbəʊfət] [skeɪl]
普郎特对数法则 Prandtl logarithmic law [ˌlɒgəˈrɪðmɪk]
普郎特叶尖损耗因子 Prandtl's tip loss factor

谱隙 spectral gap

Q

漆膜测厚仪 paint film thickness gauge [geɪdʒ]

起动风速 start wind speed

起动力矩；起动转矩 starting torque [tɔːk]

起动力矩系数 starting torque coefficient [ˌkəʊɪ'fɪʃənt]

起动信号 starting signal

起伏 undulate ['ʌndjʊleɪt]

起始风速 onset wind speed ['ɒnset]

气垫 air cushion ['kʊʃ(ə)n]

气垫；空气弹簧 air-spring

气动泵 pneumatic pump [njuː'mætɪk]

气动粗糙度 aerodynamic roughness

气动粗糙度参数 aerodynamic roughness parameter

气动粗糙度长度 aerodynamic roughness length

气动弹性 aeroelastic/aeroelasticity [ˌeərəʊɪ'læstɪk]

气动弹性不稳定性 aeroelastic instability

气动弹性传递函数 aeroelastic transfer function

气动弹性导数 aeroelastic derivative [dɪ'rɪvətɪv]

气动弹性刚度 aeroelastic stiffness

气动弹性力学 aeroelastics

气动弹性模拟 aeroelastic modeling

气动弹性模态 aeroelastic mode

气动弹性模型 aeroelastic model

气动弹性特性 aeroelastic characteristics [ˌeərəʊɪ'læstɪk]

气动弹性稳定性 aeroelastic stability

气动弹性响应，气动弹性反应 aeroelastic response

气动弹性效应 aeroelastic effect

气动弹性运动方程 aeroelastic equation of motion [ɪ'kweɪʒ(ə)n]

气动弹性载荷 aeroelastic load

气动弹性振动 aeroelastic vibration/oscillation [ˌɒsɪ'leɪʃən]

气动导纳 aerodynamic admittance [ˌerəʊdaɪ'næmɪk] [əd'mɪt(ə)ns]

气动导数 aerodynamic derivative [dɪ'rɪvətɪv]

气动导数系数 aerodynamic derivative coefficient

气动刚度 aerodynamic stiffness

气动功率 aerodynamic power

气动光滑度 aerodynamic smoothness

气动力（空） aerodynamic force

气动力边界层 aerodynamic boundary layer

气动力导数 aerodynamic force derivative [dɪ'rɪvətɪv]

气动力端部效应 aerodynamic end effect

气动力矩 aerodynamic torque [tɔːk]

气动力矩峰值 peak aerodynamic torque [ˌeərəʊdaɪˈnæmɪk]
气动力扰动 aerodynamic disturbance [dɪˈstɜːb(ə)ns]
气动力失稳 aerodynamic destabilizing
气动力数据 aerodynamic data
气动力系数 aerodynamic force coefficient
气动力效应 aerodynamic effect
气动扭转 aerodynamic twist
气动扰流器，气动扰流板 aerodynamic spoiler (device) [ˈspɒɪlə]
气动实度 aerodynamic solidity [səˈlɪdɪtɪ]
气动特性 aerodynamic characteristics
气动尾流 aerodynamic wake
气动稳定性 aerodynamic stability
气动弦线 aerodynamic chord of airfoil [ˌeərəʊdaɪˈnæmɪk]
气动相似性 aerodynamic similarity [sɪməˈlærətɪ]
气动效率比 aerodynamic efficiency
气动性能 aerodynamic behavior
气动翼型 aerodynamic airfoil [ˈeəfɒɪl]
气动阴影效应 blanketing effect
气动载荷 aerodynamic loads
气动噪声 aerodynamic noise
气动制动系统 aerodynamic braking system
气动中心 aerodynamic centre
气动转子 aerodynamic rotor
气动装置 aerodynamic device
气动阻力制动器 air resistance brake
气动阻尼 aerodynamic damping
气功效率 aerodynamic efficiency
气候 climate
气候变动 climatic fluctuation [ˌflʌktjʊˈeɪʃən]
气候带 climate zones
气候分类 climatic classification
气候风速 climatic wind speed
气候环境 climatic environment
气候控制 climatic control
气候图 climatic chart
气候稳定度 climatic stability
气候效应；气候影响 climatic effect
气候学特性 climatological characteristics [ˌklɑɪmətəˈlɒdʒɪkəl]
气候要素 climatic element
气候噪声 climatic noise
气候资料 climatic data
气流 aerial/air current；air stream
气流分离 flow separation [ˌsepəˈreɪʃən]
气流畸变 flow distortion [dɪsˈtɔːʃən]
气流结构 air flow structure
气流流速；气流流量 air flow rate
气流流速表 aerodromometer
气流脉动 airflow pulsation[pʌlˈseɪʃən]
气流偏角 flow angularity [ˌæŋgjʊˈlærətɪ]
气幕 air curtain
气泡击穿 bubble breakdown
气体绝缘 gaseous insulation [ˈgæsɪəs]
气体绝缘变电站 gas insulated substation (GIS) [ˈɪnsəˌleɪtəd]
气隙 air gap

气象测量塔 meteorological tower
气象风洞 meteorological wind runnel [ˈrʌn(ə)l]
气象符号 meteorological symbol
气象塔 meteorological tower
气象学 meteorology [ˌmiːtɪəˈrɒlədʒɪ]
气象预报 meteorological forecast [ˌmiːtɪərəˈlɒdʒɪkəl]
气象站 meteorological station
气压表，压力计 manometer [məˈnɒmɪtə]
气压计；晴雨表 barometer [bəˈrɒmɪtə]
汽封 labyrinth packing [ˈlæbərɪnθ]
千分尺 dial micrometer
千斤顶 jack
牵引板 towing plate
前扰流板 front spoiler [ˈspɒɪlə]
前缘 leading edge
前置机 front end processor [ˈprəusesə]
潜在故障 latent/dormant failure [ˈdɔːmənt]
欠速 under speed
强电控制 strong current control
强度刚度比 strength-to-stiffness ratio
强度重量比 strength-to-weight ratio
强风（六级风） strong breeze
强风；大风 gale [geɪl]
强风暴 severe storm [sɪˈvɪə]
强风边界层 strong-wind boundary layer
强力通风 sharp draft [drɑːft]
强龙卷 violent tornado [tɔːˈneɪdəu]
强迫流动 forced flow
强阵风 severe gust
强制对流 forced convection [kənˈvekʃ(ə)n]
强制环流 forced circulation [sɜːkjʊˈleɪʃ(ə)n]
跷板角 teetering angle
跷板铰链 teeter hinge [hɪn(d)ʒ]
跷板式风轮 teetering rotor
跷板式桨毂 teetering hub
跷板式结构 teeter
跷板式制动器 teeter brake [breɪk]
跷跷板固有频率 teeter natural frequency [ˈfrɪkwənsɪ]
跷跷板角 teeter angle
跷跷板铰链 teeter hinge
跷跷板式轮毂 teetered hub
跷跷板稳定 teeter stability
跷跷板载荷 teeter loads
撬棒；螺丝钻 tommy bar
撬棒技术 crowbar
切边风 shear wind
（风）切变发生器 shear generator [ˈdʒenəreɪtə]
切出风速 cut-out speed
切换；通断 switching
切换运行 switching operation
切入风速 cut-in wind speed
切向风 tangential wind
切向力 tangential force [tænˈdʒenʃəl]
切向流 tangential flow
切向诱导速度 tangential induced velocity
切向诱导因子 tangential flow in-

duction factor

侵蚀（海上） corrosion（offshore）[kə'rəʊʒən]

轻风（二级风） light breeze/wind；slight breeze

轻微旗状（Ⅱ级植物风力指示） slight flagging

倾翻力矩 overturning moment [əʊvə'tɜːnɪŋ]

倾覆风速 overturning wind speed

倾覆力矩 overturning moment；tilting moment

倾角 tile angle；angle of inclination [ɪnklɪ'neɪʃ(ə)n]

倾角设计 tile angle design

倾斜 tilting

曲柄轴箱 crankcase

驱动电机 drive motor

驱动控制 drive control

驱动力系 drive train

驱动器 driver

趋势分析 trend analysis

取样频率 frequency of sampling

全尺寸测量 full-scale measurement ['meʒəm(ə)nt]

全尺寸雷诺数 full-scale Reynolds number ['renəldz]

全尺寸模型 full-scale model

全尺寸时间 full-scale time

全尺寸原型物 full-scale prototype ['prəʊtətaɪp]

全范围桨距控制 full-span pitch control

全身式安全带 harness

全向风 omni-directional wind

权重函数 weighting functions

确定性模型 deterministic models [dɪˌtɜːmɪ'nɪstɪk]

确定载荷 deterministic loads

确定转子载荷 deterministic rotor loads

群控制器（控制风机群的） cluster controller

R

扰动强度 turbulence intensity ['tɜːbjʊləns] [ɪn'tensətɪ]

扰动中心 center of disturbance

扰流板 spoiler

扰流带；噪音条 turbulator strip

扰流发生器 vortex generator ['vɔːteks]

绕线式转子感应发电机 wound rotor induction generators

绕线转子 wound rotor

绕组系数 winding factor

人工（形成）速度梯度 artificial velocity gradient [vɪ'lɒsɪtɪ] ['greɪdɪənt]

人工催雨 artificial precipitation stimulation [ˌstɪmjə'leʃən]

人工降雨 artificial precipitation/rainfall [prɪˌsɪpɪ'teɪʃ(ə)n]

人工气候 artificial climate [ɑːtɪ'fɪʃ(ə)l] ['klaɪmət]

人工通风 artificial draft/ventilation [drɑːft] [ˌventɪˈleɪʃ(ə)n]
人为扰动 artificial disturbance [dɪˈstɜːb(ə)ns]
人造风 artificial wind
人造云 artificial cloud
日变幅 daily amplitude [ˈæmplɪtjuːd]
日变化 diurnal variation [daɪˈɜːnəl] [veərɪˈeɪʃən]
日极端值 daily extremes
日平均温度 mean daily temperature
日平均值 daily mean value [ˈdeɪlɪ]
日最高温度 daily maximum temperature
容积式泵 positive displacement pump
容库法 Runge-Kutta method [ˈmeθəd]
容器；蓄水池；储藏处；贮存器 reservoir [ˈrezəvwɑː]
容许极限 allowable limit [əˈlaʊəbl]
容许应力 allowable stress
溶媒；溶剂；解决方法 solvent [ˈsɒlvənt]
柔性塔 soft towers
柔性塔架；软塔架 compliant tower (soft tower) [kəmˈplaɪənt]
入流 inflow
入流角 inflow angle (flow angle)
入射角 incidence angle
软启动 soft start
软切入 soft cut in
软塔 soft tower
软制动 soft brake system
瑞利分布 Rayleigh distribution [ˈreɪlɪ]

S

扫掠角 sweep angle
扫掠面积 swept area
扫描电镜 scanning electron microscope
刹车 brake duty
刹车（风电机）；制动器 brake (for wind turbines)
刹车风速 shut down wind speed; shut-off wind speed
刹车块 brake block
刹车盘 brake disc
刹车片 brake block/pads
刹车尾舵 braking vane
刹车位置 brake position
刹车油 brake fluid [ˈflu(ː)ɪd]
刹车状态 motor brake status
山地空气 mountain air
山地障碍物 mountain barrier [ˈbærɪə]
山风 mountain breeze/wind
山谷风 mountain and valley breeze
山谷气流 valley flow
山坡风 slope wind
山坡气流 slope flow
闪变，闪烁 flicker [ˈflɪkə]
闪变阶跃系数 flicker step factor
闪变系数 flicker coefficient
闪变值 severity of flicker

闪点 flashpoint [ˈflæʃˈpɒɪnt]

闪电 lightning [ˈlaɪtnɪŋ]

闪络 flashover [ˈflæʃˌəʊvə]

上吹效应 upwind effect

上风/迎风 upwind

上风吹程 upwind fetch

上风风向 upwind orientation [ˌɔːrɪenˈteɪʃən]

上风面 windward [ˈwɪndwəd]

上风式风力机，迎风机组 upwind turbine

上风式风轮 upwind rotor

上风式结构 upwind configuration [kənˌfɪgjʊˈreɪʃən]

上风向 upwind

上风向的，迎风向的 windward (upwind)

上坡风 slope up-wind; upslope wind; anabatic wind

上升（扬压）系数 uplift coefficient [ˌkəʊɪˈfɪʃ(ə)nt]

上升空气 ascending air [əˈsendɪŋ]

上升流 upflow

上升气流 anabatic flow; anaflow; ascendant current [ˌænəˈbætɪk] [əˈsend(ə)nt]

上升气流；向上通风 updraft (updraught) [ˈʌpdrɑːft]

上网 grid access

上网频率 frequency

上吸风载 uplift wind load

上洗（流） up wash

上下盖（升降梯轿厢入/出口） top and bottom hatches

上游 upstream

上游风速 upstream windspeed; upstream speed

上游影响 upstream influence [ˈɪnflʊəns]

梢根比；尖削比 taper ratio [ˈreɪʃɪəʊ]

设备故障信息 equipment failure information

设备线夹 terminal connector

设定压力 setting pressure

设定值 set value

设计参数 design parameter

设计风况 design wind conditions

设计风速 design wind speed

设计风载 design/factored wind load

设计风阻 factored resistance [rɪˈzɪst(ə)ns]

设计工况 design situation [ˌsɪtjʊˈeɪʃən]

设计荷载 design load

设计极限 design limits

设计尖速比 design tip speed ratio

设计升力系数 design lift/pressure coefficient

设计寿命 design lifetime

设计原则；设计原理 design philosophy [fɪˈlɒsəfɪ]

升降梯 service lift

升力 lift (force)

升力和阻力 lift and drag (L/D)

升力角 lift angle

升力面 lifting surface

升力面理论 lifting plane/surface theory

升力曲线斜率 lift curve slope;

slope of lift curve [kɜːv]

升力系数 lift coefficient [ˌkəʊɪˈfɪʃənt]

升力线 lifting line

升力线理论 lifting line theory [ˈθɪərɪ]

升力效应 lift effect

升力型风力机 lift type wind machine

升力中心 center of lift

升起；起重机（台、架等）；固定器 hoist [hɒɪst]

升压变压器 step-up transformer

升致阻力 lift-dependent drag

升阻比 lift (to) drag ratio; lift-drag ratio [ˈreɪʃɪəʊ]

生产过剩 overproduction

生产准备费 operation start-up expense

生存风速（拒用），安全风速 survival wind speed

生态评价 ecological assessment [ˌiːkəˈlɒdʒɪkəl]

声的基准风速 acoustic reference wind speed [əˈkuːstɪk]

声功率级 sound power level (L_W)

声级 weighted sound pressure level; sound level [ˈweɪtɪd]

声级计 sound level meter

声压级 sound pressure level (L_P)

声阻尼效应 acoustic damping effect [əˈkuːstɪk]

盛行风 prevailing wind

盛行风向 most frequent wind direction [dɪˈrekʃ(ə)n]

失灵 malfunction [mælˈfʌŋkʃən]

失速；使失速 stall

失速颤振 stall flutter [ˈflʌtə]

失速迟滞激励 stall hysteresis excitation [ˌhɪstəˈriːsɪs] [ˌeksɪˈteɪʃ(ə)n]

失速调节 stall regulation

失速调节机组 stall regulated machines

失速规律 stall regular

失速角 burble angle; stall angle

失速控制 stall control

失速气流 stalled flow

失速区 stalled area

失速式限速 stall regulation [regjʊˈleɪʃ(ə)n]

失速条 stall-strips

失速停机 wind turbine furling/stalling

失速延迟（滞后） stall hysteresis [ˌhɪstəˈriːsɪs]

失速迎角；临界迎角 stalling angle of attack

失效 failure [ˈfeɪljə]

失效—安全 fail-safe

失效系统 fail-safe-system

施工辅助工程 temporary works

湿度计 hygrometer [hɑɪˈgrɒmɪtə]

石油化学工业 petrochemical industry [ˌpetrəʊˈkemɪkəl]

时域 time-domain [dəʊˈmeɪn]

实度损失 solidity losses [səˈlɪdɪtɪ]

实际输出功率 actual extracted power

实时 real time

实时电价 spot power price

史密斯—普特南大型风轮机 Smith-Putnam wind turbine

使用风地图集 using wind atlases [ˈætləs]
使用极限状态 serviceability limit states [ˌsəːvɪsəˈbɪlɪtɪ]
使用寿命 service life
使用条件 service condition
事故频率 failure rate
视风 apparent wind
视觉冲击 visual impact
视觉冲击区 zones of visual impact (ZVI) [ˈɪmpækt]
视觉影响 visual impact
视速度 apparent velocity
视迎角 apparent angle of attack [əˈpær(ə)nt]
视在功率 apparent power [əˈpærənt]
视在声功率级 apparent sound power level [saʊnd]
视重力 apparent gravity
试验场地 test site
试验数据 test data
试验台 test-bed
室内气候 indoor climate [ˈɪndɔː]
释放；去激励 deactivate [diːˈæktɪveɪt]
(风洞) 收缩段 contraction section/cone [kənˈtrækʃ(ə)n]
收缩装配 shrink-fitted
手推车 wheelbarrow [ˈhwiːlˌbærəʊ]
寿命估算 lifetime estimations
寿命试验 life test
寿命系数；使用年限因数 life factor
寿命周期 lifetime; life-cycle
寿命周期成本 life cycle costing
受风面积 wind exposure area [ɪkˈspəʊʒə]
受风区；风扫掠面积 wind swept area
售后服务 after sales service
输入 input
输入功率 input power
输入轴 input shate
束缚物；U形挂环 shackle [ˈʃækl]
树枝放电 treeing
树脂 resin [ˈrezɪn]
树脂层板 wood-epoxy laminates [ˈlæmɪneɪts]
竖轴风力机 vertical-axis wind machine
竖轴风轮 vertical-axis rotor
数据组（测试功率特性） data set (for power performance measurement)
数学模拟 mathematical simulation [ˌsɪmjʊˈleɪʃən]
数学模型 mathematical model [mæθ(ə)ˈmætɪk(ə)l]
数值模拟 numerical simulation [ˌsɪmjʊˈleɪʃən]
数值模型 numerical model
数字地形模型 digital terrain model [teˈreɪn]
数字控制 digital control
数字控制器 digital controller
数字控制系数 digital control systems
数字面板 digitizing tablet [ˈtæblɪt]

数字信号处理 digital signal processing [prəʊˈsesɪŋ]
衰减 attenuation [əˌtenjʊˈeɪʃən]
衰减系数 attenuation factor
甩负荷 load shedding
双级制动 two level braking
双卡头 double clamp
双馈 double-feds/doubly-fed
双馈发电机 doubly fed generator
双馈发电系统 doubly fed generating system
双馈感应发电机 double feed/doubly-fed induction
双馈式变速恒频风力发电机组 doubly fed variable speed constant frequency (DFVSCF) WTGS
双馈式风电机组 doubly-fed WTGS
双馈异步发电机 doubly fed asynchronous power generator
双速运行 two-speed operation
双向风标 bi-directional vane; bivane [ˈbɑɪveɪn]
双正态分布 binormal distribution
霜；霜冻；严寒；结霜 frost
霜凇 rime [rɑɪm]
水分含量 moisture content [ˈmɒɪstʃə]
水分浸入 ingress moisture [ˈɪngres]
水平风 horizontal wind
水平风场 horizontal wind field
水平风切 horizontal wind shear
水平浮力修正 horizontal buoyancy correction [hɒrɪˈzɒnt(ə)l] [ˈbɒɪənsɪ]
水平湍流扩散 horizontal turbulent diffusion [ˈtɜːbjʊl(ə)nt][dɪˈfjuːʒ(ə)n]
水平相干性 horizontal coherence [kə(ʊ)ˈhɪər(ə)ns]
水平仪 spirit level
水平轴 horizon axis [həˈrɑɪzən] [ˈæksɪs]
水平轴风力发电机 horizontal axis wind turbine (HAWT) [ˈhɒrɪˈzɒntəl]
水平轴转子 horizon axis rotor
水下电缆 insulated submarine cable/underwater cable [ˈsʌbməriːn]
顺风 upwind
顺风调向灵敏度 sensitivity of following wind
顺风观察 looking downwind
顺风向驰振 along-wind galloping [ˈgæləpɪŋ]
顺风向的 along-wind
顺风向抖振 along-wind buffeting
顺风向互相关 along-wind cross-correlation
顺风向加速度 along-wind acceleration [əkseləˈreɪʃ(ə)n]
顺风向脉动 along-wind fluctuation [ˌflʌktjʊˈeɪʃən]
顺风向挠度 along wind deflection [dɪˈflekʃ(ə)n]
顺风向位移 in line displacement [dɪsˈpleɪsm(ə)nt]
顺风向相关 along-wind correlation
顺风向响应 in-line response
顺风向振动 along-wind oscillation [ˌɒsɪˈleɪʃən]

顺桨 feather; pitching to feather [ˈfeðə]

顺桨方向 feather direction

顺桨位置 feathering position

顺流向的 along-flow

顺序；时序；数列；定序 sequence [ˈsiːkwəns]

顺压差 favourable pressure difference

顺压梯度 favourable pressure gradient [ˈgreɪdɪənt]

顺应，顺风向反应 along-wind response

顺转风 veering wind

瞬间的；即刻的；即时的 instantaneous [ˌɪnstənˈteɪnjəs]

瞬时测值 instantaneous measured [ˈmeʒəd]

瞬时点源 instantaneous point source

瞬时风速 instantaneous wind speed

瞬时功率 instantaneous power [ˌɪnstənˈteɪnjəs]

瞬时倾覆力矩 instantaneous overturning moment

瞬时体源 instantaneous volume source

瞬时湍流能量 instantaneous turbulent energy

瞬时压力 instantaneous pressure

瞬时迎角 instantaneous incidence [ˈɪnsɪd(ə)ns]

瞬时值 instantaneous value

瞬态电流 transient rotor [ˈtrænzɪənt]

瞬态载荷 transient loads

纳维—斯托克斯方程 Navier-Stokes equations

斯托克斯流 Stokes'flow

四腿钢桁架式塔架 4-legged steel truss tower

四象限 quadrant [ˈkwɒdrənt]

四象限运行 four quadrant operation

伺服电动机 servomotor [ˈsɜːvəʊˌməʊtə]

伺服电机驱动（偏航系统） servo motor drive

伺服系统 servomechanism [ˌsɜːvəʊˈmekənɪzəm]

速度波动 speed fluctuations [ˌflʌktjʊˈeɪʃəns]

速度幅值 velocity amplitude [ˈæmplɪtjuːd]

速度廓线 velocity profile

速度曲线 velocity duration curve

速度三角形 velocity triangle [ˈtraɪæŋg(ə)l]

速度时间曲线 velocity duration curve [vɪˈlɒsətɪ]

速度势 velocity potential [vɪˈlɒsɪtɪ]

速度头 velocity head

随机存储器 ram

随机塔弯矩 stochastic tower bending moments [stɒˈkæstɪk]

随机载荷 stochastic loads [stɒˈkæstɪk]

随机振动 random vibration [vaɪˈbreɪʃən]

随机转子推力波动 stochastic rotor thrust fluctuations [ˌflʌktjʊˈeɪʃəns]

损耗 loss
损坏频率 damage frequency [ˈfrɪkwənsɪ]
损伤等效载荷 damage equivalent load [ɪˈkwɪvələnt]
锁定 blocking
锁定转子 locked-rotor
锁定转子转矩 locked-rotor torque
锁定装置 locking device

T

塔的背风侧 lee side of the tower/wind shade of the tower
塔杆 pole tower
塔高 tower height
塔基临界载荷 tower base critical load case [ˈkrɪtɪkəl]
塔基前后弯矩 tower base fore-aft bending moment
塔架 tower
塔井设备 shaft for towers
塔式吊车 tower crane
塔筒 tubular tower [ˈtjuːbjʊlə]
塔筒内平台 tower platform [ˈplætfɔːm]
塔位移 tower displacement [dɪsˈpleɪsmənt]
塔影 tower shadow [ˈʃædəʊ]
塔影响效应 influence by the tower shadow [ˈʃædəʊ]
台架试验 test on bed
太阳常数 solar constant
太阳辐射 solar radiation [ˌreɪdɪˈeɪʃən]
太阳光谱 solar spectrum [ˈspektrəm]
太阳轮 sun gear
泰奥多森函数 Theodorsen's function
泰勒假说 Taylor's hypothesis [hɑɪˈpɒθɪsɪs]
炭刷 carbon brush [ˈkɑːbən]
碳化硅 silicon carbide [ˈkɑːbaɪd]
碳纤维复合材料 CFRP
特高频 VHF (Very High Frequency)
特高压 Ultra-high voltage (UHV)
特性曲线 characteristic curve [ˌkærəktəˈrɪstɪk]
特性系数 coefficient of performance [ˌkəʊɪˈfɪʃənt] [pəˈfɔːməns]
特征波长 characteristic wavelength
特征参数 characteristic parameter [pəˈræmɪtə]
特征长度 characteristic length
特征尺寸 characteristic dimension [dɪˈmenʃ(ə)n]
特征方程 characteristic equation [ɪˈkweɪʒ(ə)n]
特征函数 characteristic function
特征面积 characteristic area
特征频率 characteristic frequency
特征寿命 characteristic life
特征速度 characteristic velocity
特征温度 characteristic temperature

特征相关长度 characteristic correlation length
梯度风 gradient wind [ˈgreɪdɪənt]
梯度风高度 gradient wind height
梯度风速 gradient wind speed
替代能源 alternative energy source
替代能源税收法 alternative energy tax law
天空辐射 sky radiation
天然阵风 natural gusty wind
天线；触角 antenna [ænˈtenə]
跳线线夹 jumper clamp
跳闸电路 trip circuit
跳闸线圈 trip coil [kɔɪl]
停机 shutdown/shutting down；parking
停机影响 park effect
停机制动 parking brake
停机状态 standstill
通风口 air inlet
通风装置；通风 ventilation [ˌventɪˈleɪʃən]
通过ISO体系认证 ISO certified
通过认证的产品 certified product
统计方法 statistical methods [stəˈtɪstɪkəl]
统计能量分析 Statistical Energy Analysis [əˈnæləsɪs]
统计预测 statistical prediction
投标设计 tender design
投运试验 commissioning test [kəˈmɪʃənɪŋ]
投资成本 investment cost [ɪnˈvestmənt]
透气性 air permeability [ˌpɜːmɪəˈbɪlɪtɪ]
透视图 perspective drawing [pəˈspektɪv]
突发阵风 sharp-edged gust [edʒd]
土地利用的影响 land-use impacts
湍流边界层 turbulent boundary layers [ˈtɜːbjʊlənt] [ˈbaʊndərɪ]
湍流尺度参数 turbulence scale parameter [pəˈræmɪtə]
湍流惯性负区 inertial sub-range [ɪˈnɜːʃjəl]
湍流频率 frequency of turbulence [ˈtɜːbjʊləns]
湍流强度 turbulence intensity
推或拉力系数 thrust coefficient [θrʌst]
推力负载；轴向载荷 thrust load
推力系数 thrust coefficient
推力轴承 thrust bearing
退磁；去磁 demagnetization [ˈdiːˌmægnɪtɑɪˈzeɪʃən]
退火 annealing [əˈniːlɪŋ]
褪漆剂 paint remover
脱流 vortex shedding
脱模剂 mold release agent
脱模涂层 releasing coating
陀螺力 gyroscopic force
陀螺效应 gyroscopic effect [ˌdʒɑɪərəsˈkɒpɪk]
陀螺仪；回旋装置；回转仪 gyroscope [ˈdʒɑɪərəskəʊp]
陀螺运动 gyroscopic motion
陀螺载荷 gyroscopic loads

W

外（部流）场 external field
外部动力源 external power supply
外部防雷系统 external lighting protection system [ɪkˈstɜːnəl]
外部风荷载 mean external wind loading [ɪkˈstɜːn(ə)l]
外部风载 external wind load
外部荷载 external load
外部绕流 external flow
外部条件 external conditions
外部吸力 external suction
外部压力损失 external pressure loss
外部压力系数 external pressure coefficient
外场联机试验（风电机组） field test with turbine
外齿轮 external gear
外联机试验 field test with turbine
外推功率曲线 extrapolated power curve [ekˈstræpəleɪt]
弯度 degree of curvature [ˈkɜːvətʃə]
弯度函数 curvature function of airfoil [ˈeəfɒɪl]
弯接头 angular connection
弯矩；挠矩 bending moment；moment of deflection [dɪˈflekʃ(ə)n]
弯扭颤振 flexure torsion flutter [ˈtɔːʃ(ə)n]
弯曲；挠曲 flexure [ˈflekʃə]
弯曲刚度 flexural rigidity
弯曲固有频率 bend natural frequency [ˈfrɪkwənsɪ]
弯曲和扭转载荷 flexural and torsional loads [ˈflekʃərəl] [ˈtɔːʃənəl]
弯曲力矩 bending moment [ˈbendɪŋ]
弯曲模态 bending mode
弯曲挠度 bending deflection
弯曲频率 bending frequency
弯曲位移 bending displacement
弯曲响应；弯曲反应 bending response
弯曲应力 bending stress
弯曲振动 bending/flexural oscillation [ˌɒsɪˈleɪʃən]
完全顺桨 full feather [ˈfeðə]
万用表；多用途计量器 multimeter [mʌlˈtɪmɪtə]
网侧防雷 grid-side lightning protection
网格化方法 gridding method [ˈmeθəd]
网格塔架 lattice tower [ˈlætɪs]
往返空间 shuttle [ˈʃʌtl]
威布尔比例因子 Weibull scale factor
威布尔分布 Weibull distribution
威布尔概率函数 Weibull probability function [ˌprɒbəˈbɪlətɪ]
威布尔函数 Weibull function
微动开关 microswitch
微观选址 micrositing

微机程控 minicomputer program
维恩图 Venn Diagram
维护 preventive maintenance [ˈmeɪntənəns]
维护吊车 service crane
维护试验 maintenance test [ˈmeɪntənəns]
维护试验 maintenance test
维修间 maintenance building
尾车 tail wagon [ˈwæg(ə)n]
尾舵 tail-vane [veɪn]
尾舵（尾翼） tail vane
尾流模型 wake models
尾流扰动 wake turbulence [ˈtɜːbjʊləns]
尾流损失 wake losses
尾流效应 wake effects [ɪˈfekts]
尾流折减后发电量 energy production after wake flow
尾涡；尾流 wake
纬线 weft [weft]
位置指示器 position indicator
温度垂直梯度 lapse rate
温度范围 temperature range
温度梯度 temperature gradient [ˈgreɪdɪənt]
温度系数 temperature coefficient
温升 temperature rise
温室效应 greenhouse effect [ɪˈfekt]
文氏管，文丘里管 Venturi Tube [venˈtjʊərɪ]
紊流 turbulence flow
紊流强度 turbulence intensity [ɪnˈtensətɪ]
稳定分层 stable stratification [ˌstrætɪfɪˈkeɪʃən]
稳定负荷 steady loads
稳态电压调节 steady-state voltage regulation
稳态分析 steady state analysis
涡轮机欧拉方程 Euler's turbine equation
涡轮喷气飞机 turbojet [ˈtɜːbəudʒet]
涡强 vortex strength
涡系 vortex systems
涡旋相关 eddy correlation [ˌkɒrəˈleɪʃ(ə)n]
涡黏性 eddy viscosity [vɪˈskɒsətɪ]
无风试验 wind-off test
无风阻力系数 zero-wind drag coefficient [ˈkəʊɪˈfɪʃ(ə)nt]
无量纲的 non-dimensional [dɪˈmenʃənəl]
无人操作 unmanned operation
无限尾流 infinite wake
无限涡街 infinite vortex street
无限翼展；无限展长 infinite span [spæn]
无续流 no follow current
无液气压表；无液晴雨表 aneroid barometer [ˈænərɔɪd] [bəˈrɒmɪtə]
无因次时间 non-dimensional time
无源滤波器 passive filter [fɪltə]
无阻尼固有频率 natural undamped frequency [ˌʌnˈdæmpt]
无阻尼振荡 undamped vibration
雾 fog

X

吸附 adsorption [æd'sɔːpʃən]
吸附效应 adsorption effect
吸管式风速计 suction anemometer [ˌænɪ'mɒmɪtə]
吸力；负压；吸气 suction
吸力荷载 suction load
系统优化设计 optimum system design
下风式风力机 downwind turbine
下风式风轮 downwind rotor
下风式结构 downwind configuration [kənˌfɪgjʊ'reɪʃən]
下风向 downwind
下风向风轮 downwind rotor
下风向扇形医 downwind sector
下降气流 downwash ['daʊnwɒʃ]
下坡风 downslope winds ['daʊnsləʊp]
下坡风；出谷风 down-valley wind flow
下坡风；下降风 fall wind
下吸式风洞 suck-down wind tunnel
下游；顺流 downstream
纤维增强塑料 FRP
纤细设计 slim size
弦长 chord length
弦宽 chord width [wɪdθ]
弦线 chord line
现场 in-site
现场测量，实测 on-site measurment
现场调查；实地勘查 site investigation [ɪnˌvestɪ'geɪʃən]
现场观测，实测 on site observation
现场可靠性试验 field reliability test
现场实测 site measurement
现场数据 field data
现地控制 local control
现有数据；可用数据 available data
线化空气动力学 linearized aerodynamics [ˌeərə(ʊ)daɪ'næmɪks]
线路压降补偿 line-drop compensation (LDC)
线圈；绕组 winding ['waɪndɪŋ]
线性比例 linear scale
线性变参数 linear parameter varying (LPV)
线性变参数变增益调节技术 LPV gain scheduling techniques
线性变参数变增益控制器 LPV gain-scheduled controller ['sʃedjuːəld]
线性变参数仿射模型 LPV affine model [ə'faɪn]
线性变参数控制器 LPV controller
线性变参数模型 linearised LPV models
线性插值 linear interpolation [ɪnˌtɜːpəʊ'leɪʃən]
线性弹性 linear elasticity [elæ'stɪsɪtɪ]
线性动态响应 linear dynamic response [daɪ'næmɪk]

线性方程 linear equation
线性关系 linear relation
线性函数 linear function
线性化模型 linearized model
线性加速度法 linear acceleration method [əkˌseləˈreɪʃən]
线性矩阵不等式 linear matrix inequality (LMI) [ˌɪnɪˈkwɒlətɪ]
线性空气动力学模型 linearized aerodynamics model [ˌeərəʊdaɪˈnæmɪks]
线性能量传递 linear energy transfer
线性相关 linear correlation [ˌkɒrəˈleɪʃ(ə)n]
限幅相应 limited amplitude response [ˈæmplɪtjuːd] [rɪˈspɒns]
限速开关 limit speed switch
限位板 emergency stop plate
限位开关 limit switch
限制扭矩 torque-limiting
相对湿度 relative humidity [hjuːˈmɪdətɪ]
相对速度 relative velocity
相干函数 coherence function [kəʊˈhɪərəns]
相关长度 correlation length
相关分析 correlation analysis
相关函数 correlation function
相关系数 correlation coefficient
相关预测 measure-correlate-predict (MCP) [ˈkɔːrəˌleɪt]
相切的；正接的 tangential [tænˈdʒenʃəl]
相似理论 similarity theory
相似律 similarity law
相似准则 similarity criteria [kraɪˈtɪərɪə]
箱式变电站 box transformer/box transformer substation
响应时间 response time
向岸风 inshore wind; on shore wind
向量风场 vector wind field
项目评估 project appraisal [əˈpreɪzəl]
小尺寸涡旋 small size eddy [ˈedɪ]
小尺度湍流 small-scale turbulence [ˈtɜːbjʊləns]
小尺度阵风 small-scale gust
小扰动方法 small disturbances method [dɪˈstɜːbəns]
小扰动理论 small perturbance theory
小缩尺比模型 small-scale model
小型风洞 small scale wind tunnel
小型风能转换装置 small-scale WECS (SWECS)
斜风 skew wind; oblique wind [skjuː] [əˈbliːk]
斜坡；斜面；倾斜 slope
谐振根弯矩 resonant root bending moment
谐振规模缩减因子 resonant size reduction factor
谐振频率 resonant frequency
泄漏电流 leakage current [ˈliːkɪdʒ]
泻油 drain [dreɪn]
修正因子 correction factor
虚功 virtual work

虚弯度 virtual camber
虚弯度效应 virtual camber effect
虚翼型 virtual airfoil
虚迎角 virtual angle of attack; virtual incidence
悬式绝缘子 suspension insulator [ˈɪnsjʊleɪtə]
旋式风力机 counter-rotating wind turbine
旋翼拉力 rotor thrust [θrʌst]
旋翼理论 rotor blade theory
旋翼扭距 rotor torque [tɔːk]
旋转采样风矢量 rotationally sampled wind velocity
旋转计数器 revolution counter [ˌrevəˈluːʃən]
旋转接头 rotating union
旋转平面 plane of rotation
旋转速度 rotational speed [rəʊˈteɪʃənəl]
旋转体；旋转接头 rotating union
选址 siting; site selection
选址密度 siting density
循环载荷 cyclic loads

Y

压差阻力；压力阻力 pressure drag
压力；张力；牵力；电压 tension
压力表 pressure gauge [geɪdʒ]
压力继电器 pressure switch
压力角 pressure angle
压力控制阀 pressure control valve [ˈpreʃə]
压力头损失 loss of head
压力系数 pressure coefficient [ˌkəʊɪˈfɪʃənt]
压力下降，压强下降 pressure drop
压力中心 center of pressure
压流润滑法 pressure fed lubrication [ˌluːbrɪˈkeɪʃən]
压缩空气罐 compressed air storage
亚临界风速 subcritical wind speed
亚临界雷诺数 subcritical Reynolds number [ˈrenəldz]
亚临界流动状态 subcritical flow regime [sʌbˈkrɪtɪk(ə)l] [reɪˈʒiːm]
亚临界塔架 subcritical tower
严格执行标准 strict compliance with ISO
严重故障（风电机） catastrophic failure (for wind turbines) [ˌkætəˈstrɒfɪk]
沿边 edgewise [ˈedʒweɪz]
盐雾 salt spray (fog)
盐雾测试 salt spray test
盐雾气候环境 salty climate
验收试验 acceptance test
杨氏系数 Young's modulus
仰角（风轮轴）（rotor axis） tilt angle
仰倾；倾翻 tipping
氧化锌 zinc oxide [ˈɒksaɪd]
氧化锌避雷器 metal oxide arrester (MOA) [əˈrestə]
摇臂 rocker arm

叶端损失 tip loss
叶端损失系数 blade tip-root loss factor
叶根 blade root [bleɪd]
叶根面积 blade root area
叶根载荷 blade root load
叶尖 tip of blade
叶尖段桨距 tip section pitch [pɪtʃ]
叶尖速比 tip speed ratio
叶尖速度 tip speed; blade-tip-speed
叶尖速度系数 tip speed coefficients [ˌkəʊɪˈfɪʃənt]
叶尖损失 tip losses
叶尖损失因子 tip-loss factor
叶尖涡 blade-tip vortex; tip vortex [ˈvɔːteks]
叶尖制动 tip brakes
叶轮 impeller
叶轮扫掠面积 rotor-swept area
叶片 blade [bleɪd]
叶片安装角 blade pitch angel; setting angle of blade
叶片材料 blade material [məˈtɪərɪəl]
叶片长度 length of blade
叶片穿越频率 blade-passing frequency [ˈfrɪkwənsɪ]
叶片方位 blade azimuth angle [ˈæzɪməə]
叶片根部 root end
叶片根梢比 ratio of tip-section chord to root-section chord [kɔːd]
叶片共振 blade resonance
叶片焊接 blade bonding
叶片几何 blade geometry [dʒɪˈɒmɪtrɪ]
叶片几何攻角 angle of attack of blade [əˈtæk]
叶片几何扭转 blade geometric twist [ˌdʒɪəˈmetrɪk]
叶片计量 blade calculation [ˌkælkjuːˈleɪʃən]
叶片间距变化 blade pitch change
叶片角；桨叶角 blade angle
叶片节距控制 blade pitch control
叶片结冰 iced blade
叶片结构 blade structure
叶片结构设计 blade structural design [ˈstrʌktʃərəl]
叶片截面 blade section
叶片连接 blade connection
叶片扭角 twist of blade
叶片扭曲（连续） blade twist (continued) [twɪst]
叶片偏转；叶片挠度 blade deflections
叶片平均弦长 mean blade chord
叶片剖面桨距 blade section pitch
叶片失速 blade stall
叶片实度 blade solidity [səˈlɪdɪtɪ]
叶片数 number of blades
叶片损失 blade losses
叶片—塔架间隙 blade-tower clearance [ˈklɪərəns]
叶片投影面积 projected area of blade
叶片形状 blade shape
叶片修补 blade repair
叶片旋转直径 rotor diameter
叶片翼型；叶片轮廓 blade profile

[ˈprəʊfaɪl]
叶片迎角 blade angle of attack
叶片运行角 blade orbital angle [ˈɔːbɪt(ə)l]
叶片展弦比 aspect ratio [ˈæspekt]
叶片振动 blade vibration [vaɪˈbreɪʃən]
叶片重量 blade weight
叶片轴 blade axis [ˈæksɪs]
叶片轴承 blades bearing
叶片组装 blade assembly [əˈsemblɪ]
叶素 blade element
叶素—动量方法 Blade Element Momentum (BEM) method [ˈmeθəd]
叶素—动量理论 Blade Element - Momentum Theory [ˈθɪərɪ]
叶素方法 Blade Elements Method
叶素理论 blade element theory
叶弦 blade chord [kɔːd]
叶弦优化分布 optimized chord distribution [ˈɒptɪmaɪzd]
液动弹性模型 hydroelastic modelling
液压泵 hydraulic pump [haɪˈdrɔːlɪk]
液压变桨轴 hydraulic pitch axis
液压舱口系统 hydraulic hatch system
液压传动 hydraulic transmission
液压吊车系统 hydraulic crane system
液压缸 hydraulic cylinder [ˈsɪlɪndə]
液压过滤器 hydraulic filter [fɪltə]
液压活塞 hydraulic ram
液压联动机构 hydraulic power pack
液压马达 hydraulic motor
液压软管 hydraulic hoses
液压系统 hydraulic system
液压油 hydraulic fluid [ˈfluː(ː)ɪd]
液压执行器 hydraulic actuator [haɪˈdrɔːlɪk]
液压制动器 hydraulic brake [haɪˈdrɔːlɪk]
液压制动系 hydraulic braking system
液压阻尼器 hydraulic damper
液压作功筒 hydraulic cylinder [ˈsɪlɪndə]
一维动量矩理论 1-D momentum theory [məʊˈmentəm] [ˈθɪərɪ]
一维流动 one dimensional flow [dɪˈmenʃənəl]
移动式吊车 travelling crane [kreɪn]
异常风 extraordinary wind [ɪkˈstrɔːd(ə)n(ə)rɪ]
翼（型）弦（线） airfoil chord
翼动铰 flapping hinge
翼梁 spar
翼梢尾涡 trailing vortices [ˈvɔːtɪˌsiːz]
翼式风速计 vane-type anemometer [ˌænɪˈmɒmɪtə]
翼弦 chord
翼型 airfoil；aerofoil [ˈeərəfɒɪl]

翼型；翼剖面 airfoil section
翼型；翼型廓线 airfoil profile [ˈprəʊfɑɪl]
翼型；[车] 导风板 aerofoil/airfoil
翼型板风力机 airfoil wind machine
翼型的气动弦线 aerodynamic chord of airfoil [ˌeərəʊdɑɪˈnæmɪk]
翼型厚度 thickness of airfoil
翼型厚度函数 thickness function of airfoil
翼型廓线 airfoil contour
翼型理论 airfoil theory
翼型数据 aerofoil data
翼型特性 airfoil characteristics
翼型弯度 airfoil camber
翼型弯度函数 curvature function of airfoil
翼型相对厚度 relative thickness of airfoil [ˈeəfɒɪl]
翼型叶片 airfoil shaped blade
翼型族 aerofoil family
翼型阻力 airfoil drag
阴影 shadow
音频振动 subaudible vibrations [ˌsʌbˈɔːdəbl]
音值 tonality [təʊˈnælətɪ]
引下线 down-conductor
应变生命回归线 strain-life regression lines [rɪˈgreʃən]
应变仪 strain gauges
应力 stress
应力集中 stress concentration
应力张量 stress tensor [ˈtensə]
应用空气动力学 applied aerodynamics [ˈeərə(ʊ)dɑɪˈnæmɪks]
应用流体动力学 applied fluid dynamics
迎风；顶风；上升气流；上风向 upwind
迎风壁 windward wall
迎风机构 orientation mechanism [ˌɔːrɪenˈteɪʃən] [ˈmekənɪzəm]
迎风角；来流迎角 angle of wind approach
迎风面 face-on-attack; windward side
迎风面积 frontal area [ˈfrʌntəl]
迎风坡 windward slope
迎角 incidence angle [ˈɪnsɪdəns]
迎面风；来流 approaching wind; oncoming wind [əˈprɒtʃɪŋ]
迎面投影面积 frontal projected area
迎面阻力 frontal drag/resistance [rɪˈzɪst(ə)ns]
影闪烁 shadow flicker [ˈflɪkə]
油滴（雾滴） drizzle
油封 oil seal
油冷却器 oil cooler
油位开关 level switch
油箱容积 reservoir (fuel tank) capacity
有风试验 wind-on test
有利干扰 favourable interference [ˈfeɪvərəbl] [ɪntəˈfɪər(ə)ns]
有限元法 finite element method (FEM) [ˈfaɪnaɪt] [ˈmeθəd]
有限元分析 finite-element analysis [əˈnæləsɪs]
有效地面高度 effective ground level
有效地形高度 effective terrain

height [te'reɪn]
有效风速 effective wind speed
有效浮力 effective buoyancy ['bɒɪənsɪ]
有效功率 effective power
有效弯度 effective camber
有效弦长 effective chord length
有效迎角 effective angle of attack
有载运行 on-load operation
有载指示器 on-load indicator ['ɪndɪkeɪtə]
诱导速度 induced velocity [vɪ'lɒsətɪ]
与塔架位置关系 position relative to tower
雨流法 rainflow method ['meθəd]
雨流计数 rainflow counting
雨流周期分析 rainflow cycle counting
雨凇 glaze [gleɪz]
预加载荷 preloading
预弯叶片 preflex blade/prebend blade
预装法兰 preloaded flanges [priː'ləʊdɪd]
原动机 prime mover [praɪm]
原型风轮机 prototype turbines ['prəʊtətaɪp]
圆弧翼型 arc profile
圆锥；递减 taper
圆锥式管状钢塔 tubular steel tower ['tjuːbjʊlə]
远尾流 far wake
月风速变化 monthly variation of windspeed [veərɪ'eɪʃ(ə)n]
月平均风速 monthly average wind speed
月平均温度 mean monthly temperature ['mʌnθlɪ]
允许噪音水平 permitted noise levels
运输条件 transportation condition
运维成本 operation cost
运行包络线 operational envelope
运行风速 operational wind speed
运行管理 operation management ['mænɪdʒmənt]
运行刹车 operation brake
运行寿命 operating lifetime
运行速度 operating speed
运行维护费用 operation and maintenance costs
运行与维护 operation and maintenance ['meɪntənəns]
运行转速范围 operating rotational speed range [reɪndʒ]
运用数据 using data

Z

载波干扰比 carrier to interference ratio (C/I) [ˌɪntə'fɪərəns]
载荷路径 load paths
载荷状况 load case
再生能源 renewable energy [rɪ'njuːəbl]

在场 on site
噪声 noise
噪声敏感建筑物 noise sensitive building
噪声限值及测量方法 noise limits and measurement method
增速比 speed increasing/speed up ratio
增速箱 step-up gearing
增益调度 gain schedule [ˈʃədjuːəl]
增益裕度 gain margin [geɪn] [ˈmɑːdʒɪn]
闸衬片；刹车的衬里 brake lining [ˈlaɪnɪŋ]
闸垫 brake pad
闸瓦；刹车片 brake shoe [ʃuː]
窄幅变速 narrow range variable speed [ˈveərɪəbl]
展开角；散布角 angle of spread
展弦比 aspect ratio (AR)
占地少 small right-of-way
张应力 tensile stresses
障碍物 obstacle [ˈɒbstək(ə)l]
兆级（兆瓦级）风电机组 turbine of megawatt class/MW-class wind turbine
兆瓦级 multi-megawatt [ˈmegəwɒt]
遮蔽面积；气动阴影 blankcted area
阵风 gust [gʌst]
阵风变幅 gust amplitude [ˈæmplɪtjuːd]
阵风持续时间 gust duration [djʊˈreɪʃ(ə)n]
阵风尺寸 gust size
阵风尺度 gust scale
阵风的；疾风的 gusty
阵风递减率 gust lapse rate [læps]
阵风递减时间 gust lapse time
阵风发生器 gust generator
阵风分量 gust component [kəmˈpəʊnənt]
阵风风速计 gust (measuring) anemometer [ˌænɪˈmɒmɪtə]
阵风荷载 gust loading
阵风结构 gust structure
阵风能量因子 gust energy favor
阵风频数 frequency of gust; gust frequency [ˈfriːkw(ə)nsɪ]
阵风平均时间 gust averaging time [ˈævərɪdʒɪŋ]
阵风谱 gust spectrum [ˈspektrəm]
阵风容积 gust volume
阵风衰减时间 gust decay time [dɪˈkeɪ]
阵风速度 gust speed
阵风系数 gust/gustiness factor [ˈgʌstɪnɪs]
阵风下洗 gust downwash [ˈdaʊnwɒʃ]
阵风效应 gust/gustiness effect
阵风效应因子 gust-effect factor
阵风性 gustiness [ˈgʌstɪnɪs]
阵风因子 gustiness factor
阵风影响 gust influence [ˈɪnflʊəns]
阵风阵息 gust and lull
阵风最大风速 gust peak speed
阵列的风 array of wind
阵列型式 array pattern [ˈpæt(ə)n]
阵内间距（风轮机的） array spacing

振荡 oscillation
振荡器 oscillator
振动传感器 vibration sensor
振动等级 vibration level
振动检测仪 vibration detector
振动开关 vibration switch
振动频率 vibration frequency [vaɪˈbreɪʃən] [ˈfrɪkwənsɪ]
振动试验 vibration tests
振动中心；摆动中心 center of oscillation [ˌɒsɪˈleɪʃən]
振动阻尼器 vibration damper
整机出厂试验台 wind turbine test bench
整流 rectified
整流器 rectifiers
整流罩 nose cone
整流罩；（风洞支架）风挡；减阻装置 fairing
正常风 normal wind
正常关机/停机 normal shutdown
正常运行 normal operation
正常制动系 normal braking system
正常状态 normal condition
正齿轮 spur gear [spɜː]
正交法则 Orthogonality Condition [ˌɔːθɒgəˈnælətɪ]
正交模态 orthogonal mode shapes
正面（迎风）面积 frontal area
正平均风速 yearly average wind speed
正态分布 normal distribution
正态概率曲线 normal probability curve [kɜːv]
正弦式载荷 sinusoidal loading [ˌsɪnəˈsɔɪdl]
正压力系数 thrust coefficient
支撑结构 support structure for wind turbine [ˈstrʌktʃə]
支承摩擦 bearing friction
直角 right angle
直接太阳辐射 direct solar radiation [ˌreɪdɪˈeɪʃən]
直驱 direct-drive
直驱发电机 direct-drive generators
指向性 directivity [dɪrekˈtɪvɪtɪ]
制动机构 brake mechanism [ˌmek(ə)nɪz(ə)m]
制动盘 brake flap；braking disk
制动器 brake（for wind turbines）
制动器设定；闭合 brake setting
制动器设计 brake design
制动器释放 braking releasing
制动钳 brake-caliper [ˈkælɪpə]
制动系统 braking system
制动载荷 braking loads
制动转矩 brake torque
质量惯性 mass inertia [ɪˈnɜːʃə]
质量惯性矩 mass moment of inertia [ɪˈnɜːʃə]
质量检验 quality check
质量矩阵 mass matrix
质量流量 mass flow rate
质量流率 mass flow ratios
质量中心 center of mass
滞后角；移后角 angle of lag
滞后迎角 lagged incidence [ˈɪnsɪd(ə)ns]
中等阵风 moderate gust
（翼型）中弧线 mean（camber）line

中间轴 intermediate shaft
中跨 central span
中心极限定理 central limit theorem
重力场 gravitational field
重力风 gravity wind
重力加速度 gravitational acceleration [ˌgrævɪˈteʃənl]
重力势能 gravitational potential energy
重力载荷 gravitational loads [ˌgrævɪˈteɪʃənəl]
重力桩（海上风电） gravity foundation (offshore)
重心 center of gravity [ˈgrævɪtɪ]
重新启动 restart
周变桨距 cyclic pitch
周期系数 periodic coefficients [ˌpɪərɪˈɒdɪk]
周期性交替涡 cyclically alternating vortex [ˈsɑɪklɪklɪ] [ˈɔːltəˌneɪtɪŋ]
周期性应力极限 cyclic stress limit
周期载荷 periodic loads
周期振动 periodic vibration
周围风 ambient wind
周围风能密度 ambient power density
周围风速 ambient windspeed
周围气流 ambient windstream
周围湍流（度） ambient turbulence [ˈtɜːbjʊl(ə)ns]
周围迎风角 ambient wild angle
轴距 wheelbase
轴流 axial flow
轴流干扰因素 axial flow interference factor [ˌɪntəˈfɪərəns]
轴流式风轮机 ducted wind turbine
轴流诱导因子 axial flow induction factor
轴倾斜 shaft tilt
轴向动量方程 axial momentum equation
轴向速度 axial velocity
轴向推力 axis-thrust [θrʌst]
轴向诱导速度 axial induced velocity
轴向诱导因子 axial induction factor
轴制动 shaft brake
轴重力矩 shaft gravity moment [ˈgrævɪtɪ]
骤风载荷 gust loading
逐步动态分析 step-by-step dynamic analysis [dɑɪˈnæmɪk]
逐步求解法 step-by-step solution
主动齿轮 driving gear
主动调向风轮 yaw-active rotor
主动偏航 active (driven) yaw [jɔː]
主动失速 active stall
主动失速控制 active stall control
主风能 main wind energy
主风向 main wind direction
主风向；盛行风向 prevailing wind direction
主机架 main frame
主控柜 main box
主控系统 main control unit
主梁 main spar
主流风 main stream wind
主液压缸 master cylinder [ˈsɪlɪndə]
主载荷 main loads

主阵风 deterministic gust
主轴；心轴 spindle [ˈspɪndl]
主轴测试 shaft test
主轴承 main bearing
主轴检验 shaft test
助爬器 climbing support
驻点 stagnation point [stægˈneɪʃən]
驻压 stagnation pressure
柱塞泵 piston pump
柱状图 histogram [ˈhɪstəgræm]
转/分 rpm (revolution per minute)
转差率 slip speed
转动中心 center of rotation
转换过程 transitional processes
转矩闭环控制 torque control loop [luːp]
转矩特性 torque characteristic [ˌkærəktəˈrɪstɪk]
转矩系数；力矩系数 torque coefficient
转轮质量不平衡 rotor mass imbalance
转速计；转速器 tachometer [tæˈkɒmɪtə]
转速力矩特性曲线 speed-torque curve
转子 rotor
转子的力矩风速增益 rotor torque to wind speed gain
转子轮毂 rotor hub
转盘 rotor disc
转子锁锭 rotor locking
转子旋转模式 rotor whirl modes [hwɜːl]
桩基 pile foundation
撞锤 rammer [ˈræmə]
撞击；冲击 impinge [ɪmˈpɪndʒ]
撞击荷载 shock load
撞击系数；影响系数 impact factor
锥顶角 cone angle
锥角 coning
锥形铰链 coning hinge [hɪndʒ]
锥形转子 coning the rotor [ˈrəʊtə]
准定常空气动力学 quasi-steady aerodynamics [ˈkwɑːzɪ] [ˌeərəʊdaɪˈnæmɪks]
准静态弯曲力矩 quasistatic bending moments
准静态响应 quasistatic response [rɪˈspɒns]
准均匀场 quasi-uniform field
自励；自激 self-commutated/excitation [ˈkɒmjuːteɪt]
自耦变压器 autotransformer [ˌɔːtəʊtrænsˈfɔːmə]
自然风 natural wind
自然环境 natural environment [ɪnˈvaɪrənm(ə)nt]
自然通风 natural ventilation [ˌventɪˈleɪʃ(ə)n]
自然湍流 natural turbulence [ˈtɜːbjʊl(ə)ns]
自然湍流风 natural turbulent wind [ˈtɜːbjʊl(ə)nt]
自然危害 natural hazard [ˈhæzəd]
自然阵风 natural wind gust
自由边界 free boundary [ˈbaʊnd(ə)rɪ]
自由大气风 free atmospheric wind [ætməsˈferɪk]

自由度 degrees of freedom (DOF)
自由对流 free convection [kən'vekʃ(ə)n]
自由风；顺风 free wind
自由来流湍流（度） free stream turbulence ['tɜːbjʊl(ə)ns]
自由流边界 free stream boundary
自由流风速 free stream wind speed
自由流面 free stream surface
自由流速度 free stream velocity
自由流线 free streamline
自由偏航 free yaw [jɔː]
自由脱扣 trip-free
总谐波失真 total harmonic distortion [hɑː'mɒnɪk]
纵长的；纵向长的 lengthwise ['leŋθwaɪz]
纵横比；叶片展弦比 aspect ratio
纵梁（叶片） stringer
纵向防滑纹 lengthwise ribbing ['leŋθwaɪz]
纵向风；经向风 longitudinal wind
纵向刚模量 longitudinal stiffness modulus [ˌlɒndʒɪ'tjuːdɪnəl]
纵向湍流分量 longitudinal turbulence component ['tɜːbjʊl(ə)ns]
纵向湍流谱 longitudinal turbulence spectrum ['tɜːbjʊl(ə)ns]['spektrəm]
纵向压力梯度 longitudinal pressure gradient [ˌlɒn(d)ʒɪ'tjuːdɪn(ə)l]
纵坐标；Y 轴 Y-axis
阻抗 resistance [rɪ'zɪstəns]
阻抗电压 impedance voltage [ɪm'piːdəns]
阻抗角 impedance angles
阻力 drag (force) [dræg]
阻力板 drag spoiler
阻力差型风力机 differential drag machines
阻力风荷载 drag wind load
阻力极曲线 drag polar
阻力临界值 drag crisis
阻力刹车板 drag brake
阻力系数 drag coefficient [ˌkəʊɪ'fɪʃənt]
阻力型风杯风速计 drag cup anemometer [ˌænɪ'mɒmɪtə]
阻力型风力机 drag-type wind machine
阻力型风轮 drag-type rotor
阻力型平移式风车 drag translator
阻力中心 center of resistance；drag center [rɪ'zɪst(ə)ns]
阻尼 damping
阻尼板 spoiling flap
阻尼比 damping ratio
阻尼系数 damping coefficient [ˌkəʊɪ'fɪʃənt]
阻升比 drag lift ratio
最初起动转矩 breakaway torque
最大测量功率 maximum measured power
最大风速 maximum wind speed
最大功率 maximum power
最大极限状态；极限限制状态 ultimate limit state ['ʌltɪmət]
最大剪法 maximum shear method
最大力矩系数 maximum torque coefficient [tɔːk]
最大容许速度 maximum allowable

level [ˈmæksɪməm]

最大设计风速 maximum designed wind speed

最大升力攻角，临界攻角 angle of attack of maximum lift

最大升力系数 maximum lift coefficient [ˌkəʊɪˈfɪʃ(ə)nt]

最大输出功率风速 cut-off/cut-out wind speed

最大瞬时风速 maximum instantaneous wind speed [ˌɪnst(ə)nˈteɪnɪəs]

最大英里风速 fastest mile wind speed

最大允许功率 maximum permitted power

最大转速 maximum rotational speed

最佳场址 optimum site [ˈɒptɪməm]

最佳齿轮比 optimum gear ratio

最佳额定风速 optimum rated wind speed

最佳功率系数 optimum power coefficient [ˌkəʊɪˈfɪʃənt]

最佳仰角 optimal angle of incidence [ˈɪnsɪdəns]

最佳叶尖速比 optimum tip speed ratio

最佳叶片几何形状 optimum blade geometry [ˈɒptɪməm] [dʒɪˈɒmɪtrɪ]

最佳迎风速度 optimum speed to windward

最佳转速 optimum rotational speed [rəʊˈteɪʃənəl]

最佳转子性能 optimum rotor performance

最小调向风 minimum wind to yaw [ˈmɪnɪməm]

最小输出功率风速 cut-in wind speed

最优反馈 optimal feedback

最优叶片设计 optimal blade design [ˈɒptɪməl]

坐标系 coordinate system

太阳能发电

A

安全卸压阀 safety relief valve [rɪˈliːf]
安装的，已建立的 mounted
暗电流 dark current
暗特性曲线 dark characteristic curve

B

包装 packaging [ˈpækɪdʒɪŋ]
薄膜太阳能电池 thin-film solar cell
饱和蒸汽 saturation steam [ˌsætʃəˈreɪʃən]
备用 standby
备用状态 stand-by condition
背景 background [ˈbækgraʊnd]
背压 back pressure
被动式太阳能系统 passive solar (energy) system [ˈpæsɪv]
泵站 pump station [ˈsteɪʃən]
壁温 wall temperature
标准 criteria [kraɪˈtɪərɪə]
标准测试条件 Standard test conditions (STC)
标准大气压力 standard atmosphere [ˈætməsfɪə]
表面凝汽 surface condenser [kənˈdensə]
表面式换热器 surface heat exchanger
表面温度 surface temperature
并联电阻 shunt resistance
并网发电系统 system of making common electricity
并网光伏系统 grid-connected PV system
并网接口 utility interface
并网太阳能光伏发电系统 grid-connected PV system
波纹管 flexible connector [ˈfleksəbl]
补焊 repair welding [weldɪŋ]
不锈钢 stainless steel [ˈsteɪnlɪs] [stiːl]
部分流量 partial flow [ˈpɑːʃ(ə)l]
部件 component [kəmˈpəʊnənt]

C

采光口 aperture [ˈæpətʃə(r)]
采光面积 (A) aperture area [ˈæpətʃə(r)] [ˈeərɪə]
采光平面 aperture plane [ˈæpətʃə(r)]

槽形抛物面集热器 parabolic trough collector [ˌpærəˈbɒlɪk]
测点插孔 test wells
测量值 measured value
差压 differential pressure [ˌdɪfəˈrenʃ(ə)l]
产品构成 product composition
常规性能 general behavior [bɪˈheɪvjə]
敞开系统 open system
超导金属热管空管 super conduction metal heat pipe vacuum tube
超临界压力 supercritical pressure [suːpəˈkrɪtɪk(ə)l]
城市规划 city planning
冲孔 punching holes [ˈpʌntʃɪŋ]
冲孔成型 punching holes and flanging [ˈflændʒɪŋ]
充电 charge
充电控制器 charge controller
出口温度 exit temperature
出口压力 outlet pressure
除氧器 deaerator [diːˈeəreɪtə]
储热介质 storage medium [ˈmiːdɪəm]
储热器 thermal storage device [ˈθɜːməl] [ˈstɒːrɪdʒ] [dɪˈvaɪs]
储热容 storage capacity [ˈstɒːrɪdʒ] [kəˈpæsɪtɪ]
储水（容）量 tank capacity（V）
储水箱 storage tank [ˈstɔːrɪdʒ]
传热 heat transfer [trænsˈfɜː]
传热工质 heat transfer fluid [flʊːɪd]
串联电阻 series resistance

D

大气成分，大气组成 atmospheric composition [ætməsˈferɪk]
大气质量 air mass（AM）
带硅太阳电池 silicon ribbon solar cell
单晶硅太阳电池 single crystalline silicon solar cell
单晶硅太阳能系统 mono-crystalline silicon solar cell [ˈkrɪst(ə)laɪn] [ˈsɪlɪk(ə)n]
单膜式镜片 monolithic stretched-membrane [mɒnəˈlɪθɪk]
当量长度 equivalent length [ɪˈkwɪvələnt]
导热硅脂 silicon paste tube [ˈsɪlɪkən]
导热盲管 heat-conducting blind pipe
导热盲管括口 necking up heat-conducting blind pipe [ˈnekɪŋ]
导体 conductor [kənˈdʌktə]
等径卡套 inlet/outlet connectors [kəˈnektəz]
低成本 low cost
低压力 low pressure [ˈpreʃə]
底座 tracing
地面覆盖率 Group Cover Tatio（GCR）
碲化镉太阳能电池 CdTe solar cell
电磁阀 electromagnetism valve [ɪˌlektrəʊˈmægnətɪzəm]

电动葫芦 motor-driven hoist
电极 electrode [ɪˈlektrəʊd]
电加热 electric heater
电能表 supply meter
电网保护装置 protection device for grid
电源表 supply meter
碟阀 shut off cock
定日镜 heliostats [ˈhiːlɪə(ʊ)stæt]
独立发电系统 system of making autonomous electricity [ɔːˈtɒnəməs]
独立太阳能光伏发电系统 stand alone PV system
独立系统 island system
短路电流 short-circuit current (I_{sc})
短路电流的温度系数 temperature coefficients of I_{sc}
断路开关 circuit breaker
对流 convective flow
对流传热 convective heat transfer [kənˈvektɪv]
多反射平面集热器 faceted collector
多结太阳电池 multijunction solar cell
多晶硅太阳能电池 multi-crystalline silicon solar cell
多膜式镜片 faceted stretched-membrane [ˈmembreɪn]
多元熔盐 multi-molten salts [ˈmʌltɪ]

E

额定工作压力 nominal working pressure [ˈnɒmɪnl]
额定功率 rated power
二次聚光器 secondary concentrator [ˈkɒnsəntreɪtə]

F

发电机组 power block
发电量；发电 power generation [ˌdʒenəˈreɪʒən]
发泡 injecting foaming reagent [ɪnˈdʒektɪŋ] [rɪˈeɪdʒənt]
发泡工序 foaming process [ˈfəʊmɪŋ] [ˈprəuses]
发射率 emittance [ɪˈmɪtəns]
翻边 flanging [ˈflændʒɪŋ]
反射薄膜技术 Reflection film technology (ReflecTech)
反射表面 reflective surface
反射率 reflectance
反射系数 reflection coefficient (R)
返工 rework
方位角 azimuth angle [ˈæzɪməθ]
方阵（太阳电池方阵） array (solar cell array)
防尘圈 rubber rings [ˈrʌbə(r) rɪŋz]
防腐蚀，防腐蚀的 anticorrosion [ˌæntɪkəˈrəʊʒən]
非成像集热器 non-imaging collector
非承压太阳能热管集热器 non-

pressurized solar tube collector [preˈʃəraɪzd]

非共晶点 non-eutectic point [juːˈtektɪk]

非晶硅太阳能电池 amorphous silicon solar cell [əˈmɔːfəs]

非聚光（型）集热器 non-concentrating collector [ˈkɔnsəntreɪtɪŋ]

非选择性表面 non-selective surface [ˈsɪlˈektɪv]

非涅尔透镜 Fresnel lens

非涅尔集热器 Fresnel collector [freɪˈnel] [kəlektə]

分离式（太阳热水）系统 remote storage system

分体承压热水器 Split/ Separate Pressurized Solar Water Heater [ˈpreʃəraɪzd]

风速 wind speed (v)

风压 air pressure

封闭系统 closed system

峰瓦 watts peak

峰值电流 peak current

峰值电压 peak voltage

峰值功率 peak power

峰值功率的温度系数 temperature coefficients of P_m

氟化盐 fluoride salt [ˈflʊ (:) əraɪd]

辐射 radiation

辐射计 radiometer

辐照度 irradiance [ɪˈreɪdɪəns]

腐蚀试验 corrosion test [kəˈrəʊʒ(ə)n]

复合抛物面集热器 compound parabolic (concentrating) collector [ˈkɒmpaʊnd] [ˌpærəˈbɒlɪk]

覆膜 applying film [əˈplaɪŋ]

G

干燥箱 drying oven [ˈdraɪɪŋ] [ˈʌvən]

钢化玻璃 toughened glass

高分子聚合薄膜 polymer film [ˈpɒlɪmə]

高效传热蓄热技术 high-efficiency heat transfer and thermal storage technology

隔离二极管 blocking diode

隔热体 insulator [ˈɪnsjʊleɪtə(r)]

给排水 plumbing [ˈplʌmɪŋ]

给水泵 boiler feed pump

给水阀 feed water valve

给水流量 feed water flow

跟踪 tracking [ˈtrækɪŋ]

跟踪集热器 tracking collector

跟踪误差 tracking error

工质出口温度 fluid outlet temperature (t_0) [ˈfluːɪd]

工质进口温度 fluid inlet temperature (t_i)

工质平均温度 mean fluid temperature (t_m) [ˈtempərɪtʃə]

工作时间 up time

工作温度 working temperature

工作压力 working pressure

工作状态 in-service condition

公共连接点 point of common coupling

功率因数 power-factor

共晶点 eutectic point [juː'tektɪk]

供水箱 feed water tank

构成的化学元素 elemental composition

孤岛效应 islanding

故障时间 down time

管道 pipeline ['paɪpˌlaɪn]

管接头 pipe taps

管状的 tubular ['tjuːbjʊlə]

光电子 photo-electron

光伏 photovoltaic (PV) [ˌfəʊtəʊvɒl'teɪɪk]

光伏电池（格） photovoltaic cell

光伏系统 photovoltaic system

光伏系统功率因数 power factor of PV power station

光伏系统无功功率 reactive power of PV power station

光伏系统有功功率 active power of PV power station

光伏阵列 photovoltaic array

光伏组件 photovoltaic module ['mɒdjuːl]

光管 smooth tube [smuːð]

光洁度 degree of finish

光谱 sun spectrum ['spektrəm]

光谱响应 spectral response（spectral sensitivity）

光热太阳能发电 Solar Thermal Power（STP）

光照强度 intensity [ɪn'tensɪtɪ]

滚筋 rolling rebar ['rəʊlɪŋ] [rɪ'bɑː]

国家电力生产商协会 National Electrical Manufacturers Association

国家电气代码 National Electrical Code（NEC）

国家可再生能源实验室 National Renewable Energy Laboratory (NREL)

过热器 superheater

过热蒸汽 overheating steam [ˌəʊvər'hiːtɪŋ]

H

焓 enthalpy ['enθ(ə)lpɪ]

合成的，人造的 synthetic [sɪn'θetɪk]

赫兹 Hertz（Hz）

恒温计 thermometer [θə'mɒmɪtə]

恒温水箱 heat preservation tank [ˌprezə 'veɪʃn]

横剪 cross shearing ['ʃɪərɪŋ]

后腿（内） back holder（inside）

后腿（外） back holder（outside）

后斜拉杆 cross holder

厚度损耗 thickness losses ['θɪknɪs]

化合物半导体太阳电池 compound semiconductor solar cell

化石 fossil ['fɒsəl]

化学成分；化学组成 chemical composition

化学气相沉积 Chemical vapor deposition (CVD)
化学元素，硅 silicon
画面结构 composition of a picture
环缝焊接 girth welding [weldɪŋ]
环境（空气）温度 ambient (air) temperature (t_a) ['æmbɪənt]
环境风速 surrounding air speed
环境空气 ambient air ['æmbɪənt]
环境压力 ambient pressure
换热盘管 heat exchange coil [kɒɪl]
换热器 heat exchanger
回流口 return
回流系统 drainback system
汇流箱 combiner box [kəm'baɪnə]
混合技术 hybridization [ˌhaɪbrɪdaɪ'zeɪʃən]
混合熔盐 molten salt mixture ['mɪkstʃɜː]
火炉，熔炉，马弗炉 furnace['fɜːnɪs]

J

机械组成 mechanical composition [mɪ'kænɪk(ə)l]
机制；原理；进程；机械装置 mechanism ['mekənɪzəm]
基础设施 infrastructure ['ɪnfrəˌstrʌktʃə]
基础研究 basic research ['beɪsɪk]
集（流）管 header
集成 integret
集成系统 assembling system
集热器 collector
集热器（瞬时）效率 instantaneous (collector) efficiency (h) [ˌɪnstən'teɪnɪəs]
集热器（太阳）(solar) collector; solar thermal collector ['θɜːməl]
集热器盖板 collector cover plate
集热器列阵 collector array [ə'reɪ]
集热器流动因子 collector flow factor
集热器效率 collector efficiency [ɪ'fɪʃənsɪ]
集热器效率因子 collector efficient factor (F) ['fæktə(r)]
集热器总面积 gross collector area (Ag)
几何聚光比 geometrica concentrator ratio [dʒɪə'metrɪkəl]
记录仪表 recorder
间接系统 indirect system [ˌɪndɪ'rekt]
检修口 check valve
降低熔点 Lower Temp
降温解压阀 T/P valve [vælv]
交货期 date of delivery [dɪ'lɪvərɪ]
交钥匙工程 turn-key solution
焦距 focal lengths
脚手架 scaffold ['skæfəld]
接收器 receiver
接头 taps
接线盒 junction box
节能率 fractional energy savings ['frækʃənəl]
结构 structure ['strʌktʃə(r)]

金属薄片 foil
进口 inlet
进口温度 inlet temperature [ˈɪnlet]
进气阀 air intake valve
晶体的 crystalline [ˈkrɪst(ə)laɪn]
精确度 accuracy
净采光面积（聚光集热器） net aperture area (Ae) [ˈæpətʃə]
局部阻力 local resistance [ˈləʊkəl] [rɪˈzɪstəns]
聚光（型）集热器 concentrating collector [ˈkɒnsəntreɪtɪŋ]
聚光比 concentration ratio
聚光集热器单元 solar collector elements (SCE)
聚光器或聚光镜 concentrator [ˈkɒnsəntreɪtə]
聚光式太阳能发电 concentrating solar power (CSP)
聚晶硅太阳能电池 poly-crystalline solar cell
涓流集热器 trickle collector [ˈtrɪkl]
绝对温度 absolute temperature
绝对压力 absolute pressure
竣工图 as-built drawing

K

开关 switch
开关设备 switchgear
开口系统 vented system [ventɪd]
开口直径大小 aperture diameter width
开路电压 open-circuit voltage (V_{oc})
开路电压的温度系数 temperature coefficients of V_{oc}
抗冲击能力 surge withstand capability [sɜːdʒ] [keɪpəˈbɪlɪtɪ]
可靠性 reliability [rɪˌlaɪəˈbɪlətɪ]
可利用率 availability
空气（型）集热器 air type collector
空气预热器 air preheater
空气质量 air mass (AM)
控制器 controller [kənˈtrəʊlə(r)]
控制室 control room
控制室管理系统 CRMS (Control room management system)
矿物燃料 fossil fuels [ˈfɒs(ə)l]
矿物组成，矿物成分 mineral composition [ˈmɪn(ə)r(ə)l]
扩散，漫射 diffusion [dɪˈfjʊːʒən]

L

拉铝尾条 bottom frame [ˈbɒtəm] [freɪm]
拉伸膜 stretched-membrane
拉条 strapes [stræps]
冷却 DSC 曲线 cooling DSC curve
冷却水温度表 Coolant Temperature Gauge
冷水入口 inlet of cold water

离心式泵 centrifugal type pumps [sen'trɪfjʊg(ə)l]

理论分析 theoretical analysis [ˌθɪə'retɪkəl] [ə'næləsɪs]

理论效率 theoretical efficiency [ɪ'fɪʃənsɪ]

两端补料 filling reagent to both sides filtrate valve ['fɪltreɪt]

两端扣盖 lidding up both sides

临界温度 critical temperature

临界压力 critical pressure

灵敏度 sensitivity [sensɪ'tɪvɪtɪ]

流动状态 flow condition

流量计 flow meter ['mɪːtə(r)]

流量开关 flow switch

螺线管 solenoid ['səʊlənɒɪd]

铝外壳 aluminum alloy casing [ə'luːmɪnəm] ['ælɒɪ]

氯化盐 chloride salt ['klɔːraɪd]

M

闷晒 stagnation irradiation [stæg'neɪʃən] [ɪˌreɪdɪ'eɪʃn]

密度 density ['densətɪ]

密封垫 gasket

密封装置 sealing arrangement

模块 module

摩擦阻力 frictional resistance ['frɪkʃənəl] [rɪ'zɪstəns]

摩擦损失 friction losses

目标负荷 target load ['tɑːgɪt]

N

内胆 inner tank ['ɪnə tæŋk]

内胆翻边 flanging side of inner tank ['flændʒɪŋ]

内胆覆膜 applying film to inner tank

内胆盖 inner tank lid

能量偿还期 energetic amortization period [əˌmɔːtaɪ'zeɪʃən]

能量审核 energy audit ['ɔːdɪt]

能量生产率（比能率） specific energy yield [jiːld]

能量输出 energy yield

逆变器 inverter [ɪn'vɜːtə]

逆变器变换效率 inverter efficiency

凝固；团结；浓缩 solidification [ˌsəlɪdɪfɪ'keɪʃən]

黏度 viscosity [vɪ'skɒsətɪ]

暖通 HVAC，Heating Ventilation Air Condition

P

排放系统 draindown system

排管 tube bank

排空罐 vent pot
排污阀 drain valve
旁路二极管 by-pass diode [ˈdaɪəʊd]
旁路阀 bypass valve
抛物槽 parabolic trough
抛物线的 parabolic [ˌpærəˈbɒlɪk]
配电箱 distribution box
膨胀罐 expansion vessel [ɪksˈpænʃən] [ˈvesəl]
膨胀罐接口 connection for expansion vessel [ɪksˈpænʃən]
平板（型太阳）集热器 flat plate (solar) collector
平滑性 flatness
平均日效率 average daily efficiency
平均温度 average temperature
平台屋顶系统 flat-roof system
平屋顶支架 frame for flat roof

Q

其他零配件 other spare parts
起筋压平 pressing and ribbing [ˈrɪbɪŋ]
气体组分，气体成分 gas composition
汽发生器 steam generator
汽轮机室 turbine room
汽压 steam pressure
器件 element part [ˈelɪmənt]
前横拉杆 cross braces
前腿 side frame
强化传热 enhanced heat transfer [ɪnˈhɑːnst]
切割下料 cutting stock
氢 hydrogen [ˈhaɪdrədʒ(ə)n]
倾（斜）角 tilt angle
球阀 ball cock valve
曲线修正系数 curve correction coefficient
取水流量 water draw rate
取水温度 draw-off temperature
全集热真空管 evacuated heat pipe vacuum tube [ɪˈvækjʊeɪtɪd]

R

热冲击实验 thermal shock tests [ʃɒk]
热传导 heat conduction [kənˈdʌkʃən]
热管集热器 heat pipe solar tube collector
热管接收器 receiver
（净）热耗率 (net) heat rate
热交换器 heat exchanger
热力循环 thermal cycle [ˈθɜːm(ə)l]
热水出口 outlet of hot water
热特性 thermal properties [ˈprɒpətɪs]
热稳定性 thermal stability

[stəˈbɪlɪtɪ]

热物理特性 thermophysical property [ˌθɜːməʊˈfɪzɪkəl] [ˈprɒpətɪ]

热效率 thermal efficiency [ˈθɜːməl]

热效率曲线 thermal efficiency curve [ɪˈfɪʃənsɪ]

热性能 thermal behavior [bɪˈheɪvjə]

人工干预 manual intervention [ɪntəˈvenʃ(ə)n]

日负荷 load (L)

日照强度 irradiance [ɪˈreɪdɪəns]

容量 capacity

熔点 melting point [ˈmeltɪŋ] [pɒɪnt]

熔融盐 molten salt [ˈməʊltɜːn]

熔融盐传热蓄热材料 heat transfer and thermal storage materials

熔盐制备 preparation of molten salt [ˌprepəˈreɪʃən]

入口温度 inlet temperature

入口压力 inlet pressure

入库 warehousing [ˈweəhaʊzɪŋ]

入射波 incidence waves

入射角修正系数 incident angle modifier (K_0) [ˈɪnsɪdənt] [ˈmɒdɪfaɪə(r)]

S

三通 tee [tiː]

三相电压控制器 three-phase voltage control

散射辐照(散射太阳辐照)量 diffuse irradiation (diffuse insolation)

色彩构成;彩色合成 color composition [kɒmpəˈzɪʃ(ə)n]

上外壳胶圈 installing rubber rings to holes on outer casing

烧结 flaming

身体组成 body composition

生物质 biomass

时间常数 time constant (T_c)

时空 spatial-time [ˈspeɪʃəl]

实际压力 actual pressure

实验平台 experimental platform [ekˌsperɪˈmentəl] [ˈplætfɔːm]

实验系统 experimental system

市场预测 market prediction [prɪˈdɪkʃən]

试压 pressure testing

试压检漏 pressure testing for leaks

试压口 blow out opening

室,房间 chamber

室内温度 indoor temperature

室外温度 outdoor temperature

手动排气阀 vent cock with hand wheel

受热面 heating surface

输入功率 input power

输入计 feed-in meter

输送系统 distribution subsystem [ˌdɪstrɪˈbjʊːʃn] [ˈsʌbˈsɪstəm]

水力计算 hydraulic calculation [haɪˈdrɔːlɪk] [ˌkælkjəˈleɪʃən]

水栓 hydrant [ˈhaɪdrənt]

水位探测器 water level sensor [ˈsensə(r)]

水预热器 water preheater

水嘴管 water nozzle pipe

水嘴焊接 welding water nozzle

说明书 instruction manual [ɪnˈstrʌkʃən] [ˈmænjʊəl]

塑料端盖 plastic end lid

损失，损耗 losses

T

太阳（能）加热系统 solar heating system

太阳电池 solar cell

太阳电池的伏安特性曲线 I-V characteristic curve of solar cell

太阳电池温度 solar cell temperature

太阳电池组件 module（solar cell module）

太阳辐射 solar radiation [ˌreɪdɪˈeɪʃən]

太阳跟踪控制器 sun-tracking controller

太阳能保证率 solar fraction（f）[ˈfrækʃn]

太阳能倍数 solar multiple

太阳能的特点 characteristics of solar energy [ˌkærəktəˈrɪstɪks]

太阳能电池模块，光伏模块 solar module [ˈmɒdjuːl]

太阳能独立系统 solar independent system

太阳能发电系统 Solar electric generation systems（SEGS）

太阳能光伏建筑一体化 Building-integrated PV（BIPV）

太阳能光伏系统 solar photovoltaic（PV）system

太阳能集热器组件 Solar Collector Assembly（小 SCA）

太阳能镜场 solar field

太阳能控制器 solar controller

太阳能设计模型 Solar Advisor Model（SAM）

太阳能水箱 solar cylinder [ˈsɪlɪndə(r)]

太阳能系统 solar（energy）system

太阳能循环液 solar transfer fluid [trænsˈfɜː(r)] [ˈfluːɪd]

太阳能组件 solar energy parts

太阳热水系统效率 solar water heating system efficiency [ɪˈfɪʃnsɪ]

太阳小时 sun hours

太阳（照射）常数 solar constant

碳酸盐 carbonate salt [ˈkɑːbəneɪt]

陶瓷内胆 porcelain enamel coating [ˈpɔːsəlɪn] [ɪˈnæməl]

套管 sleeve

添加 adding into

填充因子 fill factor（curve factor）

调整，调节 modulate [ˈmɒdjʊleɪt]

铁鞋 foot pad [pæd]

停止阀 stop cock

通量聚光比 flux concentrating ratio

[ˈkɒnsəntreɪtɪŋ] [ˈreɪʃɪəʊ]
同心度 concentricity [kɒnsenˈtrɪsɪtɪ]
桶，箱，罐 tank
透射率 transmittance [trænzˈmɪt(ə)ns]
透射系数 transmission coefficient (T)
图表；图解 diagram [ˈdaɪəgræm]
涂层 coat
湍流 turbulent [ˈtɜːbjʊl(ə)nt]
退火 anneal

W

外观清理 tidying up appearance [əˈpɪərəns]
外壳 outer casing
外壳接板 connection plate
外桶缩口 necking down of outer tank [ˈnekɪŋ]
弯头 bent [bent]
微晶 crystallite [ˈkrɪst(ə)laɪt]
尾架橡胶件 rubber plugs [ˈrʌbə(r)] [plʌgz]
尾托 foot holder
温差 temperature difference [ˈdɪf(ə)r(ə)ns]
温差控制器 differential temperature controller [kənˈtrəʊlə(r)]
温度补偿 temperature compensation [kɒmpenˈseɪʃ(ə)n]
温度传感器 temperature sensors [ˈtempərɪtʃə] [ˈsensəz]
温度开关 temperature switch
温度系数 temperature coefficient
温降 temperature drops
温室气体 greenhouse gases
温室效应 greenhouse effect
稳定性 stability [stəˈbɪlɪtɪ]
稳健的 robust [rə(ʊ)ˈbʌst]
稳态 steady-state [sˈtedɪ]
污染物 pollutant [pəˈluːtnt]
屋顶倾斜度 roof inclination [ɪnklɪˈneɪʃ(ə)n]
无功损耗 reactive loss

X

吸热体 absorber [əbˈsɔːbə]
吸热体面积 absorber area
吸收 receiver [rɪˈsiːvə]
吸收峰 absorption peaks [əbˈsɔːpʃən] [piːks]
吸收率 absorptance [əbˈsɔːptəns]
吸收强度 absorptions coefficient [əbˈzɔːpʃ(ə)n]
系统的效率 system efficiency
下料缩口 necking down [ˈnekɪŋ]
线聚焦（型）集热器 line focus collector
相变介质 phase-change media [feɪz]
相变潜热 latent heat changed

[ˈleɪtənt]

相变太阳能系统 phase change solar system

相位，相阶段 phase [feɪz]

镶嵌屋顶系统 in-roof installation [ɪnstəˈleɪʃ(ə)n]

硝酸盐 nitrate salt [ˈnaɪtreɪt]

效率，效能，功效 efficiency [ɪˈfɪʃənsɪ]

斜拉杆 inclined holder [ɪnˈklaɪnd]

斜屋顶支架 frame for pitched roof [pɪtʃt]

卸模 removing die

信号管 signal pipe [ˈsɪgnəl]

性能公差 performance tolerance [ˈtɒl(ə)r(ə)ns]

性能预测 performance prediction [pəˈfɔːm(ə)ns] [prɪˈdɪkʃ(ə)n]

性能质保 performance guarantee

序号 sequence number [ˈsiːkwəns]

蓄热 thermal storage [ˈstɒrɪdʒ]

旋转抛物面集热器 parabolic dish collector [ˌpærəˈbɒlɪk]

选择性表面 selective surface [sɪˈlektɪv]

削料清理 trimming and cleaning-up

循环泵 circulation pump [ˌsɜːkjʊˈleɪʃən]

循环水管 circulating water pipe (pump) [ˈsɜːkjuleɪtɪŋ]

Y

压力表 pressure gauge with mounting valve [geɪdʒ]

压力开关 pressure switch

压力温度安全器 pressure-temperature relief device

亚临界压力 subcritical pressure [sʌbˈkrɪtɪk(ə)l]

岩棉 rock wall

研究进展 research progress [ˈprəʊgres]

野外系统 field system

液体（型）集热器 liquid type collector [ˈlɪkwɪd]

一体承压太阳能热水器 compact pressurized solar water heater [ˈpreʃəraɪzd]

一体非承压太阳能热水器系统 compact non-pressurized solar water heater system [ˈpreʃəraɪzd]

一体预热承压太阳能热水器 compact pre-heat pressurized solar water heater

仪表 measuring instrument [ˈɪnstrʊm(ə)nt]

阴极 cathode [ˈkæθəʊd]

应力消除 stress-relieved [rɪˈliːvd]

硬度，刚度 stiffness [ˈstɪfnɪs]

用户信息系统 CIS（Consumer information system）

有功损耗 active loss

有机热载体（导热油） heat transfer fluid

允许限值 allowable limit [əˈlaʊəbl]

Z

粘贴标识 attaching label and brand [ˈleɪbl]

再生性能源 renewable energy [rɪˈnjuː əbl]

兆瓦 Mega Watt (MW)

照射方向自动跟踪 setpoint tracing

折射 refracte

真空管集热器 evacuated tube collector [ɪˈvækjʊeɪtɪd]

振打装置 rapping apparatus [ˌæpəˈreɪtəs]

蒸发 evaporation [ɪˌvæpəˈreɪʃən]

蒸汽吹扫系统 steam purge system

蒸汽流量 steam flow

蒸汽汽轮发电机组 steam turbine with alternator [ˈɔːltəneɪtə]

整机检验 whole machine testing

整流器 rectifier [ˈrektɪfaɪə]

整体式太阳热水器 integral collector storage solar water heater [ˈɪntɪgrəl]

正面系统 facade system [fəˈsɑːd]

支架 bracket

支模具 inserting die [ɪnˈsəːtɪŋ]

直缝焊 longitudinal seaming [lɔndʒɪˈtjuːdɪnl]

直接产生蒸汽系统 Direct steam generation (DSG)

直接光照强度 Direct normal irradiance (DNI)

直接系统 direct system [dɪˈrekt]

直流/直流电压变换器 DC/DC converter (inverter)

直流/交流电压变换器 DC/AC converter (inverter)

直流管集热器 direct flow solar tube collector [kənˈtrəʊlə(r)]

直流式（太阳热水）系统 series-connected system; once-through system

直射辐照 direct irradiation (direct insolation)

止回阀 non-return valve [vælv]

指示仪表 indicator [ˈɪndɪkeɪtə]

制作，构成 fabrication [fæbrɪˈkeɪʃən]

智能设备 IED (Intelligent electronic devices)

滞止温度 stagnation temperature [stægˈneɪʃən] [ˈtempərɪtʃə]

中央处理模块 CPM (Central processing module)

中央控制器 CCU (Central Control Unit)

主动式太阳能系统 active solar (energy) system [ˈsəʊlə(r)]

主铜管翻边 flanging holes on main copper tube [ˈflændʒɪŋ]

主铜管缩口 necking down main copper tube

主铜管钻孔 boring holes on main copper tube

主蒸汽 MS (Main Steam)

注液清洗阀 filling and rinsing cocks [ˈrɪnsɪŋ]

专利 proprietary [prəʊˈpraɪətərɪ]

转换效率 conversion efficiency [kənˈvɜːʃ(ə)n]

装配 assembly [əˈsemblɪ]

锥丝 nipple [ˈnɪpl]

子方阵 sub-array (solar cell sub-array)

自动跟踪 automatic track

自动控制 automatic control

自动排气阀 auto air vent

自动调整 automatic adjustment

自然循环式（太阳）热水器 natural cycle solar water heater

（聚光集热器）总采光面积 gross aperture area（A_a）

总发电量 total generation

总辐射 global radiation [reɪdɪˈeɪʃ(ə)n]

总辐射度（太阳辐照度） global irradiance（solar global irradiance)

总流量 total flow

总热损系数 overall heat loss coefficient（UL）[ˌkəʊɪˈfɪʃnt]

纵剪 slitting [ˈslɪtɪŋ]

组件的电池额定工作温度 NOCT (nominal operating cell temperature)

组件的规格 type of parts

组件效率 module efficiency

最大负荷 peak load

最大功率 maximum power（P_m）

最大功率点 maximum power point

最佳工作点电流 optimum operating current（I_n）

最佳工作点电压 optimum operating voltage（V_n）

核 能 发 电

10路选择器 ten path selector

132kV电气系统 132kV electrical power system

220kV电气系统 220kV electrical power system

220kV和132kV室内配电装置HVAC系统 220kV & 132kV indoor switchgear building HVAC system

220V交流计算机电源系统 220V AC computer power supply system

220V交流重要仪表电源系统 220V AC important instrumentation power system [ˌɪnstrʊmenˈteɪʃ(ə)n]

380V交流配电和控制系统 380V AC distribution and control system

6kV厂用电源系统 6kV normal auxiliary power system [ɔːgˈzɪlɪərɪ]

O形环 O-ring

pH值 pH value

U形管 U-tubes

U形管段 crossover leg

U形管束 U-tube bundle [ˈbʌnd(ə)l]

β/α比值仪 β/α ratemeter [ˈreɪtˌmiːtə]

Ω密封 omega seal [ˈəʊmɪgə]

A

安全棒 safety rod

安全参数显示系统 safety parameter display system [pəˈræmɪtə]

安全等级 safety class

安全等级的接口 safety class interface

安全端 safe end

安全阀 safety valve

安全分析报告 safety analysis report

安全规定 safety code [kəʊd]

安全壳 reactor containment [kənˈteɪnm(ə)nt]

安全壳地坑 containment sump

安全壳隔离信号 containment isolation signal [kənˈteɪnm(ə)nt] [aɪsəˈleɪʃ(ə)n]

安全壳贯穿件 containment penetration [penɪˈtreɪʃ(ə)n]

安全壳空气净化系统 containment air cleanup system

安全壳喷淋泵 containment spray pump

安全壳喷淋热交换器 containment spray heat exchanger [spreɪ]

安全壳喷淋系统 containment spray system [kənˈteɪnm(ə)nt]

安全壳喷淋系统热交换器 contain-

ment spray system heat exchanger
安全壳喷淋信号 containment spray signal
安全壳氢气混合系统 containment hydrogen mixing system [ˈhaɪdrədʒ(ə)n]
安全壳氢气排风系统 containment hydrogen ventilation system [ˌventɪˈleɪʃ(ə)n]
安全壳清洗通风系统 containment purge ventilation system
安全壳疏排系统 containment reactor coolant drain system [ˈkuːl(ə)nt]
安全壳消氢系统 containment hydrogen recombine system [ˌriːkəmˈbaɪn]
安全壳仪表系统 containment instrumentation system [ˌɪnstrʊmenˈteɪʃ(ə)n]
安全驱动器 safety actuator [ˈæktjʊeɪtə]
安全停堆地震 safe shutdown earthquake
安全系数 safety factor
安全系统 safety systems
安全相关设备 safety-related component [kəmˈpəʊnənt]
安全照明 safety lighting
安全制动器 safety brake
安全重要物项 item important to safety
安全注射 safety injection [ɪnˈdʒekʃ(ə)n]
安全组（控制棒） safety bank
安注泵 safety injection pump [pʌmp]
安注阶段 safety injection phase [feɪz]
安注系统 safety injection system [kəmˈpəʊnənt]
安注箱 safety injection tank [ɪnˈdʒekʃ(ə)n]
安注信号 safety injection signal
安装完工报告 end-of erection state report
安装完工证书 completion of erection release [ɪˈrekʃ(ə)n]
鞍式支座 saddle support [ˈsæd(ə)l]
按钮拨号 push-button dial
凹陷 denting

B

巴基斯坦恰希玛核电工程 Pakistan Chashma Nuclear power Project
板块构造 plate tectonic [tekˈtɒnɪk]
半导体探测器 Semiconductor detector [ˌsemɪkənˈdʌktə]
半球形封头 hemispherical head [ˌhemɪˈsferɪkl]
半图形显示 semi-graphic display
半自动焊 semi-automatic welding
棒电源系统 CRDM power supply

system
棒行程套管 rod travel housing
棒束控制组件 rod cluster control assembly [ə'semblɪ]
棒位光点指示器 rod position LED indicator
棒位数字显示器 rod position digital display
棒位探测器 rod position detector [dɪ'tektə]
棒位指示组件 rod position indicator assembly ['ɪndɪkeɪtə]
包壳 cladding
包壳蠕变 cladding creep
包壳压扁 cladding flattening
包壳应变 cladding strain
包壳与芯块间的间隙 clad pellet gap [klæd]
薄壁（裙）筒 thin skirt
饱和沸腾 saturated boiling ['sætʃəreɪtɪd]
饱和温度 saturation temperature
饱和系数 saturation index ['ɪndeks]
饱和蒸汽 saturated steam/vapor ['sætʃəreɪtɪd] ['veɪpə]
保持磁极（可移动磁极） movable pole [pəʊl]
保持钩爪（可移动钩爪） movable gripper latch ['grɪpə] [lætʃ]
保持衔铁（可移动衔铁） movable armature ['muːvəb(ə)l] ['ɑːmətʃə]
保持线圈（移动钩爪线圈） movable gripper coil ['grɪpə] [kɒɪl]
保存分类 retention categories
保存期 retention period
保存区 retention area [rɪ'tenʃ(ə)n]
保护继电器 protection relay
保护罩 protective cover ['kʌvə(r)]
保健物理系统 health physics system
报警（输出）组件 alarm (output) module ['mɒdjuːl]
报警窗 alarm window
报警信号 alarm signal
暴雨强度 rainfall intensity
暴雨水量 storm runoff ['rʌnɒf]
爆破盘 rupture disk ['rʌptʃə]
倍增周期 double period
泵特性曲线 pump characteristic curve [ˌkærəktə'rɪstɪk] [kɜːv]
比放 specific radioactivity [ˌreɪdɪəʊæk'tɪvɪtɪ]
比功率 specific power
比活度 specific activity
比计数管 proportional counter
比例尺，结垢 scale [skeɪl]
比例喷雾阀 proportional spray valve [prə'pɔːʃ(ə)n(ə)l] [spreɪ]
比例组电加热器 proportional heater
比燃耗 specific burnup ['bəːnʌp]
比释功能 kerma ['kɜːmə]
比重 specific gravity ['grævɪtɪ]
闭路电视系统 closed circuit TV system
闭路循环冷却水 closed-circulating cooling water
闭式循环冷却水系统 closed cycle cooling water system

闭锁继电器 latching relay ['ri:leɪ]
闭锁装置 latching device [lætʃ] [dɪ'vaɪs]
编码控制器 code controller
编线盒 marshalling box ['mɑ:ʃəlɪŋ]
编译程序 compiler [kəm'paɪlə]
变压变频调速 variable voltage & variable frequency speed regulation
标准编制说明 description for drawing up standard [dɪ'skrɪpʃ(ə)n]
标准待批稿 draft standard for approval
标准贯入试验 standard penetration test [penɪ'treɪʃ(ə)n]
标准实施日期 date of standard implementation [ˌɪmplɪmen'teɪʃən]
标准送审稿 draft standard for examination
表决逻辑组件 voting logic module
表面放射性污染 surface radioactivity contamination [ˌreɪdɪəʊæk'tɪvɪtɪ] [kənˌtæmɪ'neɪʃən]
表面活度 surface activity
表面剂量率 surface dose rate [dəʊs]
表面漏电距离 creepage distance
表面渗碳 surface carburization [ˌkɑ:bjuraɪ'zeɪʃən]
表面污染测量仪 surface contamination survey meter [kənˌtæmɪ'neɪʃən]
表面污染控制水平 control level of surface contamination
并列 side-by-side
波动管接管 surge nozzle [sɜ:dʒ]
波动箱 surge tank
玻璃剂量计 glass dosimeter [dəʊ'sɪmɪtə]
玻璃纤维 fiberglass
补偿 compensation [kɒmpen'seɪʃ(ə)n]
补偿电缆 compensation cable
补偿电离室 compensated ionization chamber [ˌaɪənaɪ'zeɪʃən]
补给水 make up water
补给水泵 make-up water pump
补给水系统 make up water system
补给水箱 make-up water tank
补焊 repair welding
补强板 reinforcing plate
不符合项审查 non-conformance review
不间断电源 uninterruptible power supply
不可接近的设备 inaccessible equipment [ˌɪnæk'sesəbl]
不利因子 disadvantage factor
不利于质量条件 condition adverse to quality
不凝性气体 non-condensable gas [kən'densəbəl]
不熔化电极 non-consumable electrode [kən'sju:məb(ə)l] [ɪ'lektrəʊd]
不受经费和进度的约束 independence from cost and schedule
步距 step length
部分焊透焊缝 partial penetration welding ['pɑ:ʃ(ə)l]

C

采风口，进风口 air intake
采购规格书 procurement specification [prəˈkjʊrmənt]
采暖通风及空调 heating ventilation and air conditioning [ˌventɪˈleɪʃn]
参考水平，基准液位 reference level
参考值，基准值 reference value
残余变形，偏移，失调 offset
残余样品 residual sample
操作规程，工作细则 working instruction [ɪnˈstrʌkʃ(ə)n]
操作过电压 operating overvoltage [ˈəuvəˈvəultɪdʒ]
操作平台 operating platform
操作器（显示） display operation station
操作员站 operator station
测量棒位 measured rod position [rɒd]
测量管 instrumentation tube [ˌɪnstrəmenvˈteɪʃn]
测量箱 measurement box
测试程序，试验大纲 testing program
测压管 pressure tap
层间温度 interpass temperature [ɪntɜːˈpɑːs]
插棒 rod insertion
插入式断路器 plug-in breaker
查错 debug
差动保护 differential protection
差压变送器 differential pressure transmitter
差压计 differential pressure gauge [ˌdɪfəˈrenʃ(ə)l]
差压调节器 differential pressure controller
拆卸按钮 disconnect button
拆卸杆 disconnect rod
柴油发电机 diesel generator
柴油发电机组 diesel generator set
产地检查 source inspection
产地验证 source verification [ˌverɪfɪˈkeɪʃ(ə)n]
产品分析 product analysis
常闭触点 normally-closed contact [ˈkɒntækt]
常规岛 conventional island（CI）[kənˈvenʃ(ə)n(ə)l]
常规岛直流和交流电源系统 conventional island DC & AC UPS system
常规监测 routine monitoring
常开触点 normally-open contact
厂房总布置图 general arrangement [ˈdʒenərəl] [əˈreɪndʒmənt]
厂内电源 onsite source
厂区保安系统 site security system [sɪˈkjʊərɪtɪ]
厂区环境辐射监测系统 in-site controlled area environment radiation monitoring system
厂区人员 site personnel [pɜːsəˈnel]

厂区消防水分配系统 yard fire protection water distribution system [dɪstrɪ'bjuːʃ(ə)n]
厂外电源 offsite power source
厂用电系统（常规岛） plant auxiliary electrical system [ɒg'zɪljərɪ]
场地管理 housekeeping
场地勘察 site investigation [ɪnˌvestɪ'geɪʃ(ə)n]
超功率 excess power
超前/滞后补偿组件 lead/lag compensation module [kəmpen'seɪʃ(ə)n] ['mɒdjuːl]
超热中子 epithermal neutron [ˌepɪ'θəːməl] ['njuːtrɒn]
超声波去污槽 ultrasonic decontaminating tank [ʌltrə'sɒnɪk] [diːk(ə)n'tæmɪneɪt]
超声波液位计 ultrasonic liquid level meter
超声检验 ultrasonic examination [ʌltrə'sɒnɪk]
超声去污 ultrasonic decontamination ['diːkənˌtæmɪ'neɪʃən]
超时 time out
超速安全制动器 overspeed safety brake
超速保护开关 overspeed protecting switch
超压保护 overpressure protection
沉淀 sedimentation
沉积腐蚀 deposit corrosion
沉积腐蚀产物 deposited corrosion product [kə'rəʊʒ(ə)n]
沉积量 deposition [ˌdepə'zɪʃ(ə)n]
沉降缝 settlement joint
沉砂池 desilting basin
成本效益分析 cost benefit analysis
成套的，整体式的 self-contained
成组显示 group display
承压边界 pressure-retaining boundary ['baʊnd(ə)rɪ]
承压热冲击 pressurized thermal shock ['preʃəraɪzd]
程控水位 programmed water level
程序鉴定试验 procedure qualification test [prə'siːdʒə] [ˌkwɒlɪfɪ'keɪʃ(ə)n]
程序鉴定试验报告 procedure qualification report [prə'siːdʒə] [ˌkwɒlɪfɪ'keɪʃ(ə)n]
澄清，净化 clarification [ˌklærɪfɪ'keɪʃ(ə)n]
澄清槽 clarifier; mud settler ['klærɪfaɪə]
持续时间 duration time
齿轮减速器 gear reducer [riː'djuːsə]
齿轮联轴器 gear coupling ['kʌplɪŋ]
冲击 punching
冲击（电压）保护 surge protection
冲击试验 shock test
冲击载荷 shock load
冲洗泵 flush pump
冲洗水 flushing water
抽气泵 air pump [pʌmp]
抽气器 ejector
抽头切换开关 tap changer
抽芯组件 pull-out unit
出口接管保温层 outlet nozzle insulation
出口温度 outlet temperature

出线断路器 outgoing breaker

初步安全分析报告 preliminary safety analysis report [prɪ'lɪmɪn(ə)rɪ]

初步处理 primary treatment

初级辐射 primary radiation

初级绕组 primary winding

初级中子源 primary source

初级中子源组件 primary neutron source assembly ['njuːtrɒn]

初始功率 initial power [ɪ'nɪʃəl]

除碘效率 iodine removal efficiency ['aɪədiːn] [ɪ'fɪʃənsɪ]

除锂床 delithium demineralizer [diː'mɪnərəlaɪzə]

除硼床 deborating demineralizer

除盐水 demineralized water

除盐水泵 dimineralized water pump

除盐装置 demineralizer

除氧水 deoxygenated water [diː'ɒksɪdʒɪneɪtɪd]

除氧水泵 deoxygenated water pump

除氧水冷却器 deoxygenated water cooler [diː'ɒksɪdʒɪneɪtɪd]

除氧水箱 deoxygenated water tank

储存设施 storage facility [fə'sɪlɪtɪ]

储水池 storage reservoir/tank ['rezəvwɑː(r)]

氚取样器 tritium sampler ['trɪtɪəm] ['sɑːmplə]

穿晶应力腐蚀 transgranular stress corrosion cracking [træns'grænjulə]

穿墙布线 knob-and--tube wiring

穿墙套管 penetration sleeve

传递磁极 stationary pole ['steɪʃ(ə)n(ə)rɪ]

传递线圈 stationary gripper coil ['grɪpə] [kɒɪl]

传送带 conveyer belt [kən'veɪə]

床后过滤器 demineralizer after-filter

床前过滤器 demineralized pre-filter

吹扫气体 purge gas

垂直加速度 perpendicular acceleration [ˌpɜːp(ə)n'dɪkjʊlə]

磁带 magnetic tape

磁带读出器 tape reader

磁带机 magnetic tape unit

磁带录音机 tape recorder

磁粉检验 magnetic particle examination ['pɑːtɪk(ə)l]

磁盘机 magnetic disk unit [mæg'netɪk]

磁脱扣断路器 magnetic trip breaker

次级辐射 secondary radiation

次级中子源 secondary neutron source ['njuːtrɒn]

次级中子源组件 secondary neutron source assembly

次临界堆芯 subcritical core [sʌb'krɪtɪk(ə)l] [kɔː]

从动计数译码组件 slave cycler count and decoder [sleɪv] [ˌdiː'kəʊdə]

从动输入组件 slave cycler input module ['mɒdjuːl]

从动循环器 slave cycler

粗滤器 strainer ['streɪnə]

粗砂 coarse sand [kɔːs]

存储程序控制 stored program control

错装料事故 fuel mispositioning accident

D

打接线 jumper
大气辐射监测装置 atmospheric radiation monitoring apparatus [ætməs'ferɪk] [reɪdɪ'eɪʃ(ə)n] [ˌæpə'reɪtəs]
大气腐蚀 atmospheric corrosion [ætməs'ferɪk] [kə'rəʊʒ(ə)n]
大气过电压 atmospheric overvoltage [ˌəʊvə'vəʊltɪdʒ]
大气弥散因子 atmospheric dispersion factor [dɪ'spɜːʃ(ə)n]
大气稳定度 atmospheric stability
大气压力 atmospheric pressure
大容积空气取样器 large volume air sampler ['sɑːmplə]
代表性样品 representative sample [reprɪ'zentətɪv]
带厂用电运行 house load operation；service power operation
带灯按钮 illuminated pushbutton [ɪ'luːmɪneɪtɪd] [pʊʃ'bʌtn]
带灯开关 illuminated switch [ɪ'ljʊmɪneɪtɪd]
带电部分 live part
带钥匙开关 key-locked switch
单独基础 spread footing
单股线 solid conductor
单轨电动葫芦 electric monorail hoist ['mɒnə(ʊ)reɪl]
单面焊 welding by one side
单能的 mono energetic ['mɒnəʊ]
单稳态组件 monostable module ['mɒnəʊˌsteɪbl]
单线图 single line diagram ['daɪəgræm]
单相流 single phase flow
单一故障 single failure
单一故障准则 single failure criteria [kraɪ'tɪərɪə]
淡水 fresh water
弹棒 rod ejection
弹棒事故 rod ejection accident [ɪ'dʒekʃən]
蛋篓型格架 egg crate grid
氮气覆盖层 nitrogen blanket ['naɪtrədʒ(ə)n] ['blæŋkɪt]
氮气系统 nitrogen system
当地加速度 site acceleration [əkselə'reɪʃ(ə)n]
当量直径 equivalent diameter [ɪ'kwɪv(ə)l(ə)nt]
挡火墙 fire stopping wall
刀口腐蚀 knife corrosion [kə'rəʊʒn]
导出极限 derived limit [dɪ'raɪvd]
导出空气浓度 derived air concentration [kɒns(ə)n'treɪʃ(ə)n]
导阀 pilot valve [vælv]
导流筒 guide cylinder ['sɪlɪndə(r)]
导式卸压阀 pilot operated pressure relief valve [rɪ'liːf]
导向管 (thimble) guide tube

[ˈθɪmbl]
导向轨道 guide rail
倒班 shift (operation personnel)
等电压连接 equipotential connection [ˌiːkwɪpəˈtenʃ(ə)l]
等同采用 adoption by equivalent [əˈdɒpʃ(ə)n] [ɪˈkwɪv(ə)l(ə)nt]
等效满功率天 effective full power days
等震线 isoseisms [aɪˈsəʊsaɪzəmz]
低本底β探测装置 low-background β detector [dɪˈtektə]
低放α，γ探测器 low activity α，γ detector
低放惰性气体β监测仪 low-level noble gas β monitor [ˈnəʊb(ə)l] [ˈmɒnɪtə]
低放废水监测器 low-level waste liquid monitor
低合金钢 low alloy steel [ˈælɒɪ]
低量程电离室γ监测器 low range ionization chamber area γ monitor
低浓缩铀 slightly enriched uranium [ɪnˈrɪtʃt] [jʊˈreɪnɪəm]
低载，轻载 underload
地层 stratigraphy [strəˈtɪgrəfɪ]
地角螺栓 anchor bolt [ˈæŋkə]
地面污染监测仪 floor contamination monitor [kənˌtæmɪˈneɪʃən]
地下配线 underground wiring
地下水 ground water
地下水位 groundwater level
地震风险 seismic risk
地震构造区 seismotectonic province [ˌsaɪzmətekˈtɒnɪk]
地震仪表系统 seismic instrumentation system
地震载荷 seismic load
第二级处理 secondary treatment
第一堆芯 first core
典型栅元 typical cell
点燃，触发 ignition [ɪgˈnɪʃn]
点蚀 pitting corrosion [kəˈrəʊʒ(ə)n]
碘-131 监测仪 iodine-131 monitor [ˈaɪədiːn] [ˈmɒnɪtə(r)]
碘吸附盒 iodine cartridge [ˈɑɪədiːn] [ˈkɑːtrɪdʒ]
电厂火灾探测及报警系统 plant fire detection and alarm system
电厂运行管理部门 plant management
电厂总平面布置图 plant layout drawing [ˈleɪaʊt]
电磁流量计 electromagnetic flowmeter
电磁线圈 solenoid [ˈsəʊlənɒɪd]
电磁液压制动器 electric-hydraulic actuating brake
电磁制动器 electromagnetic braker [ˈbreɪkə]
电动单梁悬挂起重机 motor-driven single beam suspending crane [biːm] [səˈspend] [kreɪn]
电动机控制中心 motor control center
电度表 watt-hour meter
电加热器 electric heater
电喇叭 electric horn
电缆沟槽 trough duct [trɒf]

电缆通道 raceway
电离室 ionization chamber [ˌaɪənaɪˈzeɪʃən] [ˈtʃeɪmbə]
电离线载波 power line carrier
电力调度网 electric dispatch network [dɪsˈpætʃ]
电流波形图 current curve versus time [ˈvɜːsəs]
电流密度 current density
电炉 electric furnace [ˈfɜːnɪs]
电路图 circuit diagram [ˈsɜːkɪt]
电气厂房 electric building
电气厂房给排水系统 electric building water supply and drainage system
电气厂房排烟系统 electric building smoker exhaust system [ɪgˈzɔːst]
电气厂房通风系统 electric building ventilation system [ˌventɪˈleɪʃ(ə)n]
电气厂房循环冷风机组系统 electric building cooler system
电气贯穿件 electrical penetration [penɪˈtreɪʃ(ə)n]
电气配电装置及主要电气设备消防系统 switchgear & main electrical equipment fire protect system
电气箱 electrotechnical box [ɪˌlektrəuˈteknɪkəl]
电声元件 electroacoustic component [ɪˌlektrəuəˈkuːstɪk]
电网通话网络 grid telephone network
电位器 potentiometer [pəˌtenʃɪˈmɪtɪə(r)]
电压互感器 potential transformer
电源线 power cord
电渣重熔 electro-slag remelting
电站计算机系统 plant computer system
电站运行记录日志 power plant operation log
电子式袖珍剂量计 electronic pocket dosimeter
电阻温度探测器 resistance temperature detector
电阻箱 resistance box [rɪˈzɪst(ə)ns]
垫层 cushion [ˈkʊʃ(ə)n]
垫片 gasket [ˈgæskɪt]
吊兰部件存放架 lower internals storage stand
吊篮筒体法兰 core barrel flange
吊装孔 handling opening
碟形封头 dished head
碟形芯块 dished pellet
顶盖 closure head
顶盖法兰保温层 head flange insulation [flændʒ] [ˌɪnsjʊˈleɪʃn]
顶盖球冠保温层 head dome insulation [ˌɪnsjuˈleɪʃn]
定滑轮组 head block assembly [əˈsemblɪ]
定期检查 periodic inspection [ˌpɪərɪˈɒdɪk] [ɪnˈspekʃn]
定期试验 periodic/regular test
定位（距）管 localing tube [tjuːb]
定位部件 positioning part
定位格架 grid spring；spacer grid
定位格架栅元 grid cell
定位焊 tack welding [ˈweldɪŋ]
定位器 positioner [pəˈzɪʃənə]

定位系统 positioning system
定位销 aligning pin
定子 stator [ˈsteɪtə]
动滑轮组 load block assembly [əˈsemblɪ]
动力操作卸压阀 power operated relief valve
动力学不稳定性 dynamic instability [ɪnstəˈbɪlɪtɪ]
动量守衡 conservation of momentum [kɒnsəˈveɪʃ(ə)n]
动密封环 rotating seal ring
动平衡动压头 dynamic balance [daɪˈnæmɪk]
动态参数 dynamic parameter [pəˈræmɪtə]
动态分析 dynamic analysis
动态平衡 dynamic equilibrium [ˌiːkwɪˈlɪbrɪəm]
动压头 dynamic head
陡斜性能曲线 steep performance curve [kɜːv]
独立检查 independent review
读出和控制柜 readout and control cabinet
镀银 silver-plating
端梁 end truck
端塞 end plug
端子盒 terminal box
短路电压，阻抗电压（变压器） impedance voltage（transformer） [ɪmˈpiːdəns] [ˈvəʊltɪdʒ]
短寿命裂变产物 short-lived fission product [ˈfɪʃ(ə)n]
短寿命同位素 short life isotope [ˈaɪsətəʊp]
断连，断路 disconnect
断裂，破裂 rupture [ˈrʌptʃə]
断裂力学 fracture mechanics [ˈfræktʃə]
断裂强度 rupture strength
断裂试验 rupture test
断流阀 cut-off valve
断续焊 intermittent welding [ˌɪntəˈmɪtənt]
锻粗 upsetting
堆焊层下裂纹 underclad cracking
堆内构件 reactor internals
堆内构件存放台 internals lifting support stand [ɪnˈtɜːnlz]
堆内构件吊具 reactor internals lifting device
堆内密实 in-pile densification [densɪfɪˈkeɪʃən]
堆内下部支撑 lower core support structure
堆腔 reactor cavity
堆腔冷却系统 reactor cavity cooling system
堆外测量 ex-core instrumentation [ˌɪnstrʊmenˈteɪʃ(ə)n]
堆芯 in-core
堆芯布置 core configuration [kənˌfɪgəˈreɪʃ(ə)n]
堆芯测量 in-core instrumentation [ˌɪnstrəmenˈteɪʃn]
堆芯测量管 in-core instrumentation thimble [ˈθɪmbl]
堆芯等效直径 equivalent core diameter [ɪˈkwɪv(ə)l(ə)nt]

堆芯吊篮组件 core barrel cylinder [ˈsɪlɪndə]
堆芯活性高度 active core height
堆芯交混分因子 core mixing subfactor
堆芯流量分因子 core flow subfactor
堆芯裸露 core uncover
堆芯平均热通量 core average heat flux (density)
堆芯区域 core area
堆芯热工设计 core thermal design [ˈθɜːm(ə)l]
堆芯上板 upper core plate
堆芯上部构件 upper core support structure
堆芯完整性 core integrity [ɪnˈtegrɪtɪ]
堆芯温度测量系统 in-core temperature monitoring system [ˈmɒnɪtɪrɪŋ]
堆芯物理 core physics
堆芯下板 lower core plate [kɔː] [pleɪt]
堆芯卸料 core unloading
堆芯仪表室 in-core instrumentation room [ˌɪnstrəmenˈteɪʃn]
堆芯再淹没 core reflooding
堆芯支承块 core support pad
堆芯中子通量测量系统 in-core neutron flux monitoring system [ˈnuːtrɒn]
堆芯最大热通量 core maximum heat flux
对称截面 symmetrical section [sɪˈmetrɪk(ə)l] [ˈsekʃ(ə)n]
对地电阻 resistance to earth
对话键盘 dialog keyboard
对流沸腾 convection boiling [kənˈvekʃ(ə)n]
墩式基础 pier foundation
钝边 root face
钝化 passivation [pæsɪˈveɪʃən]
多层焊 multi-layer welding [ˈmʌltɪ] [ˈweldɪŋ]
多道能谱分析器 multichannel spectrum analyzer [ˈspektrəm]
多普勒系数 Doppler coefficient
多普勒效应 Doppler effect
多区堆芯 multizone core [ˈmʌltɪzəun] [kɔː]
多芯插座 multi-pin socket [ˈsɒkɪt]
多重测量 redundant measurement
惰性气体 noble gas [ˈnəub(ə)l]
惰性气体取样器 noble gas sampler

E

恶劣 degradation
二次屏蔽 secondary shielding [ˈʃiːldɪŋ]
二次蒸汽 secondary steam
二回路辅助厂房通风 secondary auxiliary building ventilation system [ɔːgˈzɪlɪərɪ] [ˌventɪˈleɪʃ(ə)n]

F

发电机和主变压器系统 generator & main transformer system

发射天线 transmittion antenna [æn'tenə]

乏燃池净化泵 spent fuel pool purification pump [ˌpjʊərɪfɪ'keɪʃən]

乏燃冷却器 spent fuel pool cooling water heat exchanger

乏燃料 depleted fuel

乏燃料操作工具 spent fuel handling tool

乏燃料池 spent fuel pool [fjʊəl]

乏燃料池冷却泵 spent fuel cooling pump

乏燃料池冷却及净化系统 spent fuel pool cooling and cleanup system

乏燃料储存池 spent fuel storage pool

乏燃料储存格架 spent fuel storage rack

乏燃料检测装置 spent fuel examination facility

乏燃料组件操作工具 spent fuel assembly handling tool [ə'semblɪ]

乏燃料组件运输容器 spent fuel assembly shipping cask [kɑːsk]

阀盖 valve bonnet

阀杆 valve stem

阀杆密封 stem seal [siːl]

阀盘 valve disc

阀位指示器 valve position meter

阀芯 valve trim

阀座 valve seat

法定计量单位 legal metrological unit [ˌmetrəu'lɒdʒɪkəl]

法兰连接 flanged connection

反射层 reflector

反射层节省 reflector saving

反射率 albedo [æl'biːdəʊ]

反向电压 reverse voltage [rɪ'vɜːs] ['vəultɪdʒ]

反向流 counter flow

反应厂房照明系统 reactor building lighting system

反应堆保护系统 reactor protect system

反应堆厂房给排水系统 reactor building water supply and drainage system

反应堆功率控制系统 reactor power control system

反应堆冷却剂泵 reactor coolant pump

反应堆冷却剂泵轴封水系统 reactor coolant pump seal water system

反应堆冷却剂疏水箱 reactor coolant drain tank

反应堆冷却剂系统 reactor coolant system

反应堆容器顶盖吊具 reactor vessel head lifting device

反应堆水化学 reactor water chem-

istry [ˈkemɪstrɪ]
反应堆压力容器放气系统 reactor pressure vessel vent system
反应性变化 reactivity change
反应性单位 pcm (pour cent mille)
反应性反馈 reactivity feedback
反应性阶跃引入 step insertion of reactivity [ɪnˈsɜːʃ(ə)n]
反应性控制组件 reactivity control component
反应性事故 reactivity accident
反应性停堆裕量 reactivity shutdown margin
反应性系数 reactivity coefficient [ˌkəʊɪˈfɪʃ(ə)nt]
反应性线性引入 ramp insertion of reactivity
反应性引入 insertion of reactivity [ɪnˈsɜːʃn]
反应性引入组件 reactivity insertion rate
返洗强度 intensity of back washing [ɪnˈtensətɪ]
返修焊 rewelding
泛光灯，探照灯 flood light
方案设计 conceptual design [kənˈseptjʊəl]
（堆芯）防断底板 secondary support base plate [pleɪt]
防断支撑板 secondary support plate
防断组件 secondary support assembly
防护服，气衣 ventilated suit [ˈventɪleɪtɪd]
防护屏 protective barrier
防护手套 protective glove [ɡlʌv]
防护鞋套 protective shoe cover
防护衣具 protective clothing
防火阀 fire damper
防火墙 fire break
防火设施 fire proof installation [ɪnstəˈleɪʃ(ə)n]
防泡剂 antifoam reagent [ˈæntɪˌfəʊm]
防水材料 water proofing material
防锈漆 anti-rust agent
防咬润滑剂 anti-seizer lubricant [ˈluːbrɪk(ə)nt]
防振条 anti-vibration bar [vaɪˈbreɪʃ(ə)n]
防震缝 aseismic joint [eɪˈsaɪzmɪk]
放化分析 radiochemical analysis [reɪdɪəʊˈkemɪkəl]
放气阀 vent valve
放射化学 radio-chemistry [ˈkemɪstrɪ]
放射性 radioactivity
放射性沉降物 radioactive fallout [ˌreɪdɪəʊˈæktɪv]
放射性废物固化系统 radioactive waste solidification system [ˌsəlɪdɪfɪˈkeɪʃən]
放射性含氢排出液 radioactive hydrogenated effluent
放射性核素 radio-nuclide [ˈnjuːklaɪd]
放射性活度 activity level
放射性浓度 activity/ radioactive concentration [kɒns(ə)nˈtreɪʃ(ə)n] [ˌreɪdɪəʊˈæktɪv]
放射性同位素 radioisotope [ˌreɪdɪəʊˈaɪsətəʊp]

放射性污染 radioactivity contamination [kənˌtæmɪˈneɪʃən]
放射性物质 radioactive material
飞溅 spatter
非安全相关的 nonsafety-related
非弹性散射 inelastic scattering [ɪnɪˈlæstɪk] [ˈskætərɪŋ]
非金属夹杂物 non-metallic inclusion [mɪˈtælɪk]
非裂变俘获 non-fission capture [ˈfɪʃ(ə)n] [ˈkæptʃə]
非随机效应 non-stochastic effect [stəˈkæstɪk]
非重要冷冻水冷却水系统 non-essential chilled water cooling water system [ɪˈsenʃ(ə)l] [tʃɪld]
非重要冷冻水系统 non-essential chilled water system [ɪˈsenʃ(ə)l]
废气后置冷却器 waste gas after-cooler
废气前置冷却器 waste gas pre-cooler
废气取样箱 waste gas sampling box
废气压缩机 waste gas compressor [kəmˈpresə]
废树脂收集箱 spent resin collection tank
废物取样箱 waste sampling box
废物收集箱 waste collecting tank; waste collector tank
废物衰变箱 waste decay tank
废物暂存箱 waste hold-up tank
废液处理系统 liquid waste treatment system
废液接收箱 liquid waste receiver tank [tæŋk]
分包商 sub-contractor [ˈkɒntræktə]
分辨，辨别，分解，甄别 resolution
分辨时间 resolution time
分布式计算机系统 distributed computer system
分布式网络 distributed network
分发清单 distribution list
分隔间 subcompartment
分光光度法 spectrophotometric method [ˌspektrəufəuˈtɒmɪtrɪk]
分接头，抽头 tap
分卷 subsection [ˈsʌbsekʃ(ə)n]
分离器 separator
分流器 shunt
分流制 separate system [ˈsep(ə)rət]
分配柜 distribution cabinet [ˈkæbɪnɪt]
分配系数 partition coefficient [ˌkəuɪˈfɪʃ(ə)nt]
分配因子 partition factor
分区 zoning
分线盒，电缆接头 splice box [splaɪs]
粉砂 silty sand [ˈsɪltɪ]
风阀 damper
风管 duct
风机 fan
封闭式电机，防水电动机 enclosed motor
封头 head
峰值地面加速度 peak ground acceleration (PGA)
缝隙腐蚀 crevice corrosion [ˈkrevɪs] [kəˈrəuʒ(ə)n]

服务程序 service program
浮充电 float charge
浮点 floating point
浮子流量计 float type flowmeter
符合要求 compliance with requirements [kəm'plaɪəns]
辐板（堆芯围板的径向支撑板） former
辐射本底 radiation background [reɪdɪ'eɪʃ(ə)n]
辐射防护 radiation protection
辐射防护监督人 radiation supervisor ['suːpəvaɪzə]
辐射防护评价 assessment of radiation protection [ə'sesmənt] [reɪdɪ'eɪʃ(ə)n]
辐射防护最优化 optimization of radiation protection [ˌɒptɪmaɪ'zeɪʃən]
辐射分区图 radiation map
辐射腐蚀 radiation corrosion [kə'rəʊʒ(ə)n]
辐射工作许可证 radiation work permit
辐射化学 radiation chemistry
辐射监测 radiation monitoring
辐射监测通道 radiation monitoring channel
辐射监测系统 radiation monitoring system
辐射监测仪 radiation monitor [reɪdɪ'eɪʃ(ə)n]
辐射监测专用计算机 special computer for radiation monitoring [reɪdɪ'eɪʃ(ə)n]
辐射量 radiation quantity ['kwɒntɪtɪ]
辐射漏束 radiation streaming
辐射屏蔽 radiation shielding
辐射区 radiation zone
辐射事故 radiation accident [reɪdɪ'eɪʃ(ə)n] ['æksɪdənt]
辐射探测器 radiation detector
辐射危险标志 radiation precaution sign [prɪ'kɔːʃ(ə)n]
辐射效应 radiation effect
辐照 irradiation [ɪˌreɪdɪ'eɪʃ(ə)n]
辐照脆化 irradiation embrittlement [em'brɪtəlmənt]
辐照感生生长 irradiation induced growth [ɪˌreɪdɪ'eɪʃn]
辐照监督管 irradiation surveillance capsule [sə'veɪl(ə)ns] ['kæpsjuːl]
辐照前 pre-irradiation [ɪˌreɪdɪ'eɪʃ(ə)n]
辐照损伤 irradiation damage
腐蚀产物 corrosion product
腐蚀介质 corrosion medium
腐蚀疲劳 corrosion fatigue [kə'rəʊʒ(ə)n] [fə'tiːg]
腐蚀深度 depth of corrosion
腐蚀速率 corrosion rate
负反应性 negative reactivity [ˌrɪæk'tɪvətɪ]
负反应性裕量 negative reactivity margin ['mɑːdʒɪn]
负荷冲击 load surge [sɜːdʒ]
负荷跟踪 load follow
负荷阶跃变化 step load change
负荷阶跃降低 step load reduction

负荷阶跃升高 step load increase [ɪn'kriːs]
负荷因子 load factor
负压 negative pressure
负载变化 load change
复用水 reusing water
复用水泵 reusing water pump
复用水池 reusing water pool
复用水箱 reusing water tank
富集度 degree of enrichment [ɪn'rɪtʃmənt]
富集燃料 enriched fuel [ɪn'rɪtʃt]

G

盖革/弥勒管 Geiger-Muller counter ['kaʊntə(r)]
感生放射性 induced radioactivity [ɪn'djuːst]
干沉降 dry deposition
干涸 dryout
干球温度 dry bulb temperature
干式变压器 dry-type transformer
干式绝缘变压器 dry-insulated transformer
干预水平 intervention level [ˌɪntə'venʃən]
干燥箱，烘干箱，烘箱，干燥炉 drying oven
刚架 rigid frame
（格架的）刚凸 rigid dimple ['rɪdʒɪd] ['dɪmp(ə)l]
钢绞线 strand [strænd]
钢筋束 tendon
钢支撑梁 steel support beam
岗位培训 on(-the)-job training
高纯锗 high-purity germanium ['pj'ʊərɪtɪ] [dʒɜː'meɪnɪəm]
高-低压加热器疏水系统 HP-heater & LP heater drain system
高放惰性气体 γ 监测仪 high-level noble gas gamma monitor ['gæmə 'mɒnɪtə(r)]
高能管道 high energy pipe
高效过滤器 high efficient filter ['fɪltə(r)]
高压电离室 high-pressure ionization chamber [ˌaɪənaɪ'zeɪʃən] ['tʃeɪmbə]
高压电源 high voltage power supply ['vəʊltɪdʒ]
高压取样冷却器 HP sampling cooler
锆-4 合金 zircalloy-4
锆合金包壳管 zircalloy cladding tube ['zəːkəlɒɪ]
锆-水反应 zirconium-water reaction [zɜː'kəʊnɪəm]
戈瑞 Gray/Gy [greɪ]
隔板 partition plate [pɑː'tɪʃ(ə)n]
隔间，小室 compartment [kəm'pɑːtm(ə)nt]
隔离（风）阀 isolation damper ['dæmpə]
隔离电话亭 sound limiting booth [buːð, buːθ]
隔离阀 isolating valve ['aɪsəleɪtɪŋ]

隔离放大器 isolating amplifier [ˈaɪsəleɪtɪŋ] [ˈæmplɪfaɪə]
隔离开关 isolating switch
隔膜阀 diaphragm valve [ˈdaɪəfræm]
隔墙 partition wall
隔热片(块) insulator [ˈɪnsjuleɪtə(r)]
葛莱码 Gray code [ɡreɪ kəʊd]
葛莱码整形器 Gray code shaper
个人剂量计 personal dosimeter [dəʊˈsɪmɪtə]
给水 feed water
给水泵 feed water pump
给水环管 feeder ring
给水加热器 feed water heater
给水进口管嘴 feedwater inlet nozzle [ˈnɒz(ə)l]
给水系统 water supply system
给水箱 feed water storage tank
根部间隙 root gap
根阀 root valve
跟踪/储存组件 track/store module
跟踪球 tracer ball [ˈtreɪsə]
工程公司 engineering company
工程焓升因子 engineering enthalpy rise factor [ˈenθ(ə)lpɪ]
工程经理 construction manager [kənˈstrʌkʃ(ə)n]
工程热通量热点因子 engineering heat flux hot spot factor
工程热通量热管因子 engineering heat flux hot channel factor
工程师站 engineer station
工程手册 engineering manual [ˈmænjʊ(ə)l]
工程因子 engineering factor
工业废水 industrial wastewater [ɪnˈdʌstrɪəl] [ˈweɪstwɔːtə]
工艺辐射监测系统 process radiation monitoring system
工艺规范书 process specification
工艺流程及表计图 piping and instrumentation diagram [ˈdaɪəˌɡræm]
工艺疏水 process drain
工作负荷 workload
工作检查 inspection of work [ɪnˈspekʃn]
工作介质 working medium
工作进度表 work schedule [ˈʃedjuːl]
工作线圈 operation coils [kɒɪlz]
工作许可(证) working authorization [ɔːθəraɪˈzeɪʃ(ə)n]
工作指令 work order
工作状态 on-position
公共点 common point
公共配电盘 common switchboard
公用底板 common base plate
公用电话网 public telephone network
功率表 wattmeter [ˈwɒtmiːtə]
功率分布 power distribution [dɪstrɪˈbjuːʃ(ə)n]
功率峰值因子 power spike factor [spaɪk]
功率空间振荡器 spatial power osciliation
功率亏损 power defect
功率量程 power range

功率密度 power density
功率通道比较装置 power channel comparison facility [ˈtʃænl]
功率效应 power effect
功率因子 power factor
功率运行 power operation
功能键 function key
功能试验 function test
共频单工操作 dual-frequency simplex operation
共享文件 shared files
共享资源 shared resource
共振 resonance [ˈrez(ə)nəns]
共振俘获 resonance capture
共振积分 resonance integral
共振宽度 resonance width
共振区 resonance region
共振吸收 resonance absorption [ˈrez(ə)nəns] [əbˈzɔːpʃ(ə)n]
共振中子 resonance neutron [ˈnjuːtrɒn]
供电 power supply
供方评价 supplier evaluation [ɪˌvæljuˈeɪʃn]
供方选择 selection of supplier
供气管道 air delivery pipe [dɪˈlɪv(ə)rɪ]
钩爪组件 latch assembly [əˈsemblɪ]
骨架 skeleton [ˈskelɪt(ə)n]
钴基合金 cobalt-base alloy
固定点 fixed point
固定套管 fixed tube
固化 solidification [ˌsəlɪdɪfɪˈkeɪʃən]
固化厂房通风系统 radioactive waste solidification building ventilation system
固体废物处理和存系统 solid waste treatment and storage system [ˈstɔːrɪdʒ]
固有安全设备 intrinsically safe equipmen [ɪnˈtrɪnzɪklɪ]
故障检测元件 failure detector module [ˈmɒdjuːl]
关闭扬程 shutoff head
关键核素 critical nuclide [ˈnjuːklaɪd]
关键照射途径 critical exposure pathway [ɪkˈspəʊʒə]
观察窗 viewing window
管板 tube sheet
管程 tube stream
管道防甩装置 anti-whip device
管道贯穿件 piping penetration [penɪˈtreɪʃ(ə)n]
管道系统 piping system
管道支撑件 piping support
管壳式热交换器 shell-and-tube heat exchanger
管廊 tube lane
管理导则 regulatory guide
管理机构要求 regulatory requirement [ˈregjulətərɪ]
管束 tube bundle [ˈbʌnd(ə)l]
管网 duckwork
管子间距 tube spacing (or pitch)
管子支承板 tube support plate
贯穿辐射 penetrating radiation [ˈpenɪtreɪtɪŋ]
惯性流量 coast down flow
惯性流量下降 flow coastdown

[ˈkəustdaʊn]
光电倍增管 photomultipliler [ˌfəʊtəʊˈmʌltəˌplaɪə]
光源 light source
规程编制 preparation of procedures [ˌprepəˈreɪʃ(ə)n]
规定限值 prescribed limits [prɪˈskraɪbd]
规范案例 code case
规格书，设计说明书 specification
轨道梁 runway track
柜式机组 package unit
滚球 track ball
国际标准化协会 International Standardization Association (ISA)
国际标准化组织 International Organization for Standardization (ISO)
国际电工委员会 International Electrotechnical Commission [kəˈmɪʃn]
国际原子能机构 International Atomic Energy Agency (IAEA)
国家标准 national standard
国家标准机构 national standard body
过程监测系统 process monitoring system
过程控制 process control
过渡段 transition ring
过冷度 degree of subcooling
过冷喷放 subcooled blowdown [ˌsʌbˈkuːld] [ˈbləudaʊn]
过流继电器 overcurrent relay [ˈəʊvəkʌrənt]
过滤器 filter
过热度 degree of superheat
过剩反应性 excess reactivity
过剩下泄 excess letdown
过剩下泄热交换器 excess letdown heat exchanger
过压继电器 overvoltage relay
过载 overload

H

氦气检漏 helium leak check [ˈhiːliːəm]
含汽量，（蒸汽）干度 quality
含水量 water content
焓升不确定因子 enthalpyrise uncertainty factor [enˈθælpɪ]
焓升工程热管因子 enthapy rise engineering hot channel factor
焓升因子 enthalpy rise factor
涵洞 culvert
焊缝 weld seam
焊缝代号 welding symbols [ˈsɪmbəls]
焊缝金属 weld metal
焊缝正面 face of weld
焊工资格评定 welder qualification [ˌkwɒlɪfɪˈkeɪʃ(ə)n]
焊后热处理 post weld heat treatment
焊剂垫焊 welding with flux backing
焊接程序鉴定 welding procedure qualification [ˌkwɒlɪfɪˈkeɪʃ(ə)n]
焊接方向 direction of welding

焊接工艺 welding technology [tek'nɒlədʒɪ]

焊接工艺参数 welding parameter [pə'ræmɪtə]

焊接过程 welding process

焊接技术 welding technique

焊接顺序 welding sequence ['siːkw(ə)ns]

焊接条件 welding condition

焊接位置 position of welding

焊趾 toe of weld [weld]

行业标准 occupation standard

号筒扬声器 horn loudspeaker [hɔːn] ['laʊdˌspiːkə]

耗蚀 wastage

合格试验 pass test

合理可行尽量低 as low as reasonable achievable (ALARA)

合闸（断路器） closing (circuit breaker)

核安全检查 nuclear safety inspection

核不确定因子 nuclear uncertainty factor

核测仪表系统 nuclear instrumentation system [ˌɪnstrʊmen'teɪʃ(ə)n]

核岛 nuclear island

核岛、常规岛消防水分配系统 nuclear island and conventional island fire protection water distribution system

核岛 220V 交流仪表电源系统 nuclear island 220V AC instrumentation PSS

核岛 220V 直流电源系统 nuclear island 220V DC power supply system

核岛 24V 直流电源系统 nuclear island 24V DC power supply system

核岛 380V 交流电源系统 nuclear island 380V AC power system

核岛厂房楼梯间正压送风系统 nuclear island building stair positive pressure supply system ['pɒzɪtɪv]

核岛消防水系统 NI fire protection system

核电厂 nuclear power plant

核电厂堆芯安全设计 design for reactor core safety in NPP

核辅助厂房照明系统 nuclear aux building lighting system [ˌɪnstrʊmen'teɪʃ(ə)n]

核管理委员会 Nuclear Regulatory Commission (NRC) ['regjulətərɪ] [kə'mɪʃ(ə)n]

核焓升因子 nuclear enthalpy rise factor [en'θælpɪ]

核数据通信线 nuclear data link

恒流电源 constant current power supply

恒压电源 constant voltage power supply

桁架 truss [trʌs]

横（焊）缝 transverse seam [trænz'vɜːs] [siːm]

横管 horizontal pipe [ˌhɒrɪ'zɒntəl]

横焊 horizontal position welding

横向刚度 transverse stiffness [trænz'vɜːs] ['stɪfnɪs]

横向流 cross flow

横向运动 lateral movement [ˈlæt(ə)r(ə)l]

宏观截面 macroscopic cross section [ˌmækrə(ʊ)ˈskɒpɪk] [ˈsekʃ(ə)n]

虹吸滤池 siphon filter [ˈsaɪf(ə)n]

后处理工艺过程 post-processing

后热 postheat

后信息处理器 post-processor

后续行动 follow-up action

后置过滤器 after filter (or post-filter)

呼叫通话系统 page-party system

呼吸面具 respirator [ˈrespɪreɪtə]

弧光灯 arc-lamp

互层 interbedding [ɪntɜːrˈbedɪŋ]

滑道 slide way [slaɪd]

滑环电刷组件 slip ring brush assembly

滑轮 sheave [ʃiːv]

化粪池 septic tank [ˈseptɪk]

化学水源水厂区管路系统 preliminary treatment water for dimineralized water workshop yard pipe network system [dɪˈmɔːrəˌlaɪzd]

环缝焊接 girth welding

环境监测 environmental monitoring [ɪnvaɪrənˈment(ə)l]

环境监测车 environment monitoring vehicle

环境鉴定 environment qualification

环境评价 environmental assessment (EA)

环境质量 environmental quality (EQ)

环梁 ring girder [ˈgɜːdə]

环形吊车 polar crane [kreɪn]

环形轨道 circular railway

环形支架 ring support

缓冲段 dashpot section

缓冲落棒时间 dashpot drop time

缓冲区 dashpot region

缓冲特性 dashpot characteristics [ˌkærəktəˈrɪstɪks]

缓冲作用 dashpot action

缓发临界 delayed critical [ˈkrɪtɪk(ə)l]

缓发中子 delayed neutron [ˈnjuːtrɒn]

缓发中子份额 delayed neutron fraction

缓蚀剂 corrosion inhibitor [ɪnˈhɪbɪtə]

缓蚀剂添加箱 corrosion inhibitor addition tank [ɪnˈhɪbɪtə]

换料 refueling; reload

换料操作 refueling operation

换料浓度 reload enrichment

换料水池 refueling pool

换料水池底部密封结构 sealing structure on bottom of the refueling pool

换料水箱 refueling water tank

换气率 air change rate

回风机 return fan

回火 tempering

混合结构 composite structure [ˈstrʌktʃə]

混合树脂 mixed resin [ˈrezɪn]

混合系数 mixing coefficient

[ˌkəʊɪˈfɪʃ(ə)nt]

活度 activity

活化 activation [ˌæktɪˈveɪʃən]

活节拉杆 tension link

活塞泵 piston pump [ˈpɪstən]

活性碳过滤器 active carbon filter

J

机械离合器 mechanical clutch [mɪˈkænɪk(ə)l] [klʌtʃ]

机械连锁 mechanical interlock [ɪntəˈlɒk]

机械密封 mechanical seal [siːl]

机械强度 mechanical strength

机械性能 mechanical property [ˈprɒpətɪ]

机械载荷 mechanical load [mɪˈkænɪk(ə)l]

积分价值 integrated worth [ˈɪntɪgreɪtɪd]

积分热导率 conductive integral [kənˈdʌktɪv] [ˈɪntɪgr(ə)l]

积分通量 integrated flux [flʌks]

积剂量当量 committed dose equivalent [ɪˈkwɪv(ə)l(ə)nt]

基础环 foundation ring [faʊnˈdeɪʃ(ə)n]

基础埋置深度 depth of foundation

基体材料 matrix material [ˈmeɪtrɪks] [məˈtɪərɪəl]

基准电压 reference voltage

基准试块 reference block

畸变 deformation [ˌdiːfɔːˈmeɪʃ(ə)n]

激振腐蚀 fretting corrosion [kəˈrəʊʒən]

极限工况 limited condition

极限事故 limiting fault

极性 polarity [pə(ʊ)ˈlærɪtɪ]

集油盘 drip pan

计量槽 measuring tank [tæŋk]

计数管区域监测仪 counter area monitor [ˈkaʊntə]

计数率音响系统 audible count rate signal [ˈɔːdɪb(ə)l]

计算机网络 computer network

记录保存 retention of records

记录分类 classification of records [ˌklæsɪfɪˈkeɪʃ(ə)n]

记录水平 recording level

记录仪 recorder

记录移交 transfer of records

记忆组件 memory module [ˈmɒdjuːl]

技术支援中心 technical support center

剂量当量 dose equivalent [ɪˈkwɪvələnt]

剂量当量负担 dose equivalent commitment

剂量当量限值 dose equivalent limit

剂量负担 dose commitment

剂量计读数器 dosemeter reader [ˈdəʊsˌmiːtə]

剂量率转换因子 dose rate conversion factor [kənˈvɜːʃ(ə)n]

剂量评价 dose assessment

继电器驱动器组件 relay actuator module
加强环 stiffening ring [ˈstɪfnɪŋ]
加热保温 [电] heat tracing (electric) [ˈtreɪsɪŋ]
加热盘管 heating coil [kɒɪl]
加热器 heater
加湿器 humidifier [hjuː ˈmɪdɪfaɪə(r)]
加速度损失 acceleration pressure loss [əkseləˈreɪʃ(ə)n]
加速度压降 acceleration pressure drop
夹带 entrainment
夹杂物 inclusion [ɪnˈkluːʒn]
假想事故 postulated accident [ˈpɒstʃəˌleɪtɪd]
假想事故工况 postulated accident condition
架空电缆 aerial cable [ˈeərɪəl]
间隙导热 gap conductance
监查；审计 audit [ˈɔːdɪt]
监查后会议 post-audit meeting
监查后续行动 audit follow-up
监查计划 audit plan
监查记录 audit record
监查前会议 pre-audit meeting
监督大纲 surveillance program
监督计划 surveillance plan [səˈveɪl(ə)ns]
监督区 supervised area
监督人员 surveillance personnel [səˈveɪl(ə)ns]
监控区 limited access area
检测槽 monitor tank [tæŋk]
检测装置 monitoring unit [ˈjuːnɪt]
检查大纲 inspection program [ɪnˈspekʃn]
检查计划 inspection plan
检查记录 inspection record [ɪnˈspekʃən]
检查状态标识 indication of inspection status [ˈsteɪtəs]
检索 retrieval [rɪˈtriːvl]
检索和查阅 retrieval and accessibility [rɪˈtriːvl] [əkˌsesəˈbɪlətɪ]
检修电话系统 electro-sound telephone system
减速器 speed reducer [rɪˈdjuːsə]
减压 pressure reduction
剪切弹性膜量 shear modulus [ʃɪə] [ˈmɒdjʊləs]
剪切力 shear force
剪切试验 shear test
简化的日志 simplified logbook
见证点 witness point
建造许可证 construction permit
鉴定规格书 qualification specification [ˌkwɒlɪfɪˈkeɪʃ(ə)n] [ˌspesɪfɪˈkeɪʃ(ə)n]
鉴定合格证 qualification certificate [ˌkwɒlɪfɪˈkeɪʃ(ə)n] [səˈtɪfɪkət]
鉴定试验 qualification test
降压变压器 stepdown transformer
降雨量 precipitation [prɪˌsɪpɪˈteɪʃ(ə)n]
交互方式 interactive mode [ˌɪntəˈæktɪv]
交互作用 interaction [ˌɪntərˈækʃn]
交接检查 inspection at delivery point

交钥匙承包商 turn-key contractor [kən'træktə]
胶片剂量计 film badge
接触热导 contact conductance [kən'dʌkt(ə)ns]
接地保护 earth leakage protection ['liːkɪdʒ]
接地不良 ground fault
接地导体 grounding conductor [kən'dʌktə(r)]
接地电极 grounded electrode [ɪ'lektrəʊd]
接地电阻 grounding/earthing resistance [rɪ'zɪstəns]
接地系统 earthing system
接地线 grounding wire ['waɪə(r)]
接管段 nozzle belt
接近临界曲线 criticality approach curve
接口图 interface drawing
接收槽 receiving tank
接线盒 connection box
接线图 connection diagram
节流阀 throttle valve ['θrɒt(ə)l]
节流孔板 throttling orifice ['ɒrɪfɪs]
节流装置 throttle device ['θrɒt(ə)l] [dɪ'vaɪs]
结构屏蔽 structural shield ['strʌktʃ(ə)r(ə)l]
截面 cross section
截止阀 globe valve
金属包复，金属外壳 metal clad [klæd]
金相检验 metallographic examination [mɪ,tæləu'græfɪk]
紧急硼化 emergency boration [ɪ'mɜːdʒ(ə)nsɪ]
紧急停堆 emergency shutdown
紧急停堆，跳闸 trip
紧急停堆时间 emergency trip time
紧急通知 emergency announcement [ə'naʊnsm(ə)nt]
进度计划，时间表 schedule
进口接管保温层 inlet nozzle insulation ['nɒzəl] [ˌɪnsjʊ'leɪʃn]
进口温度 inlet temperature
禁区边界 exclusion area boundary [ɪk'skluːʒ(ə)n] ['baʊnd(ə)rɪ]
晶间腐蚀 intergranular corrosion [ˌɪntə'grænjʊlə] [kə'rəʊʒən]
晶间应力腐蚀开裂 intergranular stress corrosion cracking ['krækɪŋ]
晶粒度 grain-size
精密脉冲发生器 precise-pulse generator ['presə-pɪsɪz]
警告标志 warning sign
径向峰值热点因子 radial peaking hot spot factor
径向峰值因子 radial peaking factor
径向功率倾斜 radial power tilt
径向和轴向温度分布 radial and axial temperature profiles
径向热膨胀 radial thermal expansion ['θɜːm(ə)l] [ɪk'spænʃ(ə)n]
径向翼 radial vane
净化段 purification section [ˌpjʊərɪfɪ'keɪʃən]
净正吸入压头 net positive suction head ['sʌkʃ(ə)n]
静密封环 stationary seal ring

[ˈsteɪʃ(ə)n(ə)rɪ]
静平衡 static balance [ˈstætɪk]
静态参数 static parameter [pəˈræmɪtə]
静态压力 static pressure
纠正措施 corrective action
就地机柜 local cabinet [ˈkæbɪnɪt]
就地控制室系统 local control room system
就地取样 local sampling
(可) 居留区 habitability area [hæbɪtəˈbɪlɪtɪ]
居留因子 occupancy factor
局部沸腾 local boiling
局部腐蚀 localized corrosion [kəˈrəʊʒ(ə)n]
局部屏蔽 local shielding [ˈʃiːldɪŋ]
卷边焊 flange welding
卷扬机 winch [wɪn(t)ʃ]
绝缘强度 dielectric strength; insulation level [ˌɪnsəˈleɪʃən]
绝缘子，绝缘体，隔离器 isolator [ˈɑɪsleɪtə]
均方根值 root mean square value [skweə]
均匀腐蚀 uniform corrosion
竣工图 as-built drawing

K

开放源 unsealed source
开关量输入 on/off input
开关站 switchyard
开关装置，配电装置 switchgear [ˈswɪtʃgɪə]
铠装电缆 armored cable [ˈɑːməd]
铠装屏蔽电缆 sheathing and shielding cable [ˈʃiːðɪŋ] [ˈʃiːldɪŋ]
铠装热电偶 sheathed cable [ʃiːət]
抗冲击强度 shock strength
抗辐照损伤 resistance to irradiation damage [ɪˌreɪdɪˈeɪʃ(ə)n]
抗震 I 类 seismic category I [ˈsɑɪzmɪk]
抗震分级 seismic classification
抗震能力 shock resistance [rɪˈzɪst(ə)ns]
抗震设计 seismic design
抗震支架 seismic stabilizer bracket [ˈsteɪbɪlaɪzə]
壳程 shell stream [kənˈsəʊl]
可拔出的 withdrawable [wɪðˈdrɔːəbl]
可编程的控制器 programmable logic controller
可拆接头 disconnect plug
可动部分 movable part
可焊性 weldability
可行性研究 feasibility study
可接受限值 acceptance limit
可裂变的 fissionable
可压缩流 compressible flow [kəmˈpresəbl]
可移动探测器 movable detector [ˈmuːvəb(ə)l] [dɪˈtektə]
可移动小型裂变室 movable miniature fission chamber
可移动中子探测器 movable neutron detector [ˈnjuːtrɒn] [dɪˈtektə]
客观证据 objective evidence

空泡比 void ratio
空气处理机组 air handling unit
空气动力特性 aerodynamic behavior [ˌeərəudaɪˈnæmɪks] [bɪˈheɪvjə]
空气断路器 air breaker
空气过滤机组 air filter unit
空气过滤吸附机组 air filter and absorber unit [əbˈsɔːbə]
空气浸没剂量 air submersion dose [səbˈməːʃən]
空气冷却器 air cooler
空气流量 air flow rate
空气取样设备 air sampling device
空气软管 air hose
空气自冷式 air self-cooling type
空腔，腔室 plenum [ˈpliːnəm]
空腔体积 void volume
空载损失 no-load loss
控制棒 control rod
控制棒棒位指示 control rod position indication [ɪndɪˈkeɪʃ(ə)n]
控制棒导向管 control rod guide thimble [ˈθɪmb(ə)l]
控制棒反应性价值 reactivity worth of control rod
控制棒刻度 control (rod) calibration [kælɪˈbreɪʃ(ə)n]
控制棒控制系统 control rod control system
控制棒落棒时间 control rod dropping time
控制棒驱动机构 control rod drive mechanism (CRDM)
控制棒驱动机构冷却通风系统 cooling system (CRDM)
控制棒驱动机构通风系统 ventilation system (CRDM) [ˌventɪˈleɪʃ(ə)n]
控制棒驱动线 control rod drive line
控制棒失控提升事故 uncontrolled rod withdrawal accident [wɪðˈdrɔː(ə)l]
控制棒提升 control rod withdrawal
控制棒移动速度 control rod movement speed
控制棒与阻力塞抽插机 rod cluster control assembly and thimble plug assembly changing fixture [ˈθɪmb(ə)l]
控制棒与阻力塞组件 rod cluster control assembly and thimble plug assembly [ˈθɪmb(ə)l] [plʌg]
控制棒组 control (rod) cluster
控制棒组件 control (rod) assembly
控制电缆 control cable
控制风阀 control damper [ˈdæmpə]
控制开关 control switch [swɪtʃ]
控制逻辑柜 control logic cabinet [ˈkæbɪnɪt]
控制盘 control board
控制区 controlled area
控制容积 control volume
控制台 console [kənˈsəul]
控制线对中 control rod drive line alignment [əˈlaɪnm(ə)nt]
跨距 span [spæn]
快拆接头 fast connector
快裂变因子 fast fission factor
快速开关 snap switch

快速停堆 rapid shutdown
快中子通量 fast neutron flux
快中子通量 fast neutron fluence
框架，构架 framework
扩口试验（传热管） flare test
扩容器 drain flash tank
扩散理论 diffusion theory [dɪˈfjuːʒ(ə)n]
扩散系数 diffusion coefficient [ˌkəʊɪˈfɪʃənt]

L

拉紧环 tension ring
喇叭型坡口焊 flare groove welding
累积剂量 accumulated dose [əˈkjʊmjəˌlet]
冷壁效应 cold-wall effect
冷冻水供水温度 supply chilled water temperature [tʃɪld]
冷冻水回水温度 return chilled water temperature
冷端补偿箱 cold junction box [ˈdʒʌŋ(k)ʃ(ə)n]
冷空腔 cold void volume
冷凝器 condenser [kənˈdensə]
冷凝器抽真空系统 condenser vacuum system [ˈvækjʊəm]
冷却剂比焓 coolant specific enthalpy [spɪˈsɪfɪk]
冷却剂交混 coolant mixing
冷却剂净化 purification of coolant [ˌpjʊərɪfɪˈkeɪʃən]
冷却剂流速 coolant flow rate
冷却剂水化学 coolant chemistry [ˈkemɪstrɪ]
冷却剂质量流量 coolant mass flow rate
冷却盘管 cooling coil
冷却水 cooling water
冷却塔 cooling tower
冷态启动 cold startup [stɑːtʌp]
冷态试验 cold test
冷态性能试验 cold performance test
冷停堆 cold shutdown
离线 off-line
离子交换器 ion exchanger [ˈɑɪən]
离子交换树脂 ion exchange resin [ɪksˈtʃeɪndʒ] [ˈrezɪn]
离子色谱 ion chromatograph [ˈmætəgrɑːf]
离子选择电极 specific ion electrode [ˈɑɪən] [ɪˈlektrəʊd]
力矩电机 torque motor [tɔːk]
立管 vertical pipe
例行检查 routine inspection [ruːˈtiːn] [ɪnˈspekʃn]
例行试验 routine test
砾砂 gravelly sand [ˈgrævəlɪ]
连接编辑 linkage editor
连接柄组件 spider assembly [ˈspaɪdə]
连接点 connection point
连接焊 continuous welding
连接器 connector [kəˈnektə(r)]
连接装入程序 linking loader
连续性 continuity [ˌkɒntɪˈnjuːɪtɪ]

联邦管理法规 code of federal regulation (CFR) [ˈfed(ə)r(ə)l]
联络渠道，通信线路 line of communication
联锁 interlock [ˌɪntəˈlɒk]
联氧 hydrazine [ˈhɑɪdrəziːn]
两群理论 two group theory
两相流 two phase flow
量热式质量流量计 thermal mass flowmeter [ˈfləʊmiːtə]
裂变产额 fission yield
裂变产物 fission product
裂变和腐蚀产物活度 fission and corrosion product activity
裂变截面 fission cross section
裂变链 fission chain [tʃeɪn]
裂变率 fission rate
裂变能 fission energy
裂变谱 fission spectrum
裂变气体 fission gas
裂变室（微型） fission chamber (miniature)
裂变碎片 fission fragment
裂变源 fission source
裂变中子 fission neutron
临界安全 criticality safety
临界报警系统 criticality alarm system
临界尺寸 critical size
临界键转移途径 critical transfer pathway
临界流 critical flow
临界浓度 critical concentration [kɒns(ə)nˈtreɪʃ(ə)n]
临界热流密度 critical heat flux (CHF)
临界热流密度格架修正因子 critical heat flux modified spacer factor
临界热流密度冷壁修正因子 critical heat flux modified coldwall factor
临界热流密度形状修正因子 critical heat flux modified shape factor
临界实验 critical experiment [ɪkˈsperɪm(ə)nt]
临界事故 criticality accident [ˈæksɪdənt]
临界体积 critical volume [ˈkrɪtɪkəl]
临界质量 critical mass
临界质量流量；临界质量流密度 critical mass flux
临时接线 temporary wiring [ˈtemp(ə)rərɪ]
磷酸盐处理 phosphate treatment [ˈfɒsfeɪt]
零功率 zero power
流程图（工艺） flow diagram
流动不稳定性 flow instability [ɪnstəˈbɪlɪtɪ]
流动升力 flow lift force
流量变送器 flow transmitter [trænzˈmɪtə]
流量分配 flow distribution
流量分配板 diffuser plate; flow distribution plate
流量一扬程曲线 flow-head cure
流致振动 flow induced vibration [vɑɪˈbreɪʃ(ə)n]

六氟化硫断路器 sulfur hexafiuoride breaker [ˈsʌlfə]
龙门吊 gantry crane [ˈgæntrɪ]
龙门架 gantry
楼面疏水 floor drain
楼面疏水箱 floor drain tank
路选择器 path selector
轮机厂房和水处理厂房消防系统 turbine building and water treatment building fire protection system
逻辑变量 logic variable
逻辑错误监测电路 logic error monitor
逻辑隔离组件 logic isolator module [ˈaɪsəleɪtə] [ˈmɒdjuːl]
逻辑运算 logic operation [ɒpəˈreɪʃ(ə)n]
螺孔塞装卸工具 studhole plug handling tool
螺栓拉伸机 stud tensioner [ˈtenʃənə]
螺旋形驱动电缆 helical-wrap drive cable [ˈhelɪkl]
落棒 rod drop
落棒时间 drop time
滤池 filter

M

埋弧焊 submerged arc welding [səbˈmɜːdʒd] [ˈweldɪŋ]
脉冲 pulse
脉冲编码调制 pulse-code modulation [ˌmɒdjʊˈleɪʃən]
脉冲发生器 pulser [ˈpʌlsə]
脉冲分配组件 pulse distributor module [ˈmɒdjuːl]
脉冲幅度分析器 pulse-height analyzer
脉冲监测组件 pulse monitor module
脉冲输入 pulse input
脉冲信号发生组件 pulse signal generator module
脉动 pulsation [pʌlˈseɪʃən]
满功率 full power
满载能力 full-load torque [tɔːk]
慢化 moderation [mɒdəˈreɪʃ(ə)n]
慢化剂 moderator [ˈmɒdəreɪtə]
慢化剂—燃料比 moderator-to-fuel-ratio [fjʊəl] [ˈreɪʃɪəʊ]
慢化剂温度系数 moderator temperature coefficient [ˈmɒdəreɪtə] [ˌkəʊɪˈfɪʃ(ə)nt]
美国材料与试验学会 American Society for Testing and Material
美国电气与电子工程师协会 The Institute of Electrical and Electronic Engineering
美国钢铁结构协会 American Institute of Steel Construction
美国国家标准学会 American National Standards Institute
美国核学会 American Nuclear Society

美国混凝土学会 American Concrete Institute
美国机械工程学会 American Society of Mechanical Engineers
美国土木工程师学会 American Society of Civil Engineers
门框式 γ 监测仪 portal γ monitor [ˈmɒnɪtə(r)]
门式构架，龙门架 portal frame
弥散流 dispersed flow [dɪˈspəːst]
密度波 density wave
密度计 densimeter [denˈsɪmɪtə]
密封的，防漏的 leakproof
密封盖 seal cover [siːl]
密封盒 latch housing
密封接插件 sealed plug and socket [ˈsɒkɪt]
密封壳 seal housing
密封源 sealed source [siːld]
密实化 densification [ˌdensɪfɪˈkeɪʃən]
面板 face plate
灭火 fire extinction
灭火剂（器） fire extinction agent
民用建筑 civil architecture
名义功率 nominal power [ˈnɒmɪn(ə)]
明配线 open wiring
模拟（流程）图 mimic diagram [ˈmɪmɪk] [ˈdaɪəgræm]
模拟/数字转换器 analog/digital converter
模拟程序 simulated program [ˈsɪmjʊleɪtɪd]
模拟接续线 analogue link [ˈænəlɒg]
模拟量输出 analog output
模拟量输入 analog input
模拟量通道 analog channel
模拟盘 mimic panel [ˈmɪmɪk] [ˈpæn(ə)l]
模拟信号 analog signals
膜态沸腾 film boiling
摩擦损失 friction loss
摩擦系数 friction coefficient [ˌkəʊɪˈfɪʃ(ə)nt]
磨损腐蚀 erosion corrosion [ɪˈrəʊʒən] [kəˈrəʊʒən]
磨损裕量 wear allowance
母钟 master clock
目标程序 object program
目视检查 visual inspection (or examination) [ɪnˈspekʃn]
目视检查报告 visual inspection report

N

钠蒸汽灯 sodium vapor lamp [ˈsəʊdɪəm] [ˈveɪpə]
耐腐蚀性 corrosion resistance
耐火的 fire resistant
耐火等级 fire rating
耐火焰的 flame resistant
耐酸瓷砖 acid-proof tile [ˈæsɪd]
耐压 withstand voltage [wɪðˈstænd]
耐压壳组件 pressure housing assembly [əˈsemblɪ]

挠性联轴器，挠性接头 flexible coupling
内部接口 internal interface [ɪnˈtɜːnəl] [ˈɪntəfeɪs]
内部通信网 intercom network
内污染 internal contamination [kənˌtæməˈneɪʃən]
内压 internal pressure
内在源 contained source
内照射 internal exposure [ɪksˈpəʊʒə]
能动部件 active component [kəmˈpəʊnənt]
能级 energy level
能量分辨率 energy resolution
能量守衡 conservation of energy [kɒnsəˈveɪʃ(ə)n]
能量响应 energy response
能谱 energy spectrum [ˈspektrəm]
能群 energy group
逆变装置 inverter [ɪnˈvɜːtə]
镍基合金 nickel base alloy [ˈnɪk(ə)l] [beɪs] [ˈælɒɪ]
凝结水 condensate [ˈkɒnd(ə)nseɪt]
凝结水泵 condensate pump
凝结水净化系统 condensate polishing system
凝结水冷却器 condensate cooler
凝结水输送泵 condensate transfer pump
凝聚剂 coagulant [kəʊˈægjʊlənt]
浓缩度，富集度 enrichment
浓缩液储存箱 concentrate storage tank [ˈkɒns(ə)ntreɪt]
浓缩铀 enriched uranium
浓缩铀反应堆 enriched uranium reactor

P

排出流辐射监测系统 effluent radiation monitoring system
排放泵 discharge pump
排放箱 removal tank
排风机 exhaust fan; vapor extractor [ɪgˈzɔːst] [ɪkˈstræktə]
排风净化机组 exhaust cleanup unit (ECU)
排气孔 exhaust holc
排气烟囱 vent stack
排器冷凝器 vent condenser
排水工程 sewage wastewater engineering [ˈsuːɪdʒ]
排烟风机 exhausting smoke fan
盘车装置 turning gearing (barring gear) [ˈgɪərɪŋ]
泡核沸腾 nuclear boiling
泡沫消防系统 fire fighting foam system
配电柜 distribution box
配电盘 switchboard (distribution)
配电屏 electrical panel
配水管 distribution system
喷淋泵 spray pump
喷淋管道 spray line
喷淋集管 spray header [spreɪ]
喷淋喷射器 spray ejector [ɪˈdʒektə]
喷淋热交换器 spray heat exchanger

喷淋添加剂箱 spray additive tank [ˈædɪtɪv]
喷砂 sand blastin
喷射器 ejector
喷丸（处理） peening
(稳压器) 喷雾器 sprayer [ˈspreɪə]
喷嘴 spray nozzle
膨胀环 expansion ring
膨胀结合，伸缩缝 expansion joint
膨胀水箱 expansion tank
批准，认可，审定 approval [əˈpruːv(ə)l]
偏离核态沸腾 departure from nuclear boiling (DNB)
偏离核态沸腾率 departure from nuclear boiling ratio (DNBR)
偏离膜态沸腾 departure from film boiling DFB
片筏基础 mat foundation [mæt]
漂移流密度 drift flux
频率计，周波表 frequency meter
频谱分析 frequency analysis [əˈnælɪsɪs]
平封头 flat head
平焊 downhand welding
平衡富集度 equilibrium enrichment [ɪnˈrɪtʃmənt]
平衡氙 xenon equilibrium [ˌiːkwɪˈlɪbrɪəm]
平衡中毒 equilibrium poisoning
平面图 plan
评价 assessment [əˈsesmənt]
屏蔽容器 flask；cask
（焊接）坡口 groove [gruːv]
坡口焊 groove welding
坡口角度 groove angle [ˈæŋgl]
破损燃料份额 failed fuel fraction
破损燃料足见定位检测系统 failed fuel assembly location detecting system
剖面图 sectional view [ˈsekʃ(ə)n(ə)l]
谱硬化 spectral hardening [ˈspektr(ə)l]

Q

其他 BOP 厂房给排水系统 other BOP building water supply and drainage system
其他 BOP 厂房通风系统 miscellaneous BOP building ventilation system [ˌmɪsəˈleɪnɪəs] [ˌventɪˈleɪʃ(ə)n]
企业标准 company standard
启动 start up
启动脉冲 go pulse [pʌls]
起拱线 springing line
起重 lift height
气动阀 pneumatic valve [njuːˈmætɪk]
气浮池 floatation [fləʊˈteɪʃən]
气候学；风土学 climatology [klaɪməˈtɒlədʒɪ]
气空间 gaseous space [ˈgæsɪəs]
气溶胶 aerosol [ˈeərəsɒl]
气体废物处理系统 gaseous waste treatment system [ˈgæsiːəs]

气体加热器 gas heater [ˈhiːtə(r)]
气体绝缘开关装置 gas insulated switchgear (GIS) [ˈɪnsjʊleɪtɪd] [ˈswɪtʃɡɪə]
气体累计流量计 gas accumulated flowmeter [əkˈjuːmjʊleɪtɪd] [ˈfləʊmiːtə]
气体冷却器 gas cooler [ˈkuːlə(r)]
气体取样 gaseous sample [ˈɡæsiːəs]
气体色谱 gas chromatograph [ˈkrəʊmətəɡrɑːf]
气象观测台 meteorological observatory [ˌmiːtɪərəˈlɒdʒɪkəl]
气象站 meteorological station
气象资料 meteorological data [ˌmiːtɪərəˈlɒdʒɪkəl]
气载放射性 airborne radioactivity [ˈeəbɔːn] [ˌreɪdɪəʊækˈtɪvɪtɪ]
气载粒子取样器 airborne particulate sample [pɑːˈtɪkjʊlət]
汽垫 steam cushion [ˈkʊʃ(ə)n]
汽机厂房暖通空调系统 turbine building HVAC system
汽机进汽阀 turbine inlet valve [ˈɪnlet]
汽轮机抽汽系统 turbine extraction steam system
汽轮机功率和频率调节系统 turbine power and frequency regulating system [ˈfriːkw(ə)nsɪ]
汽轮机降功率组件 turbine runback module [ˈmɒdjuːl]
汽轮机调节与控制 turbine regulation and control
汽水分离器 moisture separator [ˈmɒɪstʃə]
汽水分离再热器 moisture separator-reheator [ˈsepəreɪtə]
汽相 vapor phase
砌体结构 masonry structure [ˈmeɪs(ə)nrɪ]
铅室 lead container [kənˈteɪnə]
铅酸蓄电池 lead acid battery [ˈæsɪd]
前面板 front panel
前台 foreground
前置计算机 front computer
强迫循环 forced circulation
强制性标准 mandatory standard [ˈmændət(ə)rɪ]
桥式起重机 overhead crane
切割能 cut-off energy
切换 switch over
切换开关 transfer switch
轻油 light oil
氢脆 hydrogen embrittlement [ˈhaɪdrədʒən] [emˈbrɪtlmənt]
氢分析仪 hydrogen analyzer [ˈænəlaɪzə]
氢气测量装置 hydrogen measuring device [ˈmeʒərɪŋ]
氢气系统 hydrogen system [ˈhaɪdrədʒən]
氢致开裂 hydrogen induced cracking
倾翻机 upending device
倾斜焊 inclined positioning welding [ɪnˈklaɪnd] [pəˈzɪʃnɪŋ]
清水池 clear-water reservation
清洗剂 scavenger

[ˈskævɪn(d)ʒə]
球形壳体 spherical shell [ˈsferɪk(ə)l]
区域辐射监测系统 area radiation monitoring system [reɪdɪˈeɪʃ(ə)n]
区域选择控制器 zone-selected controller
驱动，动作 actuate [ˈæktʃʊeɪt]
驱动机构 drive unit
驱动机构顶部拉紧装置 tension structure on top of the CRDMS
驱动机构管座 CRDM adaptor
驱动机构通风管座 CRDM ventilation shroud [ˌventɪˈleɪʃ(ə)n]
驱动轴 drive shaft
驱动轴部件 drive shaft assembly
曲率 curvature [ˈkɜːvətʃə]
取水构筑物 intake structure [ˈstrʌktʃə(r)]
取水口，摄入量 intake
取样电阻 sampling resister [rɪˈzɪstə]
取样和探测装置 sampling and detecting assembly
取样冷却器 sampling cooler
取样器 sampler [ˈsɑːmplə]
取样系统 sampling system
取样箱 sampling cabinet [ˈkæbɪnɪt]
取样装置 sampling device
去污 decontamination [ˈdiːkənˌtæmɪˈneɪʃən]
去污系统 decontamination system
去污因子 decontamination factor
权重因子 weighting factor
全厂报警处理系统 plant alarm processing system
全厂失电 station blackout
全穿透焊 full penetration weld
全面腐蚀 general corrosion [kəˈrəʊʒn]
全面责任 overall responsibility [rɪˌspɒnsɪˈbɪlɪtɪ]
全日志 complete logbook
全身计数器 whole-body counter
全身污染监测仪 whole body contamination monitor [kənˌtæmɪˈneɪʃən]
全透焊逢 full penetration welding
全压 total pressure
裙式支座，筒式支座 skirt support

R

燃耗，烧乏 depletion [dɪˈpliːʃn]
燃料棒 fuel rod
燃料棒束 fuel cluster
燃料棒弯曲亏损因子 penalty factor for rod bow [ˈpen(ə)ltɪ]
燃料包壳管 fuel tube
燃料厂房 fuel building
燃料厂房给排水系统 fuel building water supply and drainage system
燃料厂房起重机 fuel building crane
燃料厂房桥架 fuel building bridge
燃料厂房通风系统 fuel building ventilation system

燃料储运和检验系统 fuel storage, transportation and inspection system

燃料管理 fuel management

燃料篮 fuel basket

燃料燃耗 fuel burnup [ˈbɜːnʌp]

燃料芯块 fuel pellet

燃料循环 fuel cycle

燃料循环寿期 fuel cycle lifetime

燃料元件 fuel element

燃料元件破碎事故 fuel element failure accident

燃料运输屏蔽 fuel transfer shielding

燃料运输容器吊具 fuel shipping container rig

燃料运输通道 fuel transfer tube

燃料运输系统 fuel transfer system

燃料肿胀 fuel swelling

燃料柱 fuel column

燃料抓取机(燃料池桥架) fuel pool bridge

燃料装量 fuel inventory [ˈɪnv(ə)nt(ə)rɪ]

燃料装卸系统 fuel handling system

燃料组件 fuel assembly [əˈsemblɪ]

燃料组件啜吸装置 fuel assembly sipping device

燃料组件目视检查装置 fuel assembly visual inspection equipment

绕组 winding

热备用 hot standby

热冲击 thermal shock

热处理 heat treatment

热导率 heat conductivity [ˌkɒndʌkˈtɪvɪtiː]

热点因子 hot point factor

热点因子 hot spot factor

热电偶 thermocouple [ˈθɜːməʊkʌp(ə)l]

热电偶导管 thermocouple column [ˈkɒləm]

热电偶导管密封 thermocouple conduit seal [ˈkɒndjʊɪt; ˈkɒndɪt]

热电偶合金 thermocouple alloy [ˈθɜːməʊkʌp(ə)l] [ˈælɒɪ]

热段 hot leg

热负荷 heat load

热管因子 hot channel factor

热继电器 thermal relay [ˈriːleɪ]

热交换器 heat exchanger

热交换系数 heat transfer coefficient [trænsˈfɜː] [ˌkəʊəˈfɪʃənt]

热井 hot sink; heat sink

热绝缘 thermal insulation [ɪnsjʊˈleɪʃ(ə)n]

热扩散 heat diffusion [dɪˈfjuːʒn]

热流量 heat flow rate

热流密度 DNB heat flux DNB

热能化 thermalization [ˌθɜːməlaɪˈzeɪʃən]

热膨胀 thermal expansion [ˈθɜːm(ə)l] [ɪkˈspænʃ(ə)n]

热膨胀系数 coefficient of thermal expansion [ˌkəʊɪˈfɪʃ(ə)nt] [ˈθɜːm(ə)l] [ɪkˈspænʃ(ə)n]

热平衡 heat balance [ˈbæləns]

热屏 thermal barrier [ˈθɜːm(ə)l] [ˈbærɪə]

热屏蔽 thermal shield [ʃiːld]

热启动 hot startup [ˈstɑːtʌp]

热容 heat capacity [kə'pæsətɪ]
热室 hot cell
热释光剂量计 thermoluminescene dosimeter [dəʊ'sɪmɪtə]
热释光探测器 thermoluminescent detector ['θəːməuˌljuːmɪ'nesənt]
热水生产与分配系统 hot water production and distribution system
热水系统 hot water system
热态试验 hot test
热态性能试验 hot performance test
热停堆 hot shutdown
热通道 hot channel
热通量 heat flux [flʌks]
热烟羽 thermal plume
热影响区 heat affected zone
热中子俘获 thermal capture ['kæptʃə]
热中子通量 thermal neutron flux
人机对话 man-machine interaction
人—机接口 man-machine interface ['ɪntəfeɪs]
人孔，检查孔 manhole
人口当量 population equivalent [ɪ'kwɪv(ə)l(ə)nt]
人口统计学 demography [diː'mɒgrəfɪ]
人员闸门 personnel air lock
日变化系数 daily variation coefficient [veərɪ'eɪʃ(ə)n]
日允许摄入量 acceptable daily intake
日志 logbook
容积变化 volume change
容控箱 volume control tank
容器支架保温层 vessel support bracket insulation ['ves(ə)l] [ɪnsjʊ'leɪʃ(ə)n]
熔断器开关 fuse switch
熔断器闸刀 fuse isolator ['ɑɪsəleɪtə]
熔敷金属 deposited metal
熔焊 fusion welding
熔炼分析 heat (or ladle) analysis [ə'næləsɪs]
冗余度，冗余性，多重性 redundancy [rɪ'dʌnd(ə)nsɪ]
柔线 flexible cord ['fleksɪb(ə)l]
柔性连接 flexible connection
蠕变 creep
软辫线连接 pigtail connection ['pɪgˌtel]

S

塞焊 plug welding
三阀组 three valves manifold ['mænɪfəuld]
三角形连接 delta connection
三维 three dimensional [daɪ'menʃənl]
三芯插座 three-pin socket ['sɒkɪt]
散热片 fin
散射 scattering ['skætərɪŋ]
散射定律 scattering law
散射截面 scattering cross section ['skætərɪŋ]
散射矩阵 scattering matrix

[ˈmeɪtrɪks]
扫描速率 scan rate
色标 color code
砂暴 sandstorm
砂滤器 sand filter [ˈfɪltə]
砂质粉土 sandy silt [sɪlt]
钐中毒 samarium poisoning [səˈmeərɪəm] [ˈpɒɪzənɪŋ]
闪光报警 flashing alarm
闪烁体 scintillator [ˈskævɪn(d)ʒə]
闪蒸 flashing
上封头 upper head
上管座 top nozzle [ˈnɒz(ə)l]
上筒体 upper shell
上游 upstream
上支撑筒 upper support column
设备冷却水 component cooling water [kəmˈpəʊnənt]
设备冷却水泵 component cooling water pump
设备冷却水波动箱 component cooling water surge tank
设备冷却水热交换器 component cooling water heat exchanger
设备冷却水系统 component cooling water system
设备明细表 equipment list
设备闸门 equipment hatch
设计报告 design report
设计变更 design change
设计工况 design condition
设计管理 design control
设计规范 design code
设计规范书 design specification [ˌspesɪfɪˈkeɪʃ(ə)n]
设计基准 design basis
设计基准事故 design basis accident (DBA) [ˈæksɪdənt]
设计基准事件 design basis event
设计基准源项 design basis source terms
设计假设 design assumption [əˈsʌm(p)ʃ(ə)n]
设计接口管理 design interface control
设计目标限值 design objective limit
设计审查 design review
设计温度 design temperature
设计压力 design pressure
设计验证 design verification [ˌverɪfɪˈkeɪʃən]
射线检验 radiographic examination [ˌreɪdɪəʊˈgræfɪk]
摄像机姿态控制器 pan-tilt control facility [fəˈsɪlɪtɪ]
伸缩套筒 telescopic tube [telɪˈskɒpɪk]
升压变压器 stepup transformer
生产废水（未污染的工业废水） non-polluted industrial wastewater [ɪnˈdʌstrɪəl]
生产用水 process water
生活给水厂区管系统 service water yard pipe network system
生活水 domestic water
生活水厂区管网系统 domestic water yard pipe network system
生活污水 domestic sewage
生活污水厂区管路系统 sanitary sewer yard pipe network system

['sænɪt(ə)rɪ]
生活污水处理系统 sanitary sewer treatment system ['su:ə]
剩余功率 after-power
剩余释热 after-heat；residual heat
失步 misalignment [mɪsə'lɑɪnmənt]
失步转矩 pull- out torque [tɔ:rk]
失火报警 fire alarm
失控释放 uncontrolled release
失去电源 power loss
失去电源 loss of power
失水事故 loss of coolant accident ['ku:l(ə)nt]
施工设计 detail design
施工质量 construction quality
湿保养（蒸发器） wet layup
湿沉降 wet deposition [ˌdepə'zɪʃ(ə)n]
湿啜吸装置 sipping device
湿球温度 wet bulb temperature
十六进制的 hexadecimal [ˌheksə'desɪml]
时变化系数 hourly variation coefficient [ˌveərɪ'eɪʃ(ə)n] [ˌkəuə'fɪʃənt]
时分多路 time division multiplex [dɪ'vɪʒ(ə)n] ['mʌltɪpleks]
时间片 time slice [slɑɪs]
时钟系统 clock system
识别，标记 identification [ɑɪˌdentɪfɪ'keɪʃn]
实时操作系统 real time operating system
实时能谱探测器 real time spectroscopy detector [spek'trɒskəpɪ]
实体分离 physical separation [sepə'reɪʃ(ə)n]
使用工况 service condition
事故处理 accident management ['æksɪdənt]
事故分析 accident analysis
事故工况 accident conditions
事故后取样 post accident sampling ['æksɪdənt]
事故缓解 accident mitigation [mɪtɪ'geɪʃ(ə)n]
事故联锁组件 accident interlocking module [ˌɪntə'lɒkɪŋ]
事故排出口 emergency outlet
事故停堆 accident shutdown ['ʃʌtdɑʊn]
事故用高量程辐射监测仪 high range gamma radiation monitor for accident
事故预防 accident prevention [prɪ'venʃn]
事故源 accident source [sɔ:s]
事故照射 accidental exposure [ɪk'spəʊʒə]
事件常数 time constant
事件顺序 sequence of events（events sequence） ['si:kw(ə)ns]
试块 coupon
试样 specimen
视在功率 apparent power
释热 heat generation [ˌdʒenə'reɪʃn]
收缩系数 contraction coefficient
收缩压力损失 contraction pressure loss [kən'trækʃ(ə)n]
手动—电动转换开关 manual-electric changeover switch

手动控制 manual control [ˈmænjʊ(ə)l]
手工焊 manual welding [ˈweldɪŋ]
手孔 hand hole
手套箱 glove box
手提式灭火器 portable extinguisher [ɪkˈstɪŋgwɪʃə]
手选取样 grab sampling [ˈsæmplɪŋ]
手足监测仪 hand and foot monitor [ˈmɒnɪtə(r)]
首次碰撞几率 first collision probability [kəˈlɪʒ(ə)n]
首发报警 first out alarm
寿命试验 life test
寿期 lifetime
寿期末 end of life (EOL)
疏水回收箱 drain recovery tank
疏水器，捕集器，存水弯管 trap [træp]
疏水箱 drain tank
输入/输出 input/output (I/O)
输入隔离组件 input isolator module [ˈaɪsleɪtə] [ˈmɒdjuːl]
输入设备 input device
鼠笼电动机 squirrel-cage motor [ˈskwɪr(ə)l]
树脂冲排泵 resin flush pump
树脂冲排水 resin flush water
树脂床 resin bed
树脂碎片 resin fine
树脂添加箱 resin addition tank
树脂再生 resin regeneration
数据采集 data acquisition [ˌækwɪˈzɪʃ(ə)n]
数据采集和处理系统 data acquisition and processing system
数据处理模块 data processing module
数据传输系统 data communication system [kəmjuːnɪˈkeɪʃ(ə)n]
数据基础 data base
数据结构 data organization
数据通道 data channel
数据文件 data file
数据总线 data bus
数字电压表组件 digital voltmeter module [ˈvəʊltmiːtə]
数字失步报警 digital misalignment alarm module [mɪsəˈlaɪnmənt]
数字输入 digital input
刷新 refresh
衰变 decay [dɪˈkeɪ]
衰变常数 decay constant
衰变链 decay chain
衰变热 decay heat
衰变箱 decay tank
甩电荷 load shedding [ˈʃedɪŋ]
双端断裂 double ended break
双工收发信机 duplex transceiver
双钩 double hook
双极单掷开关 double-pole single-throw (DPST) switch
双迹示波器 dual trace oscilloscope
双卷绕系统 double-reeling system
双脉冲发生器 double-pulse generator
双面焊 welding by both sides
双位通风柜 double fume chamber and hood
双音多频 dual-tone multifrequency

水处理 water treatment

水处理厂房暖通空调系统 water treatment building HVAC system

水封 water seal

水化学 water chemistry [ˈkemɪstrɪ]

水力阻力 hydraulic resistance [haɪˈdrɔːlɪk] [rɪˈzɪstəns]

水喷淋器 water sprinkler [ˈsprɪŋklə]

水平导向轨 horizontal guide wheel [ˌhɒrɪˈzɒntəl]

水平加速度 horizontal acceleration [ækˌseləˈreɪʃən]

水—汽平衡 water-steam equilibrium [ˌiːkwɪˈlɪbrɪəm]

水头损失 head loss

水位 water level

水下运输通道 underwater transfer tube

水下照明灯具 underwater lights

水下照明装置 underwater lighting device

水线腐蚀 water line corrosion [kəˈrəʊʒ(ə)n]

水压机 hydraulic press [haɪˈdrɔːlɪk]

水质 water quality

水装量 water inventory [ˈɪnv(ə)nt(ə)rɪ]

顺时针方向 clockwise [ˈklɒkwaɪz]

瞬发 γ 辐射 prompt gamma radiation

瞬发临界 prompt critical

瞬发中子 prompt neutron [ˈnuːtrɒn]

瞬发中子份额 prompt neutron fraction

瞬时额定值 momentary rating [ˈməʊm(ə)nt(ə)rɪ] [ˈreɪtɪŋ]

瞬时继电器 instantaneous relay [ˌɪnstənˈteɪnjəs]

瞬态 transient [ˈtrænzɪənt]

说明（书），细则 instruction [ɪnˈstrʌkʃn]

死区 dead band

伺服马达 servomotor [ˈsɜːvəʊˌməʊtə]

送风机 supply fan

速度压头 velocity pressure（or head)

酸度计 pH meter

随机效应 stochastic effect [stəˈkæstɪk]

索引图 key plan

索引文件 index file [ˈɪndeks]

锁定钮 locking button

锁紧螺母 lock nut

T

塌陷裂缝 collapse-fissure [kəˈlæps] [ˈfɪʃə]

坍塌 slumping [ˈslʌmpɪŋ]

坍塌水位 collapse level

弹性散射 elastic scattering

探测器存放管 detector storage tube

探测器孔道 detector well

探测器推拉装置 detector push-pull device [dɪˈtektə]

探测限值 detection limit

逃脱共振几率 resonance escape probability
逃脱率系数 escape rate coefficient
套管 thimble tube
特殊检查 special monitoring
特种废水混合排放系统 special waste water mixed drainage system [ˈdreɪnɪdʒ]
提棒事故 rod withdrawal accident [wɪðˈdrɔː(ə)l]
提棒速度（率） withdrawal speed (rate)
提升速度 hoist speed [hɒɪst]
提升衔铁 lift armature [ˈɑːmətʃə]
提升线圈 lift coil [kɒɪl]
体积检查 volumetric examination [ˌvɒljʊˈmetrɪk] [ɪɡˌzæmɪˈneɪʃ(ə)n]
体积平均有效应力 volume average effective stress
体积缺陷 volume defect
体剂量当量 collective dose equivalent [ɪˈkwɪv(ə)l(ə)nt]
天空回散射 sky shine
天然本底 natural background
天然放射性 natural radioactivity
填充金属 filler metal
（格架内）条带 inner strap [ˈɪnə(r) stræp]
条形基础 strip footing [strɪp]
调节棒组 control (rod) bank/group
调节阀 governor valve [ˈɡʌvənə]
调试 commissioning
跳焊 skip sequence [skɪp] [ˈsiːkw(ə)ns]
跳闸 trip (circuit breaker)
铁丝网 wire fence
停堆 shut down
停堆断路器 reactor trip breaker
停堆裕量 shutdown margin [ˈmɑːdʒɪn]
停堆组 shutdown bank/group
停工待检点 hold point
通道，入口 access [ˈækses]
通电 energization
通风防护服 ventilated protective suit [ˈventɪleɪtɪd]
通风管 vent pipe
通风柜 vent cabinet (or ventilation cabinet) [ˈkæbɪnɪt] [ˌventɪˈleɪʃ(ə)n]
通风控制室系统 ventilation control room system [ˌventɪˈleɪʃ(ə)n]
通量测量导向管 flux instrumentation guide tube [ˌɪnstrʊmenˈteɪʃ(ə)n]
通量分布 flux distribution
通量分布图 flux map
通量密度 flux density
通量梯度 flux gradient [ˈɡreɪdɪənt]
通量展平 flux flattening
通信连接 communication link
通信系统 communication system
通用标准 general standard
通用闪烁计数器 general scintillation counter
同步操作 synchronized operation [ˈsɪŋkrənaɪzd]
同位素萃取份额 isotopic stripping fraction [ɑɪsəʊˈtɒpɪk] [ˈfrækʃn]
同位素丰度 isotopic abundance [ˌɑɪsəuˈtɒpɪk] [əˈbʌnd(ə)ns]
同位素天然丰度 natural isotopic a-

bundance

同轴电缆 coaxial cable [kəʊ'æksɪəl]

筒身保温层 shell insulation [ɪnsjʊ'leɪʃ(ə)n]

筒体段，堆芯筒体 core shell

筒形壳体 cylindrical shell [sɪ'lɪndrɪkəl]

头戴式送受话器 headset ['hedset]

图册目录 drawing (volume) list

图纸 drawing

土建工程 civil works

土木建筑/工程 civil construction/engineering

土壤腐蚀 soil corrosion

[kə'rəʊʒ(ə)n]

推荐性标准 recommended standard [rekə'mendɪd]

推力轴承 thrust bearing

退役 decommissioning [ˌdiːkə'mɪʃən]

吞吐量 throughput

脱扣追忆记录 post-trip review

脱气 gas stripping ['strɪpɪŋ]

脱气塔 gas stripper ['strɪpə(r)]

脱气装置 gas stripping unit ['strɪpɪŋ]

椭圆形封头 ellipsoidal head

W

外部接口 external interface

外部设备 peripheral [pə'rɪfərəl]

外负荷 external load

外污染 external contamination

外形图 outline drawing

外压 external pressure

外照射 external exposure

完工证书 completion certificate

完好环路 intact loop [ɪn'tækt] [luːp]

万用表 multimeter ['mʌltɪmiːtə]

网格，筛孔 mesh [meʃ]

往复式上充泵 reciprocating charging pump [rɪ'sɪprəkeɪtɪŋ]

危险环境 hazardous atmosphere ['hæzədəs] ['ætməsfɪə(r)]

危险载荷 critical load

微分价值 differential worth

微分器组件 differentiator module [ˌdɪfə'renʃɪeɪtə]

微观截面 microscopic cross section [maɪkrə'skɒpɪk]

微粒-碘取样器 particulate-iodine sampler [pɑː'tɪkjʊlət] ['aɪədiːn]

微粒过滤器 particulate filter [pɑː'tɪkjʊlət]

微量元素 trace element [treɪs]

微振磨损 fretting wear

围板 outside strap [stræp]

围筒 wrapper

维修工具 maintenance tool ['meɪnt(ə)nəns]

委托者，指派者 assignor [ə'saɪnə]

未焊透 lack of penetration [penɪ'treɪʃ(ə)n]

未能紧急停堆的预计瞬态 anticipated transient without scram (ATWS)

未熔合 lack of fusion [ˈfjuːʒ(ə)n]
位置指示器 position indicator [ˈɪndɪkeɪtə(r)]
位置指示仪 position indicating system
温（磁）盘 Winchester disk
文件编制 document preparation
文件抽样 sampling of document [ˈdɒkjʊm(ə)nt]
文件递交 submission of document [səbˈmɪʃ(ə)n]
文件审查 document review
文件最终储存 final storage of document
紊流，湍流 turbulent flow [ˈtɜːbjʊl(ə)nt]
稳态 steady state
稳态运行 steady state operation
稳压电源 stabilized electrical supply [ˈsteɪbɪlaɪzd]
稳压器 pressurizer [ˈpreʃəraɪzə]
稳压器气相 pressurizer steam phase
稳压器卸压箱 pressurizer relief tank
稳压器压力控制系统 pressurizer pressure control system
稳压器液位控制系统 pressurizer water level control system
稳压器液相 pressurizer liquid phase [ˈpreʃəraɪzə] [ˈlɪkwɪd] [feɪz]
蜗轮减速器 worm gear reducer
污垢系数 fouling coefficient [ˌkəʊɪˈfɪʃ(ə)nt]
污泥 sludge [ˈslʌdʒ]
污泥焚烧 sludge incineration [ɪnˌsɪnəˈreɪʃn]
污泥干化 sludge dry
污泥浓缩 sludge thickening [ˈθɪk(ə)nɪŋ]
污泥脱水 sludge dewatering [ˈslʌdʒ] [diːˈwɔːtərɪŋ]
污泥压滤 sludge pressure filtration [fɪlˈtreɪʃn]
污染的工件 polluted item
污染的工业水 polluted industrial wastewater
污水泵 sump pump
污水量 wastewater flow [ˈweɪstˌwɔːtə]
钨铬钴合金堆焊的轴颈 stellite clad journal [klæd] [ˈdʒɜːn(ə)l]
屋顶通风器 roof ventilator [ˈventɪleɪtə]
无功功率表 wattles power meter
无级调速控制 stepless speed control
无坡口对接焊 square butt welding [ˈweldɪŋ]
无损检测 non-destructive examination [dɪˈstrʌktɪv]
无限栅格 infinite lattice [ˈɪnfɪnɪt] [ˈlætɪs]
无线固定台 fixed radio station
无线寻呼系统 radio transmission [trænzˈmɪʃ(ə)n]
无线移动通信系统 radio paging system
无延性转变温度 nil-ductility transition temperature [nɪl] [dʌkˈtɪlətɪ]
无氧废气 oxygen free waste gas [ˈɒksɪdʒ(ə)n]

物理测量室 physics laboratory [ˈfɪzɪks][ləˈbɒrət(ə)rɪ]
物项管理 control of items

X

吸氢 hydrogen pickup [ˈhaɪdrədʒən]
吸收棒 absorber rod
吸收比 absorption ratio
吸收剂量 absorbed dose rate
吸收截面 absorption cross section
吸收式制冷机 absorber chiller
吸收系数 absorption coefficient [əbˈzɔːpʃ(ə)n]
吸水井 suction well [ˈsʌkʃ(ə)n]
希弗 Sievert [ˈsiːvət]
稀释速率 rate of dilution [daɪˈluːʃn]
稀释因子 dilution factor
稀有事故 infrequent incident [ɪnˈfriːkwənt]
熄火保护装置 fire annihilation protecting equipment
洗衣房通风系统 hot laundry ventilation system [ˈlɔːndrɪ] [ˌventɪˈleɪʃn]
系统负荷 system load
系统名称 designation of system
系统软件 system software
系统生成 system generation [dʒenəˈreɪʃ(ə)n]
系统运行压力 system operation pressure
细砂 fine sand
下泄 let down
下泄节流孔板 letdown orifice [ˈɒrɪfɪs]
下泄热交换器 letdown heat exchanger
下泄通道 letdown path
下栅格板组件 lower support plate assembly [əˈsemblɪ]
先导装置 pilot device [ˈpaɪlət] [dɪˈvaɪs]
氙积累 xenon buildup [ˈzenɒn]
氙中毒 xenon poisoning
显热 sensible heat [ˈsensɪb(ə)l]
显示帧帽 display frame
现场安装 site installations [ɪnstəˈleɪʃ(ə)ns]
现场监督 site supervision
现场修改 field change
现实源项或预期源项 realistic source terms [rɪəˈlɪstɪk]
限流器 flow restrictor [rɪsˈtrɪktə]
限位开关 position limit switch
限制区边界 restricted area boundary [rɪˈstrɪktɪd] [ˈbaʊnd(ə)rɪ]
线开关 incoming breaker
线圈组件 coil stack assembly [əˈsemblɪ]
线性功率密度 linear power density [ˈdensɪtɪ]
相对生物效应 relative biological effectiveness [baɪə(ʊ)ˈlɒdʒɪk(ə)l]
相对湿度 relative humidity [hjʊˈmɪdɪtɪ]
相位差 phase difference

相位移 phase displacement
镶块 clevis insert [ˈklevɪs]
镶嵌，暗装 flush-mounted
响应时间 response time
项目工程经理 project engineering manager
项目管理 project management
项目监督 project supervision
项目经理 project manager
象限功率倾斜比 quadrant power till ratio [ˈkwɒdr(ə)nt]
橡胶缓冲器 rubber bumper
消毒 disinfection [ˌdɪsɪnˈfekʃən]
消防 fire protection
消防泵 fire extinction pump [pʌmp]
消防水 fire-fighting water
消防水龙带 fire hose
消防水系统 fire protection water system
消火栓 fire hydrant
消氢器 hydrogen recombiner [ˈhaɪdrədʒən] [ˈriːkəmˈbaɪnə]
消音器 silencer [ˈsaɪlənsə]
小车架 trolley frame [ˈtrɒlɪ] [freɪm]
小流量孔板 miniflow orifice [ˈɒrɪfɪs]
小流量循环方式 mini-flow circulation mode [səːkjʊˈleɪʃ(ə)n]
小石子 stone chipping [ˈtʃɪpɪŋ]
小型裂变室 miniature fission chamber [ˈmɪnɪtʃə] [ˈfɪʃ(ə)n] [ˈtʃeɪmbə]
泄漏，逃逸 escape
泄漏辐射 leakage radiation [reɪdɪˈeɪʃ(ə)n]
泄漏探测器 leak detector [dɪˈtektə]
卸料 discharge
卸料燃耗 discharge burnup [ˈbəːnʌp]
卸压 depressurization [diːpreʃəraɪˈzeɪʃən]
卸压阀 pressure relief valve
芯空泡份额 core void fraction [ˈfrækʃ(ə)n]
芯块与包壳相互作用 pellet-cladding interaction [ɪntərˈækʃ(ə)n]
新燃料操作工具 new fuel handling tool
新燃料储存格架 new fuel storage rack [ræk]
新燃料升降机 new fuel elevator [ˈelɪveɪtə]
新燃料运输容器 new fuel shipping container [ˈʃɪpɪŋ] [kənˈteɪnə]
新燃料运输容器支架 new fuel shipping container console [kənˈsəʊl]
新燃料组件操作工具 new fuel assembly handling tool
新燃料组件测量装置操作工具 new fuel assembly measuring handling tool
新燃料组件检查台架 new fuel assembly inspection operating [əˈsemblɪ] [ɪnˈspekʃn] [ˈɒpəreɪtɪŋ]
信号灯 semaphore [ˈseməfɔː]
行程开关 position switch
行程开关，限位开关 limit switch
行程终端限位开关 end-of travel limit switch
性能曲线 performance curve [pəˈfɔːm(ə)ns]

修正的麦加利烈度（地震术语） modified Mercalli intensity [ɪn'tensɪtɪ]

修正的棋盘式交叉布置 modified checkerboard pattern ['mɒdɪfaɪd] ['tʃekəbɔːd] ['pæt(ə)n]

修正说明 modification observations [ˌmɒdɪfɪ'keɪʃ(ə)n]

袖珍式剂量计 pocket dosimeter [dəʊ'sɪmɪtə]

许可证 permit

许可证批准程序 licensing procedure [prə'siːdʒə]

许可证申请文件 licensing document

序列 train

悬臂式起重机 jib crane [dʒɪb] [kreɪn]

悬浮物 suspended solid [sə'spendɪd] ['sɒlɪd]

悬挂式按钮控制项 pendant push-button control box

旋钮 knob [nɒb]

旋塞 plug valve

旋涡泵 whirlpool pump ['wɜːlpuːl]

旋叶式分离器 swirl vane separator

选择开关 selector switch

选择器 selector

巡回检查 field inspection [ɪn'spekʃn]

循环倍率 circulation ratio [sɜːkjʊ'leɪʃ(ə)n]

循环冷风机 circulating cooler

循环冷却水 recirculating cooling water

循环冷却水泵房系统 circulating cooling water pumping station system

循环冷却水厂区管路系统 circulating cooling water yard pipe network system

循环冷却水处理系统 circulating cooling water treatment system

循环冷却水排放口消能系统 circulating cooling water energy dispersion [dɪ'spɜːʃ(ə)n]

循环冷却水排放系统 circulating cooling water drainage system

循环冷却水吸入系统 circulating cooling water intake system

循环末 end of cycle

循环水泵 circulating water pump

循环水系统；再循环系统 recirculation system [riːˌsɜːkju'leɪʃən]

Y

压扁 collapse [kə'læps]

压盖填料 gland packing

压降 pressure drop

压紧部件存放架 upper internals storage stand

压紧弹簧环 hold down spring

压紧顶帽（导向管支撑板） guide tube support plate

压紧支撑结构 hold-down support assembly [ə'semblɪ]

压力边界 pressure boundary

压力变送器 pressure transmitter

[trænz'mɪtə]
压力表 manometer
压力平衡膜片 pressure compensation diaphragm ['daɪəˌfræm]
压力容器 pressure vessel ['vesl]
压力容器 reactor pressure vessel
压力系数 pressure coefficient [ˌkəʊɪ'fɪʃnt]
压力响应 pressure response
压缩空气系统 compressed air system
烟囱等速取样头 isokinetic sampling nozzle for stack [ˌaɪsəʊkɪ'netɪk]
烟囱监测仪 stack monitor
延迟电路组件 delay circuit module
延时电路 time delay circuit [dɪ'leɪ] ['sɜːkɪt]
严重事故 severe accident
验收报告 acceptance report
验收标准 acceptance standard
验收实验 acceptance test
验收准则 acceptance criterion [kraɪ'tɪərɪən]
验证 verification [ˌverɪfɪ'keɪʃ(ə)n]
扬液器 transfer tank
仰焊 overhead position welding
氧分析器 oxygen analyzer ['ænəˌlaɪzə]
遥控 remote control
咬边 undercut
药皮焊条 coated electrode
液滴 droplet
液滴夹带 droplet entrainment [ɪn'treɪnmənt]
液化势 liquefaction potential [pə(ʊ)'tenʃ(ə)l]
液体过滤器 liquid filter ['fɪltə]
液体空间 liquid space
液体渗透检验 liquid penetrant examination ['penɪtr(ə)nt]
液位变送器 liquid level transmitter
液位变送器 water level transmitter
液相 liquid phase [feɪz]
一次屏蔽 primary shielding['ʃiːldɪŋ]
一个半断路器 one breaker and a half type
一回路辅助厂房通风系统 primary auxiliary building ventilation system [ˌventɪ'leɪʃn]
仪表（截止）阀 instrumentation (stop) valve [ˌɪnstrəmen'teɪʃn]
仪表接头 instrumental joint [ˌɪnstrʊ'mentl]
仪表室 instrument room ['ɪnstrʊmənt ruːm]
移动式γ照射率仪 portable γ exposure rate meter
移动式空气取样器 portable air sampler
移动式灭火系统 portable extinguisher system [ɪk'stɪŋgwɪʃə]
移动速度 traveling speed
移理论 transport theory
移相触发电路 phase control and firing circuit ['sɜːkɪt]
遗传效应 genetic effect [dʒɪ'netɪk]
乙烯基漆 vinyl paint ['vaɪnɪl]
已辐照燃料 irradiated fuel [ɪ'reɪdiːeɪtɪd] ['fjuːəl]

已购物项管理 control of purchased items
异步电动机 asynchronous motor [ə'sɪŋkrənəs]
异常工况 abnormal condition [əb'nɔːm(ə)l]
抑制器 suppressor [sə'presə]
役前检查 pre-service inspection
易裂变核 fissile nucleus
易燃的 flammable
溢出 overflow
因科镍 718 合金 inconel 718 alloy [ɪn'kəʊnl]
因科镍合金 inconel alloy ['ælɒɪ]
阴离子床 anion bed ['ænaɪən]
音响信号 audible signals ['ɔːdɪb(ə)l]
银—铟—镉合金 silver-indium-cadmium alloy
引出线 conducing wire
引出线管 electrical conduit ['kɒndjʊɪt]
引漏管线 leak-off line
引漏接头 leak-off connection
引用标准 quoted standard
荧光灯 fluorescent lamp
荧屏显示器 visual display
营运单位 operating organization
应变速率 strain rate [streɪn]
应急柴油发电机通风系统 emergency diesel generator room ventilation system
应急柴油发电机系统 emergency diesel generation system
应急电源 emergency power source (or supply)
应急电源转换设备 emergency transfer equipment
应急堆芯冷却 emergency core cooling
应急给水箱 emergency feedwater tank
应急控制室 emergency control room system
应急盘式制动器 emergency disk braker
应急手动操作机构 manual emergency drive [ɪ'mɜːdʒ(ə)nsɪ]
应急疏水 emergency drain
应急油泵 emergency oil pump
应急照射 emergency exposure
应急自动电话系统 emergency telephone system
应力变化范围 stress range
应力腐蚀开裂 stress corrosion cracking [kə'rəʊʒ(ə)n]
应力集中 stress concentration [kɒns(ə)n'treɪʃ(ə)n]
应力强度 stress intensity [ɪn'tensɪtɪ]
应力松弛 stress relaxation
应力消除 stress relief [rɪ'liːf]
应力-应变曲线 load-deformation curve [ˌdiːfɔː'meɪʃ(ə)n] [kɜːv]
应用程序 application program [ˌæplɪ'keɪʃ(ə)n]
应用程序 utility program
应用软件包 application package
硬拷贝打印机 hard copy printer
永久储存 permanent storage ['pɜːm(ə)nənt]
用水量 water consumption

[kən'sʌm(p)ʃ(ə)n]
优先传呼 priority call [praɪ'ɒrətɪ]
油槽，油坑 oil pit
油箱 oil tank；oil reservoir ['rezəvwɑː(r)]
有害影响 harmful effect
有限批准 limited approval [ə'pruːv(ə)l]
有线、无线转换器 wire-wireless switching controller
有线广播系统 public address system
有效剂量当量 effective dose equivalent
有效流量 effective flow rate
有用功率 active power
淤泥质粉质黏土 very soft silty clay
余热 residual heat [rɪ'zɪdjʊəl]
余热排出泵 residual heat removal pump [rɪ'muːv(ə)l]
余热排出热交换器 residual heat removal exchanger [rɪ'zɪdjʊəl]
余热排放系统 residual heat removal system
宇宙射线 cosmic radiation ['kɒzmɪk]
预过滤器 prefilter [priː'filtə]
预埋件 embedded part
预期运行事件 anticipated operational occurrences [æn'tɪsɪpeɪt] [ə'kʌr(ə)ns]
预热 preheat
预热器 pre-heater
阈值 threshold ['θreʃəʊld]
原水 raw water
原水泵 raw water pump
原水净化系统 raw water purification system [ˌpjʊərɪfɪ'keɪʃən]
原型机，样机 prototype ['prəʊtətaɪp]
原子吸收谱 atom absorption spectroscopy [əb'zɔːpʃ(ə)n] [spek'trɒskəpɪ]
源程序 source program
源量程 source range
源文件 source file
源项 source terms
源项评价 source-term evaluation
约束点 confined point
允许电路 permissive circuit [pə'mɪsɪv]
允许电压偏差 voltage tolerance ['tɒl(ə)r(ə)ns]
允许逻辑组件 permissive logic module ['lɒdʒɪk] ['mɒdjuːl]
运行层 operation floor
运行方式 mode of operation
运行工况 operational states；operating condition
运行基准地震 operation basis earthquake（OBE）
运行记录 operational records
运行限值和条件 operational limits and conditions
运行许可证 operating license
运输容器 shipping cask
运输容器清洗池 shipping cask cleaning pool
运输容器装料池 shipping cask loading pool ['ʃɪpɪŋ]
运输通道 transfer canal/tube

[kə'næl]

运输小车 conveyer car

运输许可证 release for shipment

Z

杂质 foreign material；impurity [ɪm'pjʊərətɪ]

载荷传感器 load sensor ['sensə]

载流容量；安培流量 ampacity [æm'pæsɪtɪ]

再灌水阶段 refilling stage

再热调节阀 intercept valve [ˌɪntə'sept]

再生硼酸溶液 regenerated boric acid solution

再生热交换器 regenerative heat exchanger [rɪ'dʒen(ə)rətɪv]

再循环 recycle

再循环阶段 recirculation phase [riːˌsəːkj'leɪʃən]

在役检查 in-service inspection [ɪn'spekʃn]

暂存箱 hold-up tank

责任单位 responsible organization [rɪ'spɒnsɪb(ə)l]

增殖因子 multiplication constant [ˌmʌltɪplɪ'keɪʃ(ə)n] ['kɒnst(ə)nt]

增殖因子 multiplication factor

闸阀 gate valve

栅格间距 lattice pith ['lætɪs] [pɪθ]

窄焊缝 narrow-gap welding

张拉廊道 stretching gallery ['stretʃɪŋ] ['gæl(ə)rɪ]

长电离室 long ionization chamber [ˌaɪənaɪ'zeɪʃən] ['tʃeɪmbə]

长时间额定值 longtime rating ['reɪtɪŋ]

长寿命裂变产物 long-lived fission product ['fɪʃ(ə)n]

胀管 tube rolling (or expanding)

照明 illumination [ɪˌluːmɪ'neɪʃn]

照射评价 assessment of exposure [ɪk'spəʊʒə]

遮断容量 interrupting capacity [ˌɪntə'rʌptɪŋ] [kə'pæsətɪ]

着色渗透检查 dye penetrant examination ['penətrənt]

针型阀 tapered valve ['teɪpəd]

真空泵 vacuum pump

真空破坏阀 vacuum break valve ['vækjʊəm]

真空脱气器 vacuum degasifier [dɪ'gæsɪfaɪə]

真空吸尘器 vacuum cleaner ['vækjʊəm]

诊断程序 diagnostic program

振动频率 vibration frequencies

震陷 earthquake subsidence ['sʌbsɪd(ə)ns]

震中 epicenter

蒸发器 evaporator；steam generator

蒸发器保养系统 steam generator layup system ['leɪʌp]

蒸发器排污系统 steam generator blowdown system ['dʒenəreɪtə]

蒸发式空气冷却机组 evaporative air cooling (EAV) unit
蒸汽干度 steam quality
蒸汽干燥器 steam dryer [ˈdraɪə]
蒸汽旁排系统 steam (bypass) dump system [dʌmp]
蒸汽限流器 steam restrictor [rɪsˈtrɪktə]
蒸汽总管 steam manifold [ˈmænɪfəʊld]
整定点 setpoint
正比涂硼计数管 proportional boron-lined counter [ˈbɔːrɒn]
正常电源 normal power supply
正常条件，正常状态 normal condition
正常运行 normal operation
正压 positive pressure
政策声明 policy statement
支撑点 support point
支撑架 support bracket [ˈbrækɪt]
支承板 supporting plate [pleɪt]
支承柱 support column [ˈkɒləm]
支吊架 hanger
支座 support
直接接地 direct grounding
直接内拨 direct inward dialing (DID)
直接外拨 direct outward dialing
直流保持柜 DC hold cabinet [ˈkæbɪnɪt]
直流和交流不停电电源系统 DC&AC UPS system
直流稳压电源 DC stabilized power supply
直埋 direct burial [ˈberɪəl]
直梯 stepladder [ˈsteplædə]
职业性辐射工作人员 occupational radiation worker
职业照射 occupational exposure [ɪkˈspəʊʒə]
职责分工 functional assignment
止挡块，停止 stop
指令单 instruction sheet [ɪnˈstrʌkʃən]
指示灯，指示仪 indicating lamp; indicator lamp (indicator) [ˈɪndɪkeɪtɪŋ]
质保大纲 qualified assurance program
质保记录 quality assurance record
质量保证 quality assurance [əˈʃʊər(ə)ns]
质量计划 quality plan
质量监查 quality audit
质量检查 quality inspection
质量检查员 quality inspector
质量鉴定 quality qualification
质量流密度 mass flux [flʌks]
质量趋势分析 quality trend analysis
质量守恒 conservation of mass [kɒnsəˈveɪʃ(ə)n]
质量速度 mass velocity [vɪˈlɒsɪtɪ]
质量因子 quality factor
质量证书 quality release
滞留，保存 retention [rɪˈtenʃ(ə)n]
置信度 confidence [ˈkɒnfɪd(ə)ns]
中等频率事故 incident of moderate frequency [ˈmɒdərət] [ˈfriːkwənsɪ]

中段泵壳 stage casing

中放惰性气体 γ 监测仪 medium-level noble gas γ monitor ['nəʊb(ə)l]

中分线 trunk line

中和槽 neutralizer tank ['njuːtrəlaɪzə]

中间电压 medium voltage ['vəʊltɪdʒ]

中间量程 intermediate range [ˌɪntə'miːdjət]

中砂 medium sand

中性点 neutral point

中压取样冷却器 MP sampling cooler

中子衬垫 neutron pad

中子代时间 neutron generation time

中子毒物 neutron poison ['pɒɪz(ə)n]

中子反射率 neutron albedo [æl'biːdəʊ]

中子辐照 neutron irradiation [ɪˌreɪdɪ'eɪʃ(ə)n]

中子高通量紧急停堆 high neutron flux trip ['nuːtrɒn]

中子活化腐蚀产物 neutron activated corrosion product [kə'rəʊʒ(ə)n]

中子计数率 neutron counting rate

中子剂量当量率监测仪 neutron doserate-equivalent monitor [ɪ'kwɪv(ə)l(ə)nt]

中子密度 neutron density

中子谱 neutron spectrum ['spektrəm]

中子谱硬化 neutron hardening ['hɑːdənɪŋ]

中子热能化 neutron thermalization [ˌθɜːməlaɪ'zeɪʃən]

中子输送 neutron transport

中子探测器 neutron detector

中子通量 neutron flux [flʌks]

中子通量量程 neutron flux range [reɪn(d)ʒ]

中子通量探测器 neutron flux detector

中子温度 neutron temperature

中子吸收 neutron absorption [əb'zɔːpʃ(ə)n]

中子吸收材料 neutron absorber material [əb'sɔːbə]

中子吸收剂 neutron absorber

中子吸收截面 neutron absorption cross section [əb'zɔːpʃ(ə)n]

中子源 neutron source

中子源棒 source rod [rɒd]

中子源组件 neutron source assembly [ə'semblɪ]

终端，接线端子 terminal ['tɜːmɪn(ə)l]

钟罩形计数管 end-window counter

肿胀 swelling ['swelɪŋ]

重叠，搭接 overlap

重度 unit weight

重量百分比 percent by weight; weight percent

重量含量 content by weight

重入程序 reentrant program [riː'entrənt]

重现期 recurrence interval ['ɪntəv(ə)l]

重新校正（对中） realignment [ˌriːə'laɪnmənt]

重要厂用水泵 essential service water pump

重要厂用水泵房系统 essential service water pumping station

system

重要厂用水厂区管路系统 essential service water yard pipe network

重要厂用水取水过滤系统 essential service water intake filteration system

重要冷冻水系统 essential chilled water system

重要仪表电源 vital instrumentation power supply [ˌɪnstrʊmen'teɪʃ(ə)n]

重油 heavy oil

周向环脊 circumferential ridge [səˌkʌmfə'renʃəl]

轴封回流过滤器 seal water reflux filter ['fɪltə]

轴封回流热交换器 seal water reflux heat exchanger

轴封注水过滤器 seal water injecting filter

轴封组件 shaft seal assembly [ə'semblɪ]

轴流风机 vane axial fan

骤冷 quenching

主电路故障 mains outage ['aʊtɪdʒ]

主给水泵 main feedwater pump

主给水系统 main feed water system

主检查员 lead auditor

主控/就地切换开关 MCR/LL transfer switch

主控/应控切换开关 MCR/ECR transfer switch

主控室警报处理系统 main control room alarm processing system [ə'lɑːm]

主控室可居留区排烟（通风）系统 main control habitability ventilation system [ˌhæbɪtə'bɪlətɪ] [ˌventɪ'leɪʃ(ə)n]

主控室系统 main control room system

主梁 girder ['gɜːdə]

主凝结水泵 main condensate pump ['kɒnd(ə)nseɪt]

主起升机构 main hoist [hɒɪst]

主汽阀 main stop valve

主循环器 master cycler

主循环器输出组件 master cycler output module ['mɒdjuːl]

主蒸汽 main steam

主蒸汽及旁路蒸汽系统 main steam and turbine bypass system ['tɜːbaɪn] ['baɪpɑːs]

主蒸汽排放控制系统 main steam dump control system

注油 oil injection [ɪn'dʒekʃən]

驻极体传声器 electret microphone ['maɪkrəfəʊn]

柱塞泵 plunger pump ['plʌn(d)ʒə]

抓具 gripper ['grɪpə]

专家报告 expert's report

专家意见 expert's opinion

专业标准 specialized standard

专用自动小交换机 private automatic branch exchange

转变温度 transition temperature

转换装置 converter [kən'vɜːtə]

转子流量计 variable area flowmeter

转子流量转换器 variable area flow converter

装管 tubing

装料与卸料 fuel loading and unloading

装卸料机 refueling machine

锥形封头 conical head [ˈkɒnɪk(ə)l]

锥形过渡端 transition cone

子体产物 daughter product

子通道 sub-channel [ˈtʃæn(ə)l]

子项 sub-item

子钟 secondary clock

自动电话系统 normal telephone system

自立式包壳 free-standing cladding

自屏蔽 self-shielding

自屏蔽因子 self-shielding factor

自然地理学，地文学 physiography [ˌfɪzɪˈɒgrəfɪ]

自然通风 natural draft [ˈnætʃ(ə)r(ə)l] [drɑːft]

自然循环 natural circulation

字母数字键盘 alphanumeric keyboard [ˌælfənjuːˈmerɪk]

总峰值因子 total peaking factor [ˈpiːkɪŋ]

总功率峰值因子 total power peaking factor

总焓升因子 total enthalpy rise factor [ˈenθ(ə)lpɪ]

总配线架 main distribution frame [dɪstrɪˈbjuːʃ(ə)n]

总扬程 total head

纵（焊）缝 longitudinal seam [ˌlɒn(d)ʒɪˈtjuːdɪn(ə)l]

纵剖面 longitudinal section

阻火件 fire stop

阻火器 flame arrester

阻抗 impedance [ɪmˈpiːdns]

阻力曲线 resistance curve [rɪˈzɪst(ə)ns]

阻力塞组件 thimble plug assembly [ˈθɪmb(ə)l]

阻尼器 snubber; restraint [snʌb]

组件 module [ˈmɒdjuːl]

组选择器 group selector [sɪˈlektə(r)]

组织独立性 organization freedom

组织机构 organization structure [ˈstrʌktʃə]

最大可信事故 maximum credible accident (MCA)

最大空气浓度 maximum air concentration [kɒns(ə)nˈtreɪʃ(ə)n]

最大日供应量 maximum daily output

最大下泄流 maximum letdown flow

最大允许冷却速率 maximum allowable cooldown rate

最小使用压头 minimum service head

最终安全分析报告 final safety analysis report (FSAR) [əˈnælɪsɪs]

最终热阱 ultimate heat sink [ˈʌltɪmət]

最终热阱冷却塔储水池系统 ultimate heat sink cooling tower and water storage tank system

最终验收 final acceptance

最终验收报告 final acceptance report

水 能 发 电

“O”形环 O-ring
CO_2 保护焊 CO_2 shielding welding
CT 异常保护 CT abnormal protection
C 形夹“C” clamp
I 形坡口对边焊 square butt welding
O 型密封圈 O ring protection
X 射线 X-ray

A

安装高程 setting elevation [ˌelɪˈveɪʃ(ə)n]
安装基准圆 installation fundamental circle [fʌndəˈment(ə)l]

B

巴氏合金 Babbit metal
白点 fish eye; flake
摆动次数 times of moving
摆度 run out; displacement
半径调整 radius alignment [əˈlaɪnm(ə)nt]
半自动焊缝 semi-automatic welding
薄垫片 shim
备用泵 standby pump
备用自投 spare automatic operation
背板 back strip
背缝垫板 backing plate
背面清根 back gouging
背弯试验 root bend test
被覆熔接 clad weld
鼻端导叶 nose vane
比能 specific energy [spəˈsɪfɪk]
变压比 voltage ratio
标称尺寸 nominal dimension [dɪˈmenʃ(ə)n]
标准压力试验台 standard pressure testing table
表计压力（简称为压力） gauge pressure [geɪdʒ]
表面裂纹 surface crack
并行操作 parallel operation
铂电阻 Pt resistance (platinum)
补偿器（节） compensator
补焊 repair welding
补气阀 gulp valve
补气试验 air admission test
补气系统 air admission system
不可调式水力机械 non-regulated hydraulic machinery
不平衡 unbalance
不相容性 incompatibility
不锈钢焊接 stainless steel welding
不锈钢焊条 stainless steel welding

electrode

C

擦痕 scratch; stria; striation
操作拐臂 operation lever
操作规程 operation sequence control ['si:kw(ə)ns]
操作机构 operation mechanism
操作及维修手册 operation and maintenance manual
操作架 crosshead
操作流程图 operation flowchart
操作系统 operating system
操作须知 operation instruction
操作油管 operating oil pipe
槽衬纸 slot liner
槽底垫条 ground strip
测速装置 speed sensor
测圆架 roundness control device
测圆架中心柱 roundness measuring device pedestal ['pedɪst(ə)l]
测振测摆系统 vibration and run out measuring system
层间垫条 intermediate strip [ˌɪntə'mi:dɪət]
叉管 branch pipe
柴油发电机 diesel generator ['di:z(ə)l]
常闭触点 constant close contact
常规试验 conventional test; routine test [kən'venʃ(ə)n(ə)l]
常规维修 routine maintenance
常规作业 routine work
常开触点 constant open contact
厂房排水廊道 plant drainage gallery ['dreɪnɪdʒ]
厂用 station service
厂用电 plant service; station service power; plant electrical consumption
超声波探伤 ultrasonic testing; ultrasonic examination; ultrasonic inspection
车间 working grounding
车间装配 workshop assembly
衬垫 pack
衬套 spacer
持续电流 continuance current [kən'tɪnjʊəns]
齿盘测速 gear speed detector
冲击 impact
冲击式水轮机 impulse turbine
冲水平压管 pressure balance water pipe
充水 water filling
充油 oil filling
初生空化系数 incipient cavitation factor; cavitation inception factor; initial cavitation factor [ɪn'sɪpɪənt] [ˌkævɪ'teɪʃ(ə)n]
初始压力 initial pressure
初始转速 initial speed
储气罐 air receiver
穿墙布电线 knob-and-tube wiring
锤击 peening

瓷绝缘子 porcelain insulator [ˈpɔːs(ə)lɪn]
磁场强度 magnetic field density
磁场绕组 magnetic field winding
磁轭 magnet rim/yoke
磁轭补偿 rim compensating
磁轭堆片 rim piling
磁轭隔板 rim spacer
磁轭挂钩 rim lip
磁轭加垫 shrinking and shimming
磁轭螺杆 rim stud
磁轭螺杆扭紧 rim stud torque [tɔːk]
磁轭螺母 rim stud nut
磁轭片 rim plate
磁轭收缩 rim shrinking
磁轭调整 rim alignment [əˈlaɪnm(ə)nt]
磁粉探伤 magnetic particle inspection; magnetic particle testing
磁化 magnetization [ˌmægnɪtaɪˈzeɪʃən]
磁化电流 magnetizing current
磁化试验 magnetizing test
磁极 magnetic pole; pole
磁极挂装 pole installation
磁极环 pole ring
磁极连接 pole connection
磁极强度 pole strength
磁极绕组 pole winding
磁开关 magnetic switch
磁力钻 drilling machine with magnetic base; magnetic base drilling machine
磁通量 magnetic flux
凑合余量 fitting allowance
错边 misalignment; dislocation

D

大轴补气器 shaft air admission pipe
大轴中心补气排水管 air admission drainage pipe
单导叶接力器 individual guide vane servomotor [ɪndɪˈvɪdjʊ(ə)l]
单机容量 unit capacity
单级水泵水轮机 single stage pump-turbine
单位飞逸转速 unit runaway speed
单位功率 unit power
单位流量 unit discharge
单位水力矩 unit hydraulic torque
单位水推力 unit hydraulic thrust
单位转速 unit speed
单相接地 single phase earthing
挡风板 air shroud
挡风板环形工字钢 circular “H” steel; circular channel steel
挡风围带 strip
挡油圈 oil baffle ring
挡油裙环 thrust skirt ringI
刀闸开关 knife switch
导电膏 conduct grease
导电率 conductivity
导流板 baffle plate

导水机构 distributor system [dɪˈstrɪbjʊtə]
导水机构中心线 distributor centerline
导通角 firing angle
导向块 guide block
导叶 guide vane; wicket gate
导叶臂 guide vane lever; wicket gate lever
导叶臂杆 guide vane lever
导叶操纵杆 wicket gate actuating rod
导叶操纵环 wicket gate operating ring
导叶端面间隙 wicket gate side gap
导叶端面密封 guide vane end seal
导叶高度 guide vane height
导叶拐臂 gate lever; guide vane arm
导叶关闭接触线 closing edge
导叶过载保护装置 guide vane overload protection device
导叶环 guide vane ring
导叶接力器 guide vane servomotor
导叶开度 guide vane opening; gate-opening
导叶控制阀 guide vane control valve
导叶力特性 guide vane force character
导叶立面间隙 wicket gate closing gap
导叶立面密封 guide vane seal
导叶连杆 guide vane link; wicket gate link
导叶损失 guide blade loss
导叶填料 guide vane packing
导叶调节接力器 wicket gate adjust servomotor [ˈsɜːvəʊˌməʊtə]
导叶限位块 guide vane stop block
导叶叶柄 guide vane stem
导叶止推轴承 guide vane thrust bearing
导叶轴承 guide vane bearing
导叶轴密封 guide vane stem seal
导叶转动轴，导叶叶柄 wicket gate side gap
导叶装置 vane apparatus [ˌæpəˈreɪtəs]
导轴承 guide bearing
灯泡式水轮机 bulb turbine
灯泡体 bulb
灯泡体支柱 bulb support
低电阻阻尼绕组 low resistance damping winding [rɪˈzɪst(ə)ns]
低合金钢 low alloy steel
低水头 low head
低水位运行 low water level operation
低碳钢 low carbon steel
低压测量断面 low pressure measuring section
低压电缆箱 low voltage cable end box
低压电器 low-voltage electric appliance
低压接线端子 low voltage terminal
低压启动过流保护 low voltage overcurrent
低油位 low oil level

低油压 lower oil pressure
低阻抗保护 minimum impedance [ɪmˈpiːd(ə)ns]
底板 base plate
底环 bottom ring；bottom cover
底孔 bottom outlet
底孔导流 bottom outlet diversion
底漆 premier painting；prime painting；primer paint
底座 base
地电极 earth electrode [ɪˈlektrəʊd]
地电势 earth potential [pə(ʊ)ˈtenʃ(ə)l]
地脚螺栓 anchor bolt；barb bolt [ˈæŋkə]
电磁干扰 electromagnetic interference [ɪˌlektrə(ʊ)mægˈnetɪk] [ɪntəˈfɪər(ə)ns]
电磁铁 electromagnet [ɪˌlektrə(ʊ)ˈmægnɪt]
电磁线圈 magnetic coil
电动扳手 dynamoelectric spanner；motor spanner
电抗器 reactor
电缆沟 cable duct
电缆管 cable pipe
电缆盒 cable box
电缆夹 cable clamp；cable clip
电缆架 cable rack；cable stand
电缆廊道 cable gallery
电缆盘 cable tray
电缆切割机 cable cutter
电缆套 cable casing
电缆终端 cable terminal end
电力电缆 power cable
电力电量平衡 load generation balance
电力电容器 power capacitor [kəˈpæsɪtə]
电力馈线 power feeder
电力系统 power system
电量变送器 electric quantity transducer
电流电压特性 current-voltage characteristics [ˌkærəktəˈrɪstɪks]
电流断路器 current breaker (CB)
电流互感器 current transformer (CT)
电流换向器 inverter
电流强度 current intensity [ɪnˈtensɪtɪ]
电气照明 electric lighting
电桥 bridge
电刷 brush
电损耗 electric loss
电压闭锁 voltage blocking
电压变送器 voltage transducer [trænzˈdjuːsə]
电压放大器 voltage amplifier [ˈæmplɪfaɪə]
电压互感器 potential transformer (PT)
电压限制器 voltage limiter
电源变压器 power transformer
电源电路 power circuit
电源电阻 power resistance [rɪˈzɪst(ə)ns]
电源过载保护 over current protection
电源开关 power switch
电源切换 power switching

电晕 corona [kə'rəunə]
电晕损失 corona loss
电站空化系数 plant cavitation factor
电站排水总管 plant dewatering header pipe
电制动 electrical braking
电制动器 electrical brake
电阻温度计传感器 sensor for resistance thermometer [rɪ'zɪst(ə)ns] [θə'mɒmɪtə]
垫块 chock
垫片 shim plate
垫圈 washer; gasket
吊板 lug plate
吊耳 lifting lug
吊耳螺栓 eye bolt
叠片平台 piling platform; stacking platform
蝶形螺母 winding tools and devices
丁字柄套筒扳手 T-wrench
顶盖 head cover; top cover
顶盖测压 head cover pressure measuring
顶盖排水泵坑衬 box for head cover drainage pump
顶盖强迫排水 head cover forced drainage
顶盖强迫排水管 head cover forced drainage pipe
顶盖卸压管连接座 connection box for head cover pressure relief
顶盖卸压管支管 head cover pressure relief pipe
顶盖压力测点 head cover pressure measuring point
顶盖预留补气管 reserved head cover air admission pipe
顶盖自流排水管 head cover gravity drainage pipe
顶起螺钉 jack screw
定期检查 periodic inspection [ˌpɪərɪ'ɒdɪk] [ɪn'spekʃn]
定时限 time limit
定位 locate
定位筋 dovetail bar ['dʌvteɪl]
定位筋托板 dovetail bar strap
定位销 dowel pin; set pin
定子 stator
定子电流 stator current
定子吊装 stator lifting and installing
定子机架基础 stator frame sole plate
定子机座 stator frame
定子机座基础板 stator frame sole plate
定子基础板 stator soleplate ['səulpleɪt]
定子接地保护 stator earth fault
定子绕组 stator winding
定子绕组节距 stator winding pitch
定子上环板 stator upper ring plate
定子调整 stator adjustment
定子铁片厚度 thickness of lamination [ˌlæmɪ'neɪʃən]
定子铁芯 stator core
定子下环板 stator lower ring plate
定子线圈 stator coil
定子一点接地 stator one point earthing

定子罩 stator case

定子中环板 stator intermediate ring plate [ˌɪntəˈmiːdɪət]

定子组焊 stator assembly welding

定子组装 stator assembly

动水 flowing water

动水头 hydrodynamic head [ˌhaɪdrəʊdaɪˈnæmɪk]

动态控制 dynamic mode

动态试验 dynamic test

动作值 active value

端子箱，集线箱，接线盒 terminal box [ˈtɜːmɪn(ə)l]

短路 short-circuit

短路继电器 shortdown relay

短路开关 short-ciruit breaker

短路特性 short-circuit characteristic [kærəktəˈrɪstɪk]

短压条 short back plate

断路器 breaker

断路器失灵保护 circuit breaker failure

多级式蓄能泵 multi-stage storage pump

多级水泵水轮机 multi-stage pump-turbine

E

额定电流 rated current

额定电压 rated voltage

额定流量 rated discharge

额定水头 rated head

额定转速 rated speed

二次电路 secondary circuit [ˈsek(ə)nd(ə)rɪ]

二次绕组 secondary coil

二次线圈 secondary winding

F

发电 power generation

发电工况 electricity generation operating condition

发电机 generator

发电机 CT 断线 generator CT fault

发电机 PT 断线 generator PT fault

发电机保护屏 generator protection panel

发电机不完全差 generator incomplete differential [ɪnkəmˈpliːt]

发电机大轴 generator shaft

发电机低电压 generator low-voltage

发电机电枢 generator armature [ˈɑːmətʃə]

发电机定子 generator stator

发电机定子过负 generator stator overload protection

发电机断流器 generator breaker; generator cut-out

发电机负序过流 negative phase sequence current [ˈsiːkw(ə)ns]

发电机过电压 generator over-voltage
发电机机墩 generator support
发电机机坑 generator pit
发电机机罩 generator cover; generator shield [ʃiːld]
发电机机组 generator unit
发电机空冷器冷凝水管 generator air cooler condensating drainage
发电机库容 power storage
发电机冷凝排水 drainage canal for emptying of coolers
发电机流量 power discharge; power flow
发电机母线 generator bus
发电机频率异常保护 overfrequency/underfrequency protection
发电机升压变压器 generator step-up transformer
发电机失磁 loss of excitation [ˌeksɪˈteɪʃ(ə)n]
发电机室 generator room
发电机损失 generator loss
发电机完全差 generator complete differential [ˌdɪfəˈrenʃ(ə)l]
发电机无功功率 generator reactive power
发电机效率 generator efficiency
发电机有功功率 generator active power
发电机制动设备 generator braking equipment
发电机中性接地设备 generator neutral grounding equipment
发电机转子 generator rotor
发电水库 reservoir for power generation [ˈrezəvwɑː(r)]
发电水头 productive head
法兰，法兰面 flange
法兰密封面 flange sealing surface
法兰座 flange seat
反击式水轮机 reaction turbine
反时限 reversing time limit
反调 reverse regulation
反向滑块 reverse slide block
反应灵敏的 sensitive
反应时间 response time; reactive time
返回值 return value
方形垫圈 square washer
防尘 dust protection
防飞逸保护 runaway protection
防腐 anticorrosion [ˌæntɪkəˈrəʊʒən]
防腐保护 corrosion protection [kəˈrəʊʒ(ə)n]
防护措施 protective measure
防火布 anti-fire cloth; fire resistant cloth [rɪˈzɪstənt]
防锈漆 anti-rust paint
放电 electric discharge; discharge
放电电压 discharge voltage
飞摆 over speed pendulum [ˈpendjʊləm]
飞轮效应 flywheel effect
飞逸 runaway
飞逸工况 runaway speed operating condition
飞逸试验 runaway speed test
飞逸特性曲线 runaway speed curve
非磁性材料 non-magnetic material

非电量变送器 non-electric quantity transducer
非全相保护 non-full phase protection
分半键 split key
分瓣法兰 split flange
分瓣线 split line
分辨率 distinguishing rate; resolution [dɪ'stɪŋgwɪʃɪŋ]
分别调整 separate adjustment
分段关闭 cushioning closing ['kʊʃənɪŋ]
分段关闭管路 cushioning pipe
分降压变压器 step transformer
分界面，接触面，接口 interface
分流管 manifold
分流器 current divider
风铲 air shovel
风动扳手 air driving spanner
风闸 brake
风闸控制板 brake control board
封头曲面 ellipse head [ɪ'lɪps]
峰值 peak value
伏安特性 voltage-current characterisitic
辅助触点 auxiliary contact [ɔːg'zɪlɪərɪ]
辅助设备 auxiliary equipment
负序 negative sequence ['negətɪv] ['siːkw(ə)ns]
负载能力 load capacity
负载试验 load test
附件 accessories
腹板 wearing ring

G

感性 inductance character [ɪn'dʌkt(ə)ns]
感应线圈 induction coil
钢衬 steel liner
钢衬板 steel scaleboard ['skeɪlbɔːd]
高差 elevation difference
高程 elevation
高度 height
高度传感器 position sensor
高空作业 high-fit operation
高频 high frequency
高压侧过流保护 230kV side overcurrent protection
高压侧零序过流保护 230kV side zero-sequence overcurrent protection
高压测量断面 high pressure measuring section
高压冲水管 high pressure water pipe
高压电气设备 high-voltage equipment
高压闪络试验 flash test
高压水泵 high pressure water pump
高压油顶起系统 high pressure oil injecting system
高压油管路 high pressure oil pipeline
高油位 high oil level

隔离 isolation
隔离变压器 isolating transformer
隔离开关 disconnected switch; isolating switch
工厂装配 factory assembly
工地封焊 tight welded
工频耐压
power frequency high-voltage
工作泵 operating pump
工作接地 wiring terminal
公称直径 nominal diameter
[ˈnɒmɪn(ə)l]
功角 power angle
功率表 wattmeter [ˈwɒtmiːtə]
功率控制 power control
功率因数 power factor
功能 function
攻丝 tap
供气管 air supply pipe
钩头螺 hook bolt
固定导叶 guide vane; stay vane
固定止漏环 stationary wearing ring
[ˈsteɪʃ(ə)n(ə)rɪ]
固化剂 hardener
固有应力 inherent stress
[ɪnˈhɪər(ə)nt]
故障 fault
故障录波 wave record for faults
故障排除 clearing of fault
故障维修 breakdown maintenance
[ˈmeɪnt(ə)nəns]
挂装 hanging and installing
拐臂 crank; lever
拐点 inflexion; inflexibility
观察孔 observation hole
[ɒbzəˈveɪʃ(ə)n]
管壁厚度 pipe thickness
管夹 pipe clamp
管架 pipe support
管交叉没有连接点 pipe crossing without connection
贯流式水轮机 tubular turbine; through flow turbine
光敏电阻 light resistor
光频隔离器 optic isolator
过渡过程 transition process
[ˈtrænzɪənt]
过负荷 over load
过激磁保护 over-excitation protection
过励 over excitation
过流板 transition plate
[trænˈzɪʃ(ə)n]
过流保护 over current protection
过流继电器 over current relay
过流面 water passage
过滤精度 fineness
过滤器 filter
过速 over speed
过速飞摆 over speed pendulum
[ˈpendjʊləm]

H

毫伏 multi-voltmeter
合闸，接通（入） switching-in

合闸命令 close order
荷载传感器 load sensor
横差保护 stator interturn fault
蝴蝶阀 butterfly valve
户外作业 outdoor operation/work
环筋 circumferential rib [səˌkʌmfəˈrenʃəl]
环境湿度 ambient humidity [ˈæmbɪənt]
环境温度 ambient temperature
环境压力（或大气压） atmpspheric pressure
环筒 test ring
环筒盖 cover for test ring
环氧胶 resin
环氧树脂 epoxy resin [ɪˈpɒksɪ]
环纵焊缝 circumferential and longitudinal seam [səˌkʌmfəˈrenʃəl] [ˌlɒn(d)ʒɪˈtjuːdɪn(ə)l]
缓冲时间常数（T_d） time constant of damping
缓冲装置 buffer; buffer device
回波 echo
回复机构 regulating device
回水环管 circular return-water pipe
回缩 retraction
回填 backfilling; backfill
回油罐 oil sump tank
汇流管 manifold [ˈmænɪfəʊld]
汇流铜环 circuit ring
汇流铜环中性点 circuit ring neutral point [ˈnjuːtr(ə)l]
混流式（离心式）蓄能泵 centrifugal storage pump; mixed-flow storage pump [ˌsentrɪˈfjuːg(ə)l]
混流式水轮机 Francis turbine; mixed-flow turbine
活动扳手 adjustable spanner; monkey wrench
活动导叶 welding of turbine
活动导叶型线 guide vane profile

J

击穿电压 break down voltage
机电安装 electromechanical installation [ɪˌlektrəʊmɪˈkænɪk(ə)l]
机架 support frame
机壳 housing
机坑进人门 access of pit
机坑里衬 pit liner
机械故障 mechanical breakdown
机组 unit
机组过速 runaway speed
机组振摆 vibration control [vaɪˈbreɪʃ(ə)n]
机座分瓣 stator frame segment [ˈsegm(ə)nt]
积分平均效率 planimetric average efficiency [ˌpleɪnəˈmetrɪc]
基础 foundation
基础板 foundation plate; sole plate
基础板定位销 soleplate pin [ˈsəʊlpleɪt]
基础环 foundation ring; discharge ring
基础螺栓 anchor bolt [ˈæŋkə]

基准点 datum mark
畸变 aberrance [æ'berəns, æ'berənsɪ]
极化指数 polarization index [ˌpəʊləraɪ'zeɪʃən]
极性 polarity
棘轮扳手 ratchet spanner
集电环 slip ring
集电环室 slip ring housing
集装箱 container
加权（算术）平均效率 weighted average head
加权平均水头 wedge plate; wedge
加速时间常数（T_n） time constant of accelerating
假同期 fictive synchronize ['sɪŋkrənaɪz]
架立 erection [ɪ'rekʃ(ə)n]
架设 set up
间距 spacing
监控电路 supervisory circuit ['sju:pəˌvaɪzərɪ]
监视终端 monitor terminal ['tɜːmɪn(ə)l]
检查法兰面 flange checking
检漏 leakage inspecting; leakage checking
检修 maintenance ['meɪnt(ə)nəns]
检修密封 maintenance seal
检修密封供气 maintenance air supply
检修小车 inspection trolley [ɪn'spekʃn]
减压安全阀，降压阀 pressure relief valve; pressure reducing valve; relief valve
剪断销 shear pin
键 key
键槽 key slot
降低导电率 reduce conductivity
铰轴 hinged shaft
铰座 hinged support
脚手架 scaffold ['skæfəʊld]
接触剂 contact agent
接触器 contact; contactor
接地 grouding; earthing
接地变 earthing transformer
接地电阻 grouding resistance
接地开关 earthing switch
接地体 ground connector
接地线 ground wire
接力器 servomotor ['sɜːvəʊˌməʊtə]
接力器基础板 servomotor foundation plate
接力器坑衬 box for servomotor
接力器漏油总管 servomotor oil leakage pipe
接线板，接线盒 terminal block
接线端子 wiring drawing; wiring scheme
接线图 Winter Kennedy measuring point
节流阀 throttle; throttle valve
节流孔 throttling hole
节流片 throttle plate
节圆直径 pitch diameter
截流板 throttle plate
截止阀 shut-off valve
解列 unit disconnect
介质损耗 medium loss
紧急安全阀 pop safety valve

紧急事故停机 emergency shut down
进风槽 air passage
进口阀 intake valve
进人门 mandoor；manhole
进水 water in；water inlet
进水阀 inlet valve
进水环管 circular inlet-water pipe
进水排水阀门 valve for filling and drainage of the water
进油 oil in；oil inlet
进油阀 oil inlet valve
精度 precision；accuracy
精确定位 position accurately
径向力 radial force；lateral force
径向推力轴承 radial thrust bearing
净水头 net head
静平衡余量 balancing allowance
静水 still water
静态试验 static test
静特性 static characteristic
静压压力测量 static pressure measuring
静载试验 static loading test
镜板 rotating ring
局部变形 local distortion；local deformation
局部放电 partial discharge
局部应力 local stress
局部装置 partial discharge device
拒动 refuse to active
距离保护 distance protection
绝对压力 absolute pressure
绝缘板 insulation plate
绝缘等级 insulation grade
绝缘电阻 insulating resistance
绝缘漆 insulation paint
绝缘试验 insulation test
绝缘涂层 insulation coating
绝缘鞋 insulation shoes
绝缘子 insulator
均流 average current
均压 average voltage
均压管 balance pipe
菌形阀 mushroom valve

K

卡环 clamping ring
卡头，夹板 clamp
开度 opening
开度控制 opening control；opening limit
开放式系统 open system
开环 open circuit
开机流程 unit starting flow
开口扳，开口扳手 open spanner
开口销 split pin
抗老化性 aging resistance [rɪˈzɪst(ə)ns]
抗磨板 facing plates；wear plates
抗磨环，止漏环 wearing ring
抗磨块 wearing block
抗磨涂层 erosion resistant coating [ɪˈrəʊʒ(ə)n]
抗蚀 anti-corrosion

抗弯强度 bending strength
可锁定的三通球阀 lockable 3-way ball valve
可调式水力机械 regulated hydraulic machinery
空化 cavitation [ˌkævɪˈteɪʃ(ə)n]
空化试验 cavitation test
空化裕量 cavitation margin
空冷器 air cooler
空冷器打压 pressure test for air cooler
空冷器环管 air cooler ring pipe
空气过滤器 air filter
空气稳压罐 equalizing air tank only for blow down
空气压力罐 equalizing air tank only for blaking
空蚀 cavitation [ˌkævɪˈteɪʃ(ə)n]
空水冷却器水管 cooling water piping for heat exchanger
空压机 air compressor
空压气（压缩空气） compressed air
空载工况 no-load operating condition
空载试验 no-load test
空载特性 no-load characteristic
空载状态 no-load status
空转状态 spinning status
控制电缆 control cable
控制电路 control circuit
控制方式 control method
控制环 regulating ring；operating ring
控制环加工 operating ring machining
跨步电压 step voltage
快速接地开关 high speed earthing switch
扩散管 diffuser

L

拉杆 lifting link
拉紧器 turnbuckle
冷却水设备室 cooling water equipment room
冷却液 cooling liquid [ˈlɪkwɪd]
力特性 force character
力特性试验 force characteristic test
立筋 vertical reinforcement [riːɪnˈfɔːsm(ə)nt]
立式、卧式和倾斜式机组 vertical，horizontal and inclined unit
励磁变过负荷 excitation transformer overload protection
励磁变过流 excitation transformer overcurrent protection
励磁变速断 excitation transformer instaneous overcurrent
励磁系统故障 excitation system fault
励磁线圈 energizing coil
励磁引线 excitation lead；slip ring leads
连杆 link
连接 connect；connection

连接法兰 coupling flange

连接杆 connection shaft; connection rod

连锁装置 interlocker; interlocking

联轴法兰 shaft coupling flange

联轴螺栓 shaft coupling bolt

临界空化系数 critical cavitation factor [ˈkrɪtɪk(ə)l]

漏电 electric leakage

漏电保护 leakage protection

漏气检验器 gas leak detector

漏水 water leakage

漏油 oil leakage

轮叶 impeller blade; impeller vane

M

埋入部件 embedded component

毛水头 gross head

梅花扳手 hexagon spanner; quincuncial spanner [ˈheksəg(ə)n] [kwɪnˈkʌnʃəl]

迷宫环 labyrith ring

密封圈 O-ring

密封性试验 leak test

灭磁 excitation off [ˌeksɪˈteɪʃ(ə)n]

灭磁开关 excitation switch

灭火器 extinguisher; fire extinguisher

铭牌 name plate

模型（机） model

模型试验 model test

摩擦装置 friction device

母联充电保护 busbar coulper breaker charge protection

母联非全相保护 busbar coupler incomplete phase protection

母联过流保护 busbar coupler breaker overcurrent protection

母联继电器保护柜 busbar coupler breaker protection panel

母线 busbar

母线保护屏 busbar protection panel

母线差动 busbar differential [ˌdɪfəˈrenʃ(ə)l]

母线失灵保护 busbar coupler breaker failure

N

耐磨性 abrasion performance [əˈbreɪʒ(ə)n]

耐压（水压）试验 pressure test

耐压试验 high voltage test; withstand voltage test

内导水环 inner guide ring

内顶盖（支持盖） inner head cover; inner top cover

内螺纹 female screw thread

内支撑 inner support

内止口 female end

逆功率 reverse power; reversing power

P

排水管 drainage pipe
排油孔 oil drain hole
盘形阀 mushroom valve; hollow-cone valve; howell-Bunger valve
旁路母线 interbus
旁通阀 by-pass valve
配电盘 switchboard
配水阀 distributing valve
配重 ballast
喷针 needle
喷针接力器 needle servomotor
喷嘴 nozzle
喷嘴支管 bifurcation [baɪfə'keɪʃ(ə)n]
偏心，偏心率，偏心度 eccentricity
偏心凸轮 eccentric cam
频率限制器 frequency limiter
平板蝶阀 biplane butterfly valve; through flow butterfly valve

Q

起吊到位 lift into position
起吊孔 lifting hole
起吊梁 lifting beam
起动器 starter
起停机定子接地保 stator earth fault protection of startup condition
气密性试验 air tight test; tightness test
气体绝缘开关站 gas insulated switchyard
汽化压力 vapour pressure ['veɪpə]
千瓦小时，度（电） kilowatt-hour
潜水泵 submersible pump [səb'mɜːsɪb(ə)l]
欠励 under excitation
切向键 transversal key
球阀 ball valve; globe valve; spherical valve ['sferɪk(ə)l]
全贯流式水轮机 straight flow turbine; rim-generator unit

R

热继电器 thermal relay
热交换器 heat exchanger
热稳定性 thermal stability
热应力 thermal stress
人机对话 man-machine dialogue ['daɪəlɒg]
人机交换 man-machine communicate [kə'mjuːnɪkeɪt]
软管连接 hose connection
软接头 flexible connection
润滑管路 lubrication circuit [ˌluːbrɪ'keɪʃən]

S

三通 “T” connection; junction (tee); 3-way

三通阀，三向阀 three way valve

三相电路 three phase circuit

三相球阀（可锁定的） 3 way ball valve (lockable)

三相绕线式电动机 3-phase wind asynchronous motor [ə'sɪŋkrənəs]

三相四线制 three-phase four-wire system

上层线棒 upper bar

上导瓦 upper guide bearing segment

上导瓦温 upper guide bearing metal temperature

上导油槽 upper guide bearing oil pot

上导油冷器 upper guide bearing oil cooler

上导油温 upper guide bearing oil temperature

上导轴承 upper guide bearing

上导轴承间隙 upper guide bearing clearance

上端轴 upper shaft

上盖板基础 sole plate for top cover; top cover sole plate

上冠 crown

上环板 upper ring plate

上机架盖 upper bracket cover

上机架焊接 upper bracket welding

上机架基础板 upper bracket sole-plate

上机架拼装 upper bracket assembly

上机架支臂 upper bracket spider

上机架支柱 upper bracket pedestal

上机架中心体 upper bracket hub

上压指 upper pressure finger

上游排水沟 upstream drainage trench

上止漏环冷却供水 upper wearing ring cooling water supply

上止漏环冷却水供水管 upper wearing ring cooling water pipe

上轴向推力轴承 upper axial thrust bearing

设计水头 design head

射流入射角 jet inclined angle [ɪn'klaɪnd]

射流椭圆 ellipsed of inclined jet (when reaching the wheel vane)

射流直径 jet diameter

射流直径比 jet ratio

伸缩节 expansion joint [ɪk'spænʃ(ə)n]

失磁 de-excitation

失灵保护 breaker fail protection

湿度 humidity

时间常数 time constant

实时对多任务系统 real-time multitasking system

实时控制 real-time control

事故阀 emergency valve

事故停机流程 unit emergency shut down flow
事故照明 emergency lighting
受油器 oil head
输水阀 delivery valve
竖井贯流式水轮机 pit turbine
甩负荷 load rejection
甩负荷试验 load rejection test
双击式水轮 cross-flow turbine
双向阀 two way valve
水泵 water pump
水泵电机 water pump motor
水泵水轮机 reversible turbine; pump-turbine
水泵水轮机全特性 complete characteristics of pump-turbine
水锤 water hammer
水导油槽 turbine guide bearing oil pot
水导轴承 turbine guide bearing
水导轴承供油管路 turbine guide bearing oil supply pipe
水导轴承间隙 turbine guide bearing clearance
水导轴承冷油管 guide bearing oil pipe from cooler
水导轴承热油管 guide bearing oil pipe to cooler
水导轴承油泵坑衬 box for guide bearing oil pump
水导轴瓦 turbine guide bearing pad
水斗 bucket
水斗式水轮机 Pelton turbine
水封 seal; rubber seal
水击 water hammer
水机仪表柜 turbine instrument cubicle [ˈɪnstrʊm(ə)nt]
水接头 water connection; hydraulic connection
水接头 PT 探伤试验 water connection PT test
水接头安装 water connection installation
水接头测量 water connection measurement
水接头打压试验 water connection pressure test
水接头密封圈 water connection seal ring
水接头配置 water connection fitting
水接头铜焊 water connection brazing
水控阀 hydraulic control valve
水冷却器 water chiller
水力测量 hydraulic measuring
水力机械 hydraulic machinery
水力旋流器 hydrocyclone [ˌhaɪdrəuˈsaɪkləun]
水力旋流器排水管 hydrocyclone drainage pipe
水流量计 water flow meter
水轮发电机组安装 turbo-generator set installation
水轮发电机 hydraulic generator; hydrogenerator
水轮发电机基本参数 basic parameters of hydrogenerator
水轮发电机制动系统 braking system of hydrogenerator

水轮发电机轴 hydrogenerator shaft

水轮发电机组 hydrogenerator unit; turbine generator set

水轮发电机组保护 hydraulic generator protection

水轮发电机组运行 hydraulic generator operation

水轮发电机组自动控制系统 automatic control of hydrogenerator set

水轮机 hydraulic turbine

水轮机安装 erection of water turbine [ɪ'rekʃ(ə)n]

水轮机安装高程 turbine sitting

水轮机保护阀 turbine guard valve

水轮机比转速 specific speed of turbine

水轮机补气 turbine venting; turbine air supply

水轮机参数 parameter of hydraulic; turbine parameter

水轮机层仪表盘 instrument panel of turbine floor

水轮机出口测量断面 outlet measuring section of turbine

水轮机大轴 turbine shaft

水轮机导水机构 guide vane of hydraulic

水轮机顶盖 turbine cap

水轮机端子控制箱 turbine terminal cubicle

水轮机额定输出功率 rated output power of turbine

水轮机飞逸转速 runaway speed of turbine

水轮机辅助部件 turbine auxiliary components

水轮机辅助设备 auxiliary equipment of hydraulic turbine

水轮机功率试验 turbine output test

水轮机过渡过程 transition process of water turbine

水轮机过流部件 water passing parts of turbine

水轮机焊接 weighted (arithmetic) average efficiency

水轮机机墩外形 turbine pit contour

水轮机机坑 turbine pit

水轮机机坑里衬 turbine pit liner

水轮机机械效率 mechanical efficiency of turbine

水轮机进口测量断面 inlet measuring section of turbine

水轮机进水阀 turbine inlet valve

水轮机进水弯管 turbine inlet bend

水轮机净出力 net turbine power

水轮机空化系数 cavitation factor of turbine; cavitation coefficient of turbine

水轮机空蚀 cavitation of hydraulic turbine [ˌkævɪ'teɪʃ(ə)n]

水轮机空载流量 no-load discharge of turbine

水轮机空转 turbine idling

水轮机控制盘 turbine control panel

水轮机廊道 turbine gallery

水轮机流量 turbine discharge; turbine flow rate

水轮机磨损 abrasion of water turbine [ə'breɪʒ(ə)n]
水轮机旁通管 turbine bypass
水轮机气蚀特性 cavitation performance of hydraulic turbine [ˌkævɪ'teɪʃ(ə)n]
水轮机设计 design of hydraulic turbine
水轮机设计参数 turbine design data
水轮机事故 accident of turbine
水轮机输出功率 turbine output power
水轮机输入功率 turbine input power
水轮机水力效率 hydraulic efficiency of turbine
水轮机特性 characteristic of hydraulic turbine; turbine characteristics [kærəktə'rɪstɪk]
水轮机特性曲线 performance diagram of hydraulic turbine
水轮机调节系统 water turbine regualtion system
水轮机通气管 turbine air vent pipe
水轮机尾水管 turbine draught tube
水轮机效率 turbine efficiency
水轮机选型 turbine type selection
水轮机叶片 turbine blade
水轮机闸门 turbine gate
水轮机罩 turbine case
水轮机振动 vibration of hydraulic turbine
水轮机轴 turbine shaft
水轮机轴密封套 turbine shaft gland
水轮机主轴 turbine major axis
水轮机转轮 turbine runner
水轮机自动起转系统 automatic turbine run up system
水轮机最大瞬态压力 maximum momentary pressure of turbine
水轮机最大瞬态转速 maximum momentary overspeed of turbine
水轮机座环 turbine stay ring
水平度 level; levelness
水头测量 head measurement
水推力 hydraulic thrust
水压 water pressure
水压试验 water pressure test; hydraulic test
水压作用条件下 under water head condition
顺时针 clockwise
顺序 sequence ['siːkw(ə)ns]
瞬间故障 instant fault
瞬时电流 transient current ['trænzɪənt]
瞬时电压 transient voltage
瞬态压力变化率 momentary pressure variation ratio
瞬态转速变化率 momentary speed variation ratio
死区 dead band; dead zone
死区保护 busbar dead-zone protection
速度比能 velocity energy
速度水头 velocity head
锁定胶剂 locking glue

T

弹性变形 elastic deformation [ˌdiːfɔːˈmeɪʃ(ə)n]
弹性材料 elastic material
弹性垫层 elastic layer
弹性垫圈 spring washer
弹性极限 elastic limit
弹性轴承 elastic bearing
特性试验 characteristic test
调节保证 regulating guarantee
调节模式 regulating mode
调平垫板 support steel plate for adjusting balance
调试 commissioning [kəˈmɪʃənɪŋ]
调试试验 commissioning test
调速环 regulating ring
调速器 governor
调速器柜 governor cubicle；governor cabinet
调速轴 regulating shaft
调相 phase modulation [ˌmɒdjʊˈleɪʃən]
调相压气管路 compressed air piping for blow down
调压器 voltage regulator
跳闸命令 trip order
铁损 core lost
铁芯 core
铁芯片 core lamination [ˌlæmɪˈneɪʃən]
停机油位 oil level at standstill
通风 air ventilation [ˌventɪˈleɪʃ(ə)n]
通风保护 start cooling fans protection
通风槽片 duct lamination
通气孔 vent hole
通信，通讯 communication
同步 synchronize；synchronous
同期 synchronize
同期装置 synchronize device [əˈsɪŋkrənəs]
同心度 concentricity [kɒnsenˈtrɪsɪtɪ]
同心度偏差 concentricity deviation [diːvɪˈeɪʃ(ə)n]
同轴度 axiality [ˌæksɪˈælətɪ]
透平油 turbine oil
推拉杆 push and pull rod；connecting rod
推力头 thrust ring
推力瓦 thrust bearing pads
推力瓦弹簧 thrust bearing spring
推力轴承 thrust bearing
推力轴承高压油系统 high pressure system for lower combined bearing
脱扣 trip
椭圆度 ovality [əuˈvælɪtɪ]

W

瓦架 base ring
瓦块托架 segments bracket

瓦斯 gas
瓦斯继电器 gas relay
外导水环 outer guide ring
外径千分尺 outer micrometer；outside micrometer
外围设备 peripheral equipment [pə'rɪf(ə)r(ə)l]
弯曲半径 bend radius
网络 network
尾水层排水总管 dewatering header pipe
尾水管 draft tube
尾水管出口压力测量仪表盘 draft tube outlet pressure measuring panel
尾水管进口压力 draft tube inlet pressure
尾水管进口压力测量点 draft tube inlet pressure measuring point
尾水管进口压力测量仪表盘 draft tube inlet pressure measuring panel
尾水管进人门 draft tube mandoor
尾水管扩散段 draft tube outlet part
尾水管扩散段侧墙 draft tube diffuser side wall
尾水管扩散段出口压力测点 draft tube diffuser outlet pressure measuring point
尾水管冷却水排水管 draft tube cooling water drainage pipe
尾水管冷却水取水管 draft tube cooling water intake pipe
尾水管里衬 draft tube liner
尾水管调相压水液位监测 draft tube blow down level monitoring
尾水管压力脉动 draft tube pressure pulsation [pʌl'seɪʃən]
尾水管支墩 draft tube pier
尾水管肘管 draft tube elbow
尾水管锥管 draft tube cone
尾水门 draft tube gate
尾水排水管 draft tube drainage pipe
尾水肘管 draft tube elbow
尾水肘管压力脉动测量点 draft tube elbow pressure pulsation measuring point [pʌl'seɪʃən]
尾水锥管压力脉动测点 draft tube cone pressure pulsation measuring point
位移传感器 displacement sensor
位置比能 potential energy
位置水头 potential head
温度变送器 temperature transmitter
温度测量仪器 temperature measuring instrument
温度接触设备传感器 sensor for temperature contact instrument
温度湿度记录表 record for temperature and humidity
温升试验 heat running test
纹波 wave
涡流 vortex
蜗壳 spiral case
蜗壳包角 nose angle
蜗壳鼻端 spiral case nose
蜗壳基础 foundation of spiral case
蜗壳进口压力测点 spiral case inlet

pressure measuring point
蜗壳进口压力脉动测点 spiral case inlet pressure pulsation measuring point
蜗壳进人门 spiral case mandoor
蜗壳流量测量点 wing nut
蜗壳排水管 spiral case drainage pipe
蜗壳瓦片 spiral case segment
蜗壳压差测量 spiral case Winter-Kennedy
蜗室 spiral housing
无功 reactive power
无水试验 dry test
无损检验 nondestructive test
无损探伤 nondestructive inspection
误操作 mis-operation
误差 difference
误动作 malfunction

X

吸出高度 static suction of turbine
吸入高度 static suction of storage pump
吸水管 suction tube
下层线棒 lower bar
下导瓦 lower guide bearing segment
下导瓦间隙 lower guide bearing clearance
下导瓦调整丝顶 lower guide bearing jacking bolt
下导瓦温 lower guide bearing metal temperature
下导轴承排油雾 oil steam exhaust piping for lower combined bearing
下导轴承油管 oil piping for lower combined bearing
下导轴领 lower guide bearing journal
下环 band
下环板 lower ring plate
下机架挡油管 oil well tube
下机架焊接 lower bracket welding
下机架基础板 lower bracket sole plate
下机架油槽 lower bracket oil pot
下机架油漆 lower bracket oil painting
下机架支臂 lower bracket spider
下机架中心体 lower bracket hub
下机架组装 lower bracket assembly
下线材料 winding documentation
下线工装 winding material
下线环境 wicket gate stem
下线资料 winding condition
下压指 lower pressure finger
下止漏环冷却水供水管 lower wearing ring cooling water pipe
下轴径向推力轴承 lower axial thrust bearing
现场指令 site instruction
现地控制 local control
现地控制单元 local control unit

现地控制柜 local control cubicle
限位开关 limit switch
线棒衬纸刷胶 apply resin for the bar lining paper
线棒抽样检查 random check for the bar; spot check for the bar
线棒电接头离铁芯的距离 distance from end of bar to stator core
线棒防晕段 anti-corona section of the bar [kə'rəunə]
线棒间斜边距离 overhang distance between bars
线棒绝缘 bar insulation
线棒开箱 box opening for the bar
线棒耐压试验 bar high voltage test; HIPOT
线棒水接头支撑 water connection support for bars
线棒水接头支撑铜焊 water connection support for brazing
线间短路 line-to-line short circuit
相对效率 relative efficiency
相似工况 similar operating condition
相位角 phase angle
相序，相位 phase sequence ['siːkw(ə)ns]
相引线 phase terminal
橡皮垫 rubber shim
消除应力处理 stress relived
消防 fire protection
消防控制柜 fire fighting control cubicle
销钉 pin
校准试块 calibration block [kælɪ'breɪʃ(ə)n]
效率试验 efficiency test
楔键 taper key
楔子板 web plate
协联工况 combined condition
协联装置 combination device [kɒmbɪ'neɪʃ(ə)n]
斜击式水轮机 inclined jet turbine
斜流定桨式水轮机 fixed blade of Deriaz turbine
斜流式水轮机 diagonal turbine
斜流式蓄能泵 diagonal storage pump
斜流转桨式水轮机 Deriaz turbine
泄漏监控 leakage monitor
泄水锥 runner cone
卸负荷 load rejection
星形 star connection
行程 stroke
行程开关 overtravel-limit switch
性能试验 performance test [pə'fɔːm(ə)ns]
修改参数 parameter modify
蓄电池（组） storage battery
蓄电池充电器 battery charger
蓄电池容量 capacity of storage battery
蓄能泵 storage pump
蓄能泵［水泵水轮机的泵工况］的流量比转速 discharge specific speed of storage pump
蓄能泵［水泵水轮机的泵工况］的最大瞬态反向转速 momentary counterrotation speed of storage pump [pump-condition of pump-

turbine]

蓄能泵［水泵水轮机的泵工况］反向飞逸转速 reverse runaway speed of storage pump

蓄能泵出口测量断面 outlet measuring section of storage pump

蓄能泵的输出功率 storage pump output power

蓄能泵的输入功率 storage pump input power

蓄能泵机械效率 mechanical efficiency of storage pump

蓄能泵进口测量断面 inlet measuring section of storage pump

蓄能泵空化系数 cavitation factor of storage pump; cavitation coefficient of storage pump

蓄能泵空化余量（NPSH）（蓄能泵净吸上扬程） net positive suction head of storage pump

蓄能泵零流量功率 no-discharge input power of storage pump

蓄能泵零流量扬程 no-discharge head of storage pump

蓄能泵流量 storage pump discharge; storage pump flow rate

蓄能泵水力效率 hydraulic efficiency of storage pump

蓄能泵吸入扬程损失 suction head loss of storage pump

蓄能泵扬程 storage pump head

蓄能泵最大（最小）流量 maximum（minmum）storage pump discharge

蓄能泵最大（最小）扬程 maximum（minimum）head of storage pump

蓄能泵最大输入功率 rated input power of storage pump

蓄能泵最小输入功率 minimum input power of storage pump

蓄能器 accumulator [əˈkjuːmjʊleɪtə]

悬式绝缘子 suspended insulator

旋流器法兰 flange of hydrocyclone [ˌhaɪdrəuˈsaɪkləun]

旋转方向 direction of rotation

Y

压板 compressing plate

压差 pressure difference

压差传感器 differential pressure transducer

压紧行程 over stroke

压紧螺杆 pressing bolt; punching bolt

压力比能 pressure energy

压力变送器 pressure transducer

压力传感器 pressure sensor; pressure transducer

压力计 pressure gauge

压力开关 pressure switch

压力控制装置 pressure control device

（水轮机）压力脉动（turbine）pressure fluctuation

压力脉动测盒 pressure pulsation

[pʌl'seɪʃən]
压力脉动测量 pressure pulsation measuring
压力脉动测量座 pressure pulsation measuring seat
压力脉动试验 pressure fluctuation test
压力试验 pressure test
压力释放 pressure release
压力水头 pressure head
压力油罐 oil pressure tank; oil pressure vessel
压缩空气 compressed air
压指上翘量 finger inclination
延时 delay
延时扫描 delayed sweep
验收试验 acceptance test
叶轮 impeller
叶轮后盖 impeller back shroud
叶轮前盖 impeller front shroud
叶轮输出功率 output power of impeller
叶轮输入功率 input power of impeller
叶片 blade
叶片安放角 blade angle
叶片进（出）水边 leading（trailing）edge
叶片开口 blade opening
叶片力特性 blade force character
叶片倾角 blade tilt angle
叶片枢轴 runner blade trunnion ['trʌnjən]
叶片正（背）面 pressure（suction）side of blade
叶片轴颈圆角 radius of blade-stem
叶片转角 blade rotating angle
液位计 hydraylic level indicator
液压千斤顶 hydraulic jack
液压系统 hydraulic system
一次电路 primary circuit
仪表盘 gauge panel
移相 shift phase
异步 asynchronous [ə'sɪŋkrənəs]
溢流环盖板 overflow ring covering cap
引出盒（箱），接线盒 outlet box
引出线，电源插座 outlet
引水室（turbine）flume
应力松弛 stress relaxation
永态转差系数（b_p）permanent speed droop ['pɜːm(ə)nənt]
油泵 oil pump
油槽盖板 oil pot cover
油缸 cylinder; oil cylinder
油管 oil pipe
油过滤器 oil filter; oil filtering machine
油混水 water in oil
油混水探测器 water in oil detector
油或水管路 oil or water pipeline
油冷却器 oil cooler
油盆 oil pan
油位极低 oil level extreme low [ɪk'striːm]
油位计 oil level indicator
油位信号指示器 oil level switch
油雾吸收器 oil mist absorber [əb'sɔːbə]
油箱 oil tank
油压 oil pressure

油样 oil sample
油枕 conservator
[ˈkɒnsəˌveɪtə，kənˈsɜːvətə]
有效水头 effective head
有效值 virtual value
有载调压保护 load-ratio voltage protection
预处理 pretreatment
预留，预置 pre-set
预留顶盖强迫补气管 reserved head cover air admission pipe
预埋板 embedded plate
预埋件 embedded part
预热 pre-heat；pre-heating
预装 pre-assembly
原型（机） proto type
原型水轮机 prototype turbine
圆度调整 roundness adjustment
圆筒阀 cylindrical valve；ring gate
远方控制 remote control
远方诊断 remote diagnostics
越限 over-limit
云母板 mica plate
云母带 mica tape
允许不平衡重量 remaining rest unbalance
允许偏差 allowable variation；tolerance difference
允许相对残余不平衡力矩（转轮） relative allowable remaining unbalance
运行工况 operation condition
运行机构 travelling mechanism
运行情况 operating condition
运行油位 oil level in operation
（水轮机）运转特性曲线（turbine） performance curve
运作间隙 operating gap

Z

载荷分配 load distribution
暂态转差系数（b_t） temporary speed droop
噪声 noise
兆欧 megohm [ˈmegəʊm]
照明配电箱 lighting distribution box (board)
照明设备 lighting equipment
折向器 jet deflector
针阀 needle valve
整流 rectify
正极 positive
支撑板 support plate；retaining plate
支撑件 support
支撑柱 round bar for gauge panel
支墩安装 pedestal installation
[ˈpedɪst(ə)l]
支墩基础板 soleplate for the pedestal
支墩水平度测量 pedestal leveling check
支墩调整 pedestal adjustment
支柱绝缘 column-type insulator
直空破坏阀 vacuum break valve
直流电（DC） direct current

直流电极接负 direct current electrode negative (DCEN)
直流电极接正 direct current electorde positive (DCEP)
直流电阻 DC resistance
直流电阻测试仪 DC resistance tester
直流电阻试验 DC resistance checking
直线度 straightness accuracy
止动块 stop lock
止回阀 check valve
止推块 thrust block
制动阀 brake valve
制动及顶起系统控制柜 control and hydraulic unit for braking and lifting
制动喷嘴 brake nozzle
制动气罐 air tank for braking system
制动器 brake
制动器基础 sole plate for brake cylinder
中间位置 medium position
中间压紧 intermediate pressing
中间值 medium value
中心体安装支墩 assembly pedestal for the hub [ˈpedɪst(ə)l]
中心体转运 hub transportation [trænspɔː] [trænspɔːˈteɪʃ(ə)n]
中心线 center line
中性点 neutral point
中性点引线 neutral terminal
中压侧过流保护 33kV side overcurrent protection
中压侧零序过流保护 33kV side zero-sequence overcurrent protection
轴衬；轴套 bushing
轴承挡油环 bearing neck
轴承盖 bearing cover
轴承环，推力头 bearing collar
轴承控制 bearing control
轴承体 guide bearing housing
轴承箱 bearing box
轴承座 bearing carrier; bearing housing
轴电流 shaft current
轴电流保护 shaft current protection
轴颈 bearing journal
轴领 guide bearing collar
轴流定桨式水轮机 Propeller turbine
轴流式水轮机 axial turbine
轴流式蓄能泵 propeller storage pump; axial storage pump
轴流调桨式水轮机 Thoma turbine
轴流转桨式水轮机 Kaplan turbine; axial-flow adjustable blade propeller turbine
轴伸贯流式水轮机（S形水轮机）tubular turbine（S－type turbine）
轴瓦 bearing segment; guide bearing shoe
轴瓦，轴套 bearing bush
轴向移位 axis displacement
肘形尾水管 elbow draft tube
肘形尾水管深度 depth of elbow draft tube

主/备用切换 main/spare switch over

主变差动 main transformer differential protection

主变复合电压过流保护 main transformer overcurrent with voltage blocking

主变过负荷保护 main transformer overload protection

主变火警 transformer fire fighting

主变零序过流保护 main transformer zero-sequence overcurrent protection

主变绕组温度过高 transformer winding over temperature

主变通风保护 start cooling fans protection

主变压器 main transformer

主变油温过高 transformer oil over temperature

主阀 main valve

主梁 main girder

主令开关 master switch

主配压阀 main distribution valve

主起升电阻器 main hoisting resistor

主起升柜 main hoisting cabinet

主起升机构 main lifting system

主轴 main shaft

主轴密封 shaft seal

主轴密封供水 shaft seal water supply

主轴密封供水备用水管 shaft seal spare water supply pipe

主轴密封供水管 shaft seal water supply pipe

主轴密封加压泵 shaft seal booster pump

主轴密封检修密封供气管 shaft seal maintenance seal air pipe

主轴密封冷却供水管 shaft seal cooling water supply pipe

主轴密封冷却水供水总管 shaft seal cooling water header pipe

主轴密封排水管 shaft seal drainage pipe

主轴密封润滑水 shaft seal lubrication water

主轴密封装置 main shaft seal

转臂 rocker arm

转动部件 rotating component

转轮 runner

转轮（叶轮）公称直径 runner（impeller）diameter

转轮减压板 decompression plate

转轮密封装置 runner seal

转轮平衡 runner balancing

转轮上冠 runner upper crown

转轮室 runner chamber

转轮输出功率 output power of runner

转轮输入功率 input power of runner

转轮体 runner hub

转轮下环 runner lower band

转轮泄水锥 runner cone

转轮叶片 runner blade; runner bucket

转轮叶片接力器 runner blade servomotor

转轮止漏环 runner wearing ring

转叶机构 mechanism of runner

blade
转子 rotor
转子测圆架 rotor roundness measuring device
转子测圆架底座 rotor roundness measuring device seat
转子温度保护 rotor temperature monitoring
转子一点接地 rotor on point earthing
转子一点接地保护 rotor earth fault
转子支臂 rotor spider
转子中心体 rotor hub
装配公差 fitting allowance
装配试验 assembly test
锥销 taper pin
锥形尾水管 conical draft tube
自动操作 automatic operation
自动化元件 automatic element; automatic instrumentation
自动开关 automatic switch
自恢复 automatic-restoration [restəˈreɪʃ(ə)n]
自启动 self-start
自诊断程序 self-diagnostic [daɪəgˈnɒstɪk]
(水轮机)综合特性曲线 efficiency hill diagram; combined characteristic curve (turbine)
总间隙 clearance in diameter
总体允许不平衡量(转轮) total allowable remaining unbalance
阻力器,减震器 damper; absorber; cushion
阻尼连接 damper connection
组装把合块切割 assembly block cutting
最大(最小)水头 maximum (minimum) head
最大水头 maximal head
最佳中心 best center
最小水头 minimal head
最优比转速 optimum specific speed
最优工况 optimum operating condition
最终压紧 final pressing
座环 stay ring
座环加工 stay ring machining
座环上环板 stay ring upper deck ring
座环下法兰面 stay ring lower flange
座环下筒体 stay ring lower barrel

相近专业汉英词汇

电力系统继电保护

TA 断线 TA line-break

TV 断线 TV line-break

B

半导体 semiconductor [ˌsemɪkənˈdɒktə]

保护半径 protection radius [prəʊˈtekʃən] [ˈreɪdɪəs]

保护变压器 protective transformer [prəʊˈtektɪv] [trænsˈfɔːmə]

保护地 protective earth

保护电抗器 protection reactor [rɪˈæktə]

保护电路 protective circuit [ˈsɜːkɪt]

保护电容器 capacitor for voltage protection [kəˈpæsɪtə] [ˈvəʊltɪdʒ]

保护断路器 circuit breaker [ˈsɜːkɪt]

保护范围 protection range [reɪndʒ]

保护高度 protection height

保护合闸 close by local protection [ˈləʊkəl]

保护和接地 protection and earthing

保护继电器 protective relay

保护间隙 protective gap

保护接地 protective earthing

保护接零 protective neutralization [ˌnjuːtrəlaɪˈzeɪʃən]

保护开关 protection switch

保护帽 protective cap

保护配合时间阶段 coordination time interval [kəʊˌɔːdɪˈneɪʃən] [ˈɪntəvəl]

保护配置 protection disposition [ˌdɪspəˈzɪʃən]

保护屏 protection screen

保护屏柜 protective panel [ˈpænl]

保护设备 protection device

保护跳闸 protecting tripping [ˈtrɪpɪŋ]

保护外壳 protective casing [ˈkeɪsɪŋ]

保护网 protecting net

保护系统 protection system

保护线 protective earth (PE)

保护线路 protective link [lɪŋk]

保护性接地 protective ground

保护罩 protective cover/housing [ˈkʌvə]

保护装置 protection device

保护装置的起动电流和返回电流 starting current and returning current of protection device

报警信号 alarm signal [əˈlɑːm]

报警信号继电器 alarm relay

备用变压器 standby transformer [ˈstændbaɪ] [trænsˈfɔːmə]

备用电源 standby source [ˈstændbaɪ] [sɔːs]

备用电源自动投入 automatic

switch-on of standby power supply [ˌɔːtəˈmætɪk]

备用容量 standby capacity [ˈstændbɑɪ] [kəˈpæsətɪ]

比率差动保护 percentage differential protection [pəˈsentɪdʒ] [ˌdɪfəˈrenʃəl]

比率差动继电器 percentage differential relay

闭合电路 closed circuit [ˈsɜːkɪt]

闭锁 escapement/interlock/blocking [ɪˈskeɪpmənt]

闭锁继电器 lockout relay

闭锁重合闸 blocking auto recloser

闭锁装置 locking device

不对称负载保护装置 protection against unsymmetrical load [ˈʌnsɪˈmetrɪkəl]

不平衡电流 unbalance current [ˈkʌrənt]

C

采样保持 sampling and holding [ˈsɑːmplɪŋ]

采样同步 synchronized sampling

操作（内部）过电压 operational (internal) overvoltage [ˌəʊvəˈvəʊltɪdʒ]

差动 differential motion [ˌdɪfəˈrenʃəl]

差动保护 differential protection

差动继电器 differential relay

超压防护 protection against overpressure [ˈəʊvəˌpreʃə]

潮流计算 load flow calculations [ˌkælkjuːleɪʃən]

持续过电压 sustained overload [səˈsteɪnd]

冲击电压 surge voltage

冲击防护 surge guard [gɑːd]

冲击继电器 pulse relay/surge relay [pʌls]

触点多路式继电器 contact multiplying relay

触发器 trigger [ˈtrɪgə]

穿越故障 through-fault

传感器 sensor [ˈsensə]

瓷绝缘子 porcelain insulator [ˈpɔːsəlɪn] [ˈɪnsjʊleɪtə]

重复接地 iterative earth [ˈɪtərətɪv]

重合于故障线路加速保护动作 accelerating protection for switching onto fault [ækˈseləreɪtɪŋ]

重合闸 reclosing

重合闸后加速保护 relay acceleration after auto-reclosing [əkˌseləˈreɪʃən]

重合闸前加速保护 relay acceleration before auto-reclosing

D

导纳型继电保护装置 admittance relays [ədˈmɪtəns]

低电压保护 under voltage protection
低电压起动的过电流保护 over current relay with under voltage supervision [ˌsjuːpəˈvɪʒən]
低电压跳闸 under-voltage release (trip)
低负荷运行 under-run
低功率保护 under power protection
低功率继电器 under-power relay
低频保护 under-frequency protection [ˈfrɪkwənsɪ]
低压保护 low-voltage protection
低压继电器 low-voltage relay
低压释放继电器 low-voltage release relay
低周波保护 under-frequency protection
低阻抗继电器 under-impedance relay [ɪmˈpiːdəns]
低阻抗母线保护 low impedance busbar protection
电磁型继电器 electromagnetic relay [ɪˌlektrəumægˈnetɪk]
电磁制动系 electromagnetic braking system
电导继电器 conductance relay [kənˈdʌktəns]
电动机磁场故障继电器 motor-field failure relay
电动式继电器 dynamoelectric relay [ˌdɑɪnəməuɪˈlektrɪk]
电复位继电器 electric reset relay
电缆继电器 cable relay
电力传输继电器 power-transformer relay
电力系统继电保护 power system relaying protection
电力系统振荡 power system oscillation [ˌɒsɪˈleɪʃən]
电流闭锁的电压速断保护 instantaneous under voltage protection with current supervision [ˌɪnstənˈteɪnjəs] [ˌsjuːpəˈvɪʒən]
电流补偿式接地远距继电器 current compensational ground distance relay
电流过载 current overload
电流极化继电器 self-polarizing relay
电流平衡式差动电流继电器；差动平衡式电流继电器 current balance type current differential relay
电流平衡式继电器 current-balance relay
电流起动型漏电保护器 current actuated leakage protector
电路控制继电器 circuit control relay
电气保护 electrical protection
电容继电器 capacitance relay [kəˈpæsɪtəns]
电容接地 capacity ground
电网阻抗相角 network impedance phase angle
电压差动继电器 voltage differential relay [ˌdɪfəˈrenʃəl]
电压互感器的相角差 phase-angle of voltage transformer [feɪz]

电压控制过电流继电器 voltage-controlled over current relay

电压平衡继电器 voltage balance relay

电压响应继电器 voltage responsive relay [rɪˈspɒnsɪv]

电压选择继电器 voltage selection relay [sɪˈlekʃən]

电涌保护器 surge suppressor

电子管继电器 vacuum-tube relay [ˈvækjʊəm]

电阻继电器 Ohm relay [əʊm]

定时继电器 timing relay

定时脉冲继电器 time pulse relay [pʌls]

定时限 definite time [ˈdefɪnɪt]

定时限过电流保护 definite time over-current protection

定时限继电器 definite time relay [ˈdefɪnɪt]

定向继电器 directional relay

定子接地保护 stator earth-fault protection

动圈式继电器 moving coil relay

动作时间 operating time

短路计算 short circuit calculations

"断开"位置，"开路"位置 off-position

断路继电器 cut-off relay

断路器 circuit breaker

断路器按钮 cut-off push

断路器触点 breaker contact point

断路器故障保护装置 circuit breaker failure protection

断路器跳闸线圈 breaker trip coil

断相保护 open-phase protection

断相继电器 open-phase relay

对地漏电保护 earth-leakage protection [ˈliːkɪdʒ]

多次重合闸断路器 multiple-reclosing breaker [ˈmʌltɪpl]

多端线路保护 multi-ended circuit protection

多相补偿式阻抗继电器 multiphase compensated impedance relay [ˈmʌltɪfeɪz]

多重接地 multiple earth

E

二位置继电器 two-position relay

F

发电机保护 generator protection

发电机一变压器保护 protection of generator-transformer set

发电机的失步保护 generator out of step protection

发电机的失磁保护 field failure protection of generator [ˈfeɪljə]

发电机定子绕组单相接地保护

generators stator single phase earth fault [fɔːlt]
发电机定子绕组短路故障 generator stator winding short circuit faults
发电机断路继电器 generator cut-out relay
发电机负序电流保护 generator negative current protection ['negətɪv]
发电机横差动保护 transverse differential protection for generator turn-to-turn faults
发光二极管 light emitting diode
反馈回路 feedback loop
反馈系统 feedback system
反馈信号 feedback signal
反时限 inverse time
反时限过电流保护 inverse time over-current protection
反向制动 plugging ['plʌgɪŋ]
反转时间继电器 back-spin timer
方框图 block diagram ['daɪəgræm]
方向保护 directional protection
方向触点 directional contact
方向过流保护 directional over-current protection
方向过流继电器 directional over-current relay
方向距离继电器 directional distance relay
方向性电流保护 the directional current protection
方向纵联继电保护 directional pilot relaying [dɪ'rekʃənəl]
放大器 amplifier ['æmplɪfaɪə]
分段电抗器 bus section reactor ['sekʃən]
分段距离继电器 step-type distance relay
分段母线隔离开关 bus bar disconnecting switch [ˌdɪskə'nektɪŋ]
分段限时继电器 multi-zone relay ['mʌltɪ]
分接头 tap
分相电流差动保护 segregated current differential protection ['segrɪgeɪtɪd]
分支系数 branch coefficient [ˌkəuɪ'fɪʃənt]
辅助触点 auxiliary contacts [ɔːg'zɪljərɪ]
辅助继电器 auxiliary relay
负相位继电器 negative phase relay
负相序继电器 negative-phase sequence impendence ['siːkwəns] [ɪm'pendəns]
负载不足继电器 under-load relay
附加超速保护装置 back-up overspeed governor ['gʌvənə]

G

感应杯式继电器 induction cup relay
感应式继电器 induction type relay
感应圆盘式继电器 induction disc relay

高灵敏度继电器 high sensitive relay [ˈsensɪtɪv]
高频保护 pilot protection
高频通道 power line carrier channel [ˈtʃænəl]
高速数字信号采集系统 high speed signal acquisition system [ˌækwɪˈzɪʃən]
高速阻抗继电器 high-speed impedance relay
高压电瓷绝缘子 high tension electrical porcelain insulator [ˈtenʃən] [ˈpɔːsəlɪn][ˈɪnsjʊleɪtə]
高压继电器 high-voltage relay
高阻抗母线差动保护 high impedance busbar differential protection [ɪmˈpiːdəns]
功率方向继电器 power direction relay
功率继电器 power relay
功率因数继电器 power factor relay
固定联结式母线保护 busbar protection with fixed circuit connection [kəˈnekʃən]
故障电流 fault current
故障定位 fault location [ləʊˈkeɪʃən]
故障检测继电器 fault detecting relay [dɪˈtektɪŋ]
故障接地 fault earthing
故障切除时间 fault clearing time
过电流 over-current
过电流保护 over-current protection
过电流保护装置 over-current protective device
过电流继电器 over-current relay
过电压 over-voltage
过电压保护 over-voltage protection
过渡阻抗 transition impedance
过负荷 overload
过励磁保护 over fluxing protection
过流保护 over current protection
过流保护跳闸 over current protection trip
过热保护 thermal protection [ˈθɜːməl]
过热继电器 temperature limiting relay [ˈtempərɪtʃə]
过压 over-voltage
过压继电器 over-voltage relay
过载 over-load
过载保护 over-load protection
过载功率 over power
过载继电器 overload relay
过载跳闸 overload trip

H

合闸继电器 closing relay
合闸位置继电器 closing position relay
合闸线圈 closing coil [kɒɪl]
恒温继电器 thermostat relay [ˈθɜːməʊstæt]
横差保护 transverse differential protection [ˌdɪfəˈrenʃəl]
后备保护 back-up protection
后备电源切换 transfer of auxiliary supply [ɔːgˈzɪljərɪ]

后备继电保护 back-up system

缓动继电器 delay-action relay

缓放继电器 slow-to release relay

换流器继电器 converter relay [kən'vɜːtə]

J

机电继电器 electromechanical relay [ɪˌlektrəʊmɪ'kænɪkəl]

基础接地体 foundation earth electrode [faʊn'deɪʃən]

极化差动继电保护系统 biased differential relaying

极化继电器 electro polarized relay ['pəʊlərɑɪzd]

极限切除时间 critical clearing time

继电保护 relay protection

继电保护装置 protective relaying equipment

继电器保证启动值 relay must-operate value

继电器操作跳闸 relay act trip

继电器超限运行 relay overrun

继电器屏 relay panel

交流方向过流继电器 AC directional over current relay

交流重合闸继电器 AC reclosing relay

阶梯型时间配置 graded time settings ['greɪdɪd]

接地故障 earth fault

接地故障继电器 ground fault relay

接地装置 grounding apparatus [ˌæpə'reɪtəs]

解除闭锁信号 unblocking signal

解列重合闸 power system splitting and reclosing

晶体管 transistor

晶体管继电器 transistor relay

晶体密闭继电器 crystal can relay ['krɪstəl]

静电继电器 static relay

静态距离继电器 static distance relay ['stætɪk]

就地控制 local control

距离保护 distance protection

距离纵联保护 pilot protection using distance relay

绝缘监视 insulation supervision device [ˌsjuːpə'vɪʒən]

绝缘子串 insulator chain ['ɪnsjʊleɪtə]

绝缘子角形避雷器 insulator arcing horn

绝缘子帽 insulator cap

绝缘子闪络 insulator arc-over

K

开关站 switch station

快动缓释继电器 fast-operate slow-release relay

快释放继电器 fast-release relay

L

励磁继电器 field application relay [ˌæplɪˈkeɪʃən]

励磁涌流 inrush exciting current of transformer [ˈɪnrʌʃ]

连锁故障 cascading outages [kæˈskeɪdɪŋ]

连锁继电器 electric interlock relay

联络线 tie line

联锁装置 interlocker

两相短路故障 two-phase short circuit fault

两相接地短路故障 two-phase grounding fault

两相星形接线方式 two star connection scheme [kəˈnekʃən]

灵敏极化继电器 sensitive polarized relay [ˈpəuləraɪzd]

灵敏继电器 sensitive relay [ˈsensɪtɪv]

零序保护 zero-sequence protection [ˈsiːkwəns]

零序电流 zero-sequence current/residual current [rɪˈzɪdjuəl]

零序电流保护 zero-sequence current protection

零序电流补偿 zero sequence current compensation

零序电流构成的定子接地保护 stator ground protection based on zero sequence current [ˈsiːkwəns]

零序电流互感器 zero sequence current transducer [trænzˈdjuːsə]

零序电流继电器 residual current relay [rɪˈzɪdjuəl]

零序电压继电器 zero-sequence voltage relay

漏电断路器 residual current circuit-breaker

漏电流绝缘型（联轴节） leakage current insulating version

M

母联断路器 bus coupler CB

母线保护 bus bar protection

母线保护继电器 bus protective relay

母线电抗器 bus reactor

母线电流变压器 bus bar current transformer

母线分段断路器 bus section breaker

母线分段开关 Bus-bar sectionalizing switch [ˈsekʃənəlaɪz]

母线分离 bus separation

母线故障 bus-bar fault

母线和变压器共用保护 combined bus and transformer protection [kəmˈbaɪnd]

母线绝缘器 bus insulator
母线绝缘子 bus-bar insulator [ˈɪnsjʊleɪtə]
母线联络继电器 bus tie circuit breaker
母线联络开关 bus tie switch
母线终端故障 bus terminal fault [ˈtɜːmɪnəl]

N

内部故障 internal fault
能量方向继电器 energy directional relay
逆功率保护 inverse power protection
逆相序保护 inverse phase sequence protection [ˈɪnvɜːs]

P

偏移特性阻抗继电器 offset impedance relay [ˌɒfˈset]

Q

起动断路器跳闸 activate the breaker trip coil
欠范围允许跳闸式 permissive under reaching transfer trip scheme [pəˈmɪsɪv]
欠压继电器 under-voltage relay
轻（重）瓦斯保护 slight（severe）gas protection

R

弱电源端保护 weak power end protection

S

三端输电线保护 three terminal line protection
三相一次重合闸 three phase one shot reclosure
失步 out-of-step
失步保护 loss of synchronism protection [ˈsɪŋkrənɪzəm]
失磁继电器 excitation-loss relay
时间继电器 timer relay
时限过电流 time over-current
事故过电压 abnormal overvoltage [æbˈnɔːməl] [ˌəʊvəˈvəʊltɪdʒ]

输电线路异常运行 transmission line malfunction [mælˈfʌŋkʃən]

数字式保护 digital protection

数字信号处理 digital signal processor (DSP) [ˈprəusesə]

双母线保护 double bus bar protection

双向继电器 bi-directional relay

顺序跳闸 sequential tripping [sɪˈkwenʃəl]

瞬动继电器 instantaneous relay [ˌɪnstənˈteɪnjəs]

瞬时测值 instantaneous measured

瞬时动作 instantaneous action

瞬时性故障 temporary fault [ˈtempərərɪ]

瞬态电流 transient rotor [ˈtrænzɪənt]

瞬态响应 transient response

T

调节器 regulator

调谐器 tuner

调压器 voltage regulator

跳闸 trip off

跳闸继电器 trip relay

跳闸开关 trip switch

跳闸脉冲 tripping pulse [pʌls]

跳闸线圈 trip coil

跳闸指示灯 tripping indicating lamp

脱扣线圈 tripping coil

W

瓦斯保护 Buchholtz protecter

瓦斯保护装置 gaseous shield [ˈgæsɪəs] [ʃiːld]

微动开关 microswitch [ˈmaɪkrəuswɪtʃ]

微机继电保护 micro-processor based protective relay

无时限电流速断保护 instantaneously over-current protection [ˌɪnstənˈteɪnjəslɪ]

五防装置 fail safe interlock

误动 false tripping

X

衔铁（磁铁）吸合式继电器 attracted armature relay

限流电路 limited current circuit

限流继电器 current-limiting relay

相间故障 phase to phase fault

相位比较继电器 phase comparison relays

响应时间 response time

小电流接地系统 ground-fault of ungrounded systems [ˈʌnˈgrɑʊndɪd]

谐波制动 harmonic restraining [hɑːˈmɒnɪk] [rɪˈstreɪnɪŋ]

信号继电器 signal relay [ˈsɪgnəl]

行波继电器 traveling wave relay [ˈtrævəlɪŋ]

Y

压敏开关 pressure sensitive switch

延时 time delay

延时继电器 timing relay

延时速断 delay quick breaking

移相电容器 phase-shift capacitor

异常过载 abnormal overload [æbˈnɔːməl]

隐形故障 hidden failures [ˈhɪdən]

永久性故障 permanent fault [ˈpɜːmənənt]

远距离控制设备 remote-control apparatus [ˌæpəˈreɪtəs]

Z

匝间短路 turn to turn fault，inter turn faults

载波或导引线接受继电器 carrier or pilot-wire receiver relay

暂态保护 relay based on transient component [kəmˈpəʊnənt]

增量继电器 relay based on incremental quantity [ˌɪnkrɪˈmentəl]

增益 gain [geɪn]

振荡 oscillation [ˌɒsɪˈleɪʃən]

振荡（失步）闭锁 power swing (out of step) blocking

振荡冲击 oscillatory surge [ˈɒsɪlətərɪ]

振荡反应性扰动 oscillatory reactivity perturbation [ˌpɜːtɜːˈbeɪʃən]

振荡器 oscillator

整定 setting

整定电流 setting current

整定范围 setting range

整定值 set value

中间继电器 auxiliary relay/intermediate relay [ˌɪntəˈmiːdjət]

中心点电阻接地方式 resistance grounded neutral system

中性点电抗接地系统 reactor grounded neutral system [ˈnjuːtrəl]

中性点接地 neutral-point earthing

中性点接地自耦变压器 neutral auto-transformer

重负荷 heavy load

主保护 primary protection [ˈprɑɪmərɪ]

专用线路 private line [ˈprɑɪvɪt]

转子接地保护 rotor earth-fault protection

准稳态 quasi-steady state [ˈkweɪzaɪ]

直接接地 solidly earthing

自调节控制器 self-tuning controllers

自动重合闸 automatic reclosure

自动复归 self reset

自动准同步 automatic quasi-synchronization

自极化姆欧、导纳继电器 self-polarize mho

自适应分相方向纵差保护 adaptive segregated directional current differential protection [ˈsegrɪgeɪtɪd] [dɪˈrekʃənəl]

自适应继电保护 adaptive relay protection [əˈdæptɪv]

自持放电 sustained discharge [səˈsteɪnd]

总线请求周期 bus request cycle

纵差保护 longitudinal differential protection [ˌlɒndʒɪˈtjuːdɪnəl]

纵联保护 Pilot protection [ˈpɑɪlət]

纵联差动保护 longitudinal differential protection

纵联差动继电器 longitudinal differential relay

阻抗补偿器 impedance compensator [ˈkɒmpənseɪtə]

阻抗继电器 impedance relay

电 气 工 程

SF_6 断路器 SF_6 circuit breaker [ˈsəːkɪt]

T形线夹 T-connector [kəˈnektə]

A

安培 ampere (A) [ˈæmpeə]

安全标志 safety marking

安全隔离变压器 safety isolating transformer [ˈɑɪsəleɪtɪŋ]

安全距离 safety distance [ˈdɪstəns]

安全开关 safety switch

安全色 safety color

安全阻抗 safety impedance [ɪmˈpiːdəns]

暗敷 concealed laying [kənˈsiːld]

暗接地线 concealed earth line

暗线 concealed wire [ˈwɑɪə]

暗装 conceal mounted [ˈmɑʊntɪd]

B

饱和曲线 saturation curve [ˌsætʃəˈreɪʃən]

饱和特性 saturation characteristic [ˌkærəktəˈrɪstɪk]

饱和效应 saturation effect [ˌsætʃəˈreɪʃən]

备用电源 reserve source (stand-by source)

避雷带 lightning belt [ˈlɑɪtnɪŋ]

避雷器 arrester [əˈrestə]

避雷器的残压 residual voltage of lightning arrester [rɪˈzɪdjʊəl]

避雷网 lightning-protection net

避雷针 lightning rod, lightning conductor [kənˈdʌktə]

避雷针基础 lightning rod base

避雷针支架 lightning rod support

避雷装置 lightning protector

变比 ratio [ˈreɪʃɪəʊ]

变电站 transformer substation [ˈsʌbsteɪʃən]

变电箱 switchbox

变极电动机 pole changing generator [ˈdʒenəreɪtə]

变流器 (current) converter

变频发电机 variable-frequency generator [ˈdʒenəreɪtə]

变频交流电 variable frequency AC

变频器 frequency converter [ˈfrɪkwənsɪ] [kənˈvəːtə]

变时限 dependent time [dɪˈpendənt]

变压器 transformer

变压器额定电流 transformer rated

current

变压器额定电压 transformer rated voltage

变压器额定容量 rated power of transformer（transformer rating）

变压器风扇 transformer fan

变压器负载损耗 load loss of transformer

变压器绕组 transformer winding

变压器调压范围 transformer voltage regulation rage

变压器铁芯 transformer core

变压器外壳 transformer casing

变压器线圈 transformer coil

变压器油箱 transformer oil tank

变压器阻抗电压（变压器短路电压） impedance voltage of transformer [ɪm'piːdəns]

变阻器 rheostat ['riːəustæt]

标么值 per unit value

并励 shunt [ʃʌnt]

并励磁场 shunt field

并联 parallel connection ['pærəlel] [kə'nekʃən]

并联电容 parallel capacitance [kə'pæsɪtəns]

并联电容器 parallel capacitor

并列运行 parallel operation

并网 grid connected [grɪd]

并网点 connection point

并网风力发电机 interconnection of wind power generators [ˌɪntəkə'nekʃən]

并网功率 power of grid connected

并网时间 grid connected time

并网系统 grid connection system

波导，波导管 wave guide [gaɪd]

补偿度 degree of compensation

补偿后功率因数 power factor after compensating ['kɒmpenseɪtɪŋ]

补偿容量 compensating capacity

不带电金属外壳 non-current carrying metallic case [mɪ'tælɪk]

不平衡负荷 unbalanced load [ˌʌn'bælənst]

不停电电源 uninterrupted power supply（UPS） ['ʌnˌɪntə'rʌptɪd] [sə'plaɪ]

C

参考值 reference value ['refərəns] ['væljuː]

残压 residual voltage

残余电流 residual current ['kʌrənt]

残余电容 residual capacitance [rɪ'zɪdjʊəl] [kə'pæsɪtəns]

残余电压 residual voltage

操作电压 control voltage

操作电源 operating power source [sɔːs]

槽 solt

侧击雷 side stroke

插接装置 plug device [plʌg]

插头 plug

插座 socket；outlet ['sɒkɪt]

插座箱 socket box

差周率继电器 difference frequency relay
超高频 UHF (Ultra High Frequency)
超高压 EHV (Extra-High Voltage)
潮流 tidal current
常闭触点 normally closed contact [ˈnɔːməlɪ]
常开触点 normally opened contact
常用电源 normal source
厂用电 service power of plant
厂用负荷 load of plant
厂用供电系统 power supply system
敞开式组合电器 open-air type assembled switch gear [əˈsembəld]
充电 charge
充电室 battery-charging room
充电（阻尼）电阻 charging (damping) resistor
冲击；过电压 surge [sɜːdʒ]
冲击电流 impulse current [ˈkʌrənt]
冲击电压 impulse voltage
冲击放电电压 impulse discharge voltage
冲击负荷 shock load
抽屉式低压配电屏 drawable LT distribution panel [ˈdrɔːəbl]
抽头 tap-off
出线 outgoing line
出线屏 outgoing panel [ˈpænl]
触电 electric shock
触电电流 shock current
触头 contact
传感器 sensor [ˈsensə]
穿钢管敷设 laid in steel conduit
串励 series
串联 series connection [kəˈnekʃən]
串联变压器 series transformer
串联电抗器 series reactor
串联电路 series circuit
垂直接地极 vertical electrode [ˈvɜːtɪkəl] [ɪˈlektrəʊd]
瓷插式熔断器 plug-in fuse
磁场 magnetic field [mægˈnetɪk]
磁吹避雷器 magnetic blow-out arrester [əˈrestə]
磁吹阀式避雷器 magnetic blow-out valve type arrester
磁放大器 magnetic amplifier [ˈæmplɪfɑɪə]
磁化电抗 magnetizing reactance [rɪˈæktəns]
磁力起动器 magnetic starter
磁链 flux linkage [ˈlɪŋkɪdʒ]
磁路 magnetic circuit [mægˈnetɪk] [ˈsɜːkɪt]
从动齿轮 driven gear [gɪə]
从旋出螺丝；旋开；旋松 unscrew [ˌʌnˈskruː]

D

大功率并联电容无功功率补偿 high power parallel capacitor reactive power compensating [ˈpærəlel] [rɪˈæktɪv]

带电金属外壳 current carrying metallic case
带负荷调压变压器 on-load regulating transformer
带宽 bandwidth [ˈbændwɪdθ]
单刀双掷开关 single-pole double throw switch
单机无穷大系统 one machine-infinity bus system [ɪnˈfɪnətɪ]
单相变压器 single-phase transformer
单相短路电流 single-phase short-circuit current [feɪz]
单相二极插头 1-phase 2-pole plug
单相二线制 1-phase 2-wire system
单相接地短路 single-phase ground short circuit
单芯电缆 single-core cable
单元接线（发电机—变压器组接线） generator-transformer unit connection
刀闸(隔离开关) isolator[ˈaɪsəleɪtə]
导电性 conductivity [ˌkɒndʌkˈtɪvɪtɪ]
导体 conductor
等电位连接 equipotential bonding
等电位连接带 bonding bar
等电位连接导体 bonding conductor
低压 low voltage
低压补偿 compensating in LT side [ˈkɒmpenseɪtɪŋ]
低压电缆 LT cable
低压电器 low voltage apparatus [ˌæpəˈreɪtəs]
低压静电电容器屏 LT electrostatic capacitor panel [ɪˌlektrəˈstætɪk] [kəˈpæsɪtə]
低压配电屏 low voltage distribution panel/board [ˈpænl]
低压配电装置 low voltage switchgear installation [ˈswɪtʃɡɪə] [ˌɪnstəˈleɪʃən]
低压熔断器 LT fuse
低压室 LT room
低压无功功率补偿装置 LT reactive power compensator [ˈkɒmpənseɪtə]
低压线路 LT line
低周率继电器 low frequency relay [ˈfrɪkwənsɪ]
地下电缆 ground cable
电厂 power plant
电触头 electrical contact
电磁操动机构 magnetic control mechanism [mægˈnetɪk] [ˈmekənɪzəm]
电磁场 electromagnetic field [ɪˌlektrəumægˈnetɪk]
电磁干扰 electromagnetic interference (EMI)
电磁感应 electromagnetic induction
电磁起动器 electromagnetic starter
电磁式 electromagnetic
电导 conductance [kənˈdʌktəns]
电动操动机构 motor drived operating mechanism [ˈmekənɪzəm]
电动机 motor
电动机保护 motor protection
电动势 emf=electromotive force
电度表箱 kilowatt-hour meter box [ˈkɪləʊwɒt]

电杆长度 pole length [leŋθ]
电感 inductance [ɪn'dʌktəns]
电感器 inductor
电工学 Electrotechnics [ɪˌlektrəʊ'teknɪks]
电荷 electric charge
电弧 arc
电火花 spark
电机 electric machine
电机学 Electrical Machinery
电极 electrode
电介质 the dielectric
电抗 reactance [rɪ'æktəns]
电抗电压 reactance voltage
电抗负荷 reactive load
电抗器 reactor (shunt reactor) [rɪ'æktə]
电缆 cable
电缆槽 cable duct
电缆吊架，电缆吊杆 cable hanger ['hæŋə]
电缆分线箱 cable junction box ['dʒʌŋkʃən]
电缆沟 cable trench [trentʃ]
电缆夹 cable cleat [kliːt]
电缆接头 cable spice
电缆接线头 cable plug
电缆井 cable pit
电缆铠装 cable armouring ['ɑːmərɪŋ]
电缆盘 cable reel
电缆桥架 cable bridge
电缆人孔 cable manhole ['mænhəʊl]
电缆隧道 cable tunnel ['tʌnəl]
电缆隧道口 cable tunnel exit
电缆套 cable box
电缆托架 cable tray [treɪ]
电缆线路 cable line
电缆箱，分线盒 cable cabinet ['kæbɪnɪt]
电缆靴 cable shoes
电缆终端头 cable termination joint [ˌtɜːmɪ'neɪʃən]
电力变压器 power transformer
电力传动与控制 electrical drive and control
电力传输 power transmission
电力电缆 power cable
电力电子变流器 power electronic converter [ˌɪlek'trɒnɪk]
电力电子基础 basic fundamentals of power electronics [ˌfʌndə'mentlz] [ˌɪlek'trɒnɪks]
电力干线 main power line
电力公司 power supply company [sə'plɑɪ]
电力汇集系统（风电机组） power collection system (for WTGS)
电力网络 power network
电力系统 power system
电力系统的集中控制 centralized control of power system ['sentrəlɑɪzd]
电力系统内部过电压 past voltage within power system
电力系统稳态分析 Steady-State Analysis of Power System [ə'næləsɪs]
电力系统暂态分析 transient-state analysis of power system

[ˈtrænziənt] [əˈnæləsɪs]

电力线载波（器） power line carrier (PLC)

电力支线 power branch line

电量变送器 power transducer [trænzˈdjuːsə]

电流 electric current

电流表 ammeter, current meter [ˈæmɪtə]

电流互感器 current transformer (CT)

电流继电器 current relay

电流速断保护 current quick-breaking protection

电流脱扣，串联脱扣 series tripping

电路 electric circuit

电路参数 circuit parameters

电路元件 circuit components [kəmˈpəʊnənt]

电纳 susceptance [səˈseptəns]

电能 electric energy

电能质量 power quality

电能转换器 energy converter [kənˈvɜːtə]

电气布线 electrical wiring

电气工程师学会（英） IEE (Institution of Electrical Engineers)

电气设备 electrical device (equipment)

电气寿命 electrical endurance [ɪnˈdjʊərəns]

电气系统 electrical systems

电气与电子工程师学会（美） IEEE (Institute of Electrical and Electronics Engineers

电气元件 electrical device

电热器 heating appliance [əˈplɑɪəns]

电容 capacitance [kəˈpæsɪtəns]

电容器 capacitor

电容器室 condenser room [kənˈdensə]

电容式电压互感器 capacitor voltage transformer

电枢 armature [ˈɑːməˌtjʊə]

电枢电路 armature circuit [ˈɑːməˌtjʊə] [ˈsɜːkɪt]

电刷 brush [brʌʃ]

电网 power system

电网波动 power fluctuations [ˌflʌktjʊˈeɪʃəns]

电网毁坏；电网故障 grid failure [ˈfeɪljə]

电网连接点（风力发电机组） network connection point (for WTGS) [kəˈnekʃən]

电网频率 grid frequency [ˈfrɪkwənsɪ]

电网渗透率 grid penetration [ˌpenɪˈtreɪʃən]

电网损失 network/grid loss

电网阻抗相角 network impedance phase angle [feɪz]

电线电缆 electric wire and cable

电压 voltage

电压变化系数 voltage change factor

电压标准 voltage standard

电压表 voltmeter [ˈvəʊltˌmiːtə]

电压波动 voltage variation
电压等级 voltage grade
电压互感器 PT = potential transformer
电压继电器 voltage relay
电压降 voltage drop
电压控制母线（PV 母线） voltage control busbar [ˈbʌsbɑː]
电压控制系统 voltage control system
电压偏移 voltage deviation [ˌdiːvɪˈeɪʃən]
电压损失 voltage loss
电压调整率 voltage regulation rate
电压脱扣，并联脱扣 shunt tripping
电压稳定 voltage stability
电涌保护器 surge suppressor
电源 power source
电源进线 incoming line
电源平滑 power smoothing
电源切换箱 power change-over box
电子电机集成系统 electronic machine system
电阻 resistance [rɪˈzɪstəns]
电阻电压 resistance voltage
电阻负荷 resistance load
电阻率 resistivity [ˌriːzɪsˈtɪvətɪ]
电阻器 resistor
掉网 grid dropout [ˈdrɒpaʊt]
跌落式熔断器 drop-out fuse
叠片 lamination [ˌlæmɪˈneɪʃən]
叠片铁芯 laminated core [ˈlæmɪneɪtɪd]
定期检修 periodic maintenance [ˌpɪərɪˈɒdɪk] [ˈmeɪntənəns]
定子 stator
定子绕组 stator winding
动力电子设备 power electronics
动力负荷 power load
动力馈电屏 power feeder panel
动力配电箱 power distribution box
动态 dynamic (state) [daɪˈnæmɪk]
动态无功补偿系统 D-VAR system
动态响应 dynamic response
动作电流 action current
动作范围 action range
动作时间 action time; actuating time
端电压 terminal voltage [ˈtɜːmɪnəl]
端环 end ring
端接；终止 terminate
端子 terminal point
短路 short circuit
短路比 short-circuit ratio
短路冲击电流 short-circuit impulse current [ˈɪmpʌls]
短路点 short circuit point
短路电抗 short-circuit reactance [rɪˈæktəns]
短路电流 short circuit current
短路电流计算 calculation of short circuit current [ˌkælkjuːleɪʃən]
短路电压 short-circuit voltage
短路电阻 short-circuit resistance
短路环 short-circuiting ring [ˈsɜːkɪtɪŋ]
短路容量 short circuit capacity
短路试验 short circuit testing
短路特性 short-circuit characteris-

tic [ˌkærəktəˈrɪstɪk]

短路稳定性 short-circuit stability [stəˈbɪlɪtɪ]

短路运行 short-circuit operation

短路阻抗 short circuit impedance [ɪmˈpiːdəns]

断电；停电 blackout [ˈblækaʊt]

断开电路 open circuit

断路器 circuit breaker

对地电容 grounding capacitance [kəˈpæsɪtəns]

对地电压 voltage to earth

多隙避雷器 multigap arrester

多相（的） polyphase [ˈpɒlɪfeɪz]

多相电度表 polyphase meter [ˈpɒlɪfeɪz]

多相整流器 polyphase rectifier [ˈrektɪfɑɪə]

多油式断路器 bulk oil circuit breaker [bʌlk]

E

额定电流 rated current，nominal current [ˈnɒmɪnəl]

额定电压 rated voltage [ˈreɪtɪd]

额定负载 nominal load

额定工作电流 rated operational current

额定工作电压 rated operational voltage

额定功率 rated power

额定频率 rated frequency

额定容量 rated capacity

额定视在功率 rated apparent power

额定无功功率 rated reactive power

二次电流 secondary current [ˈsekəndərɪ] [ˈkʌrənt]

二次电压 secondary voltage

二次接线图号 secondary wiring drawing No.

二次系统图 secondary system diagram [ˈdɑɪəgræm]

二类负荷 second-class load

二线制 two-way configuration [kənˌfɪgjʊˈreɪʃən]

F

发电 generating

发电成本 energy generation cost

发电机 generator [ˈdʒenəreɪtə]

发电机安装 generator mounting [ˈmaʊntɪŋ]

发电机电压 generator voltage

发电机短路 generator shortage [ˈʃɔːtɪdʒ]

发电机主引出线 generator main outlet

发电量 energy yield [jiːld]

阀式避雷器 auto-valve arrester

法兰，垫圈 flange [flændʒ]

反向峰值电压 peak reverse voltage

反谐振频率 anti-resonance frequency

防爆开关 explosion proof switch [ɪk'spləuʒən]

防雷 lightning protection ['lɑɪtnɪŋ]

防雷变压器 lightning-proof transformer

防雷分类 classification of lightning protection [ˌklæsɪfɪ'keɪʃən]

防雷工程 lightning protection engineering

防雷接地 earthing for lightning protection

防跳 tripping prevent

放电 electric discharge

放射式 radial system ['reɪdɪəl]

非标准控制箱（柜，台）non-standard control box（cabinet，desk)

非同期重合闸 asynchronous reclosing [eɪ'sɪŋkrənəs]

分层控制系统 hierarchy control system ['hɑɪəˌrɑːkɪ]

分级转换开关 stepping switch

分接 tapping

分接头 tap

分接头位置信息 tap position information

分离轴承 split bearing

分列运行 independent operation

分裂式变压器 split winding type transformer

分路器 shunt

分压器 voltage divider [dɪ'vɑɪdə]

分压器分压比 divider ratio ['reɪʃɪəu]

封闭母线 enclosed busbar [ɪn'kləuzd] ['bʌsbɑː]

封闭式开关 closed switch

封闭式组合电器 enclosed-type assembled switch gear [ə'sembəld] [gɪə]

峰荷 peak load

峰值电压表 peak voltmeter ['vəultˌmiːtə]

峰值因数 peak factor

敷设 laying

辅助电路 auxiliary circuit [ɔːg'zɪljərɪ] ['səːkɪt]

辅助继电器 auxiliary relay

辅助设备 auxiliary equipment

辅助照明 supplementary lighting [ˌsʌplɪ'mentərɪ]

负荷电阻 load resistor

负荷电阻（水冷） load resistor (water cooled)

负荷计算 load circulation

负荷计算表 load calculation table

负荷开关 load switch

负荷率 load rate

负荷母线（PQ 母线） load busbar

负荷曲线 load-duration curves

负频率 negative frequency

负相序 negative phase sequence [feɪz] ['siːkwəns]

负相序等效电路 negative phase sequence equivalent circuit [ɪ'kwɪvələnt]

负序阻抗 negative sequence impedance ['negətɪv] ['siːkwəns] [ɪm'piːdəns]

负载饱和曲线 load-saturation curve
负载特性 load characteristic
[ˌkærəktəˈrɪstɪk]
复励 compounded [kɒmˈpaʊndɪd]
复数阻抗 complex impedance
[ɪmˈpiːdəns]

G

感应电机 induction generator
感应电流 induced current
[ɪnˈdjuːst]
感应雷 induction stroke
感应雷过电压（间接雷过电压）
induced lightning stroke over-voltage
感应式电机 induction machine
干式变压器 dry transformer
高架电线 overhead wire
高架输电线 overhead power line
高电压工程 high voltage engineering [ˌendʒɪˈnɪərɪŋ]
高顶值 high limited value
高抗 high voltage shunt reactor
高频保护 high-frequency protection
高频变压器 high-frequency transformer
高频滤波器 high-frequency filter
[fɪltə]
高性能的 high-performance
高压 high voltage
高压补偿 compensating in HT side
[ˈkɒmpenseɪtɪŋ]
高压侧 high side
高压成套开关柜 high voltage aggregated switchgear [ˈægrɪgɪtɪd]
[ˈswɪtʃgɪə]
高压电抗器 HT reactor
[rɪˈæktə]
高压电缆 HT cable
高压电器 HT equipment
高压断路器 HT circuit breaker
[ˈsɜːkɪt]
高压负荷开关 HT load switch
高压隔离开关 HT isolator[ˈɑɪsəleɪtə]
高压互感器 HT transformer
高压接触器 HT contactor
[ˈkɒntæktə]
高压静电电容器柜 HT electrostatic capacitor cabinet
[ɪˌlektrəˈstætɪk] [kəˈpæsɪtə]
[ˈkæbɪnɪt]
高压开关柜 HT switchgear
[ˈswɪtʃgɪə]
高压配电柜 HT distribution cabinet [ˈkæbɪnɪt]
高压配电装置 High voltage switchgear installation
[ˌɪnstəˈleɪʃən]
高压熔断器 HT fuse
高压室 HT room
高压无功功率补偿装置 HT reactive power compensator
[ˈkɒmpənseɪtə]
高压线路 HT line
高压真空接触器 HT vacuum contactor [ˈvækjʊəm]
高压直流断路器 high voltage direct

current circuit breaker [dɪ'rekt]
钢丝线 steel wool
钢芯铝绞线 steel-reinforced aluminum conductor [ə'lju:mɪnəm]
隔离变压器 isolating transformer ['ɑɪsəleɪtɪŋ]
隔离开关 isolating switch (disconnecting switch) ['ɑɪsəleɪtɪŋ]
工频放电电压 power frequency discharge voltage
工作变压器 operating transformer
工作电压 working voltage
工作接地 working earthing
工作特性 performance characteristic
工作线圈 operating coil
工作照明 working lighting
公用变压器 common transformer ['kɒmən]
公用电网 utility grid [ju:'tɪlətɪ]
公共供电点 point of common coupling
公共连接点 point of common coupling
功角 power angle
功角稳定 angle stability [stə'bɪlɪtɪ]
功率测量；电力参数测量 power measurement ['meʒəmənt]
功率持续曲线（负荷曲线） power duration curve
功率变化率 power ramp rate
功率方向保护 directional power protection [dɪ'rekʃənəl]
功率极限 power limit
功率控制 power control
功率流；电流量 power flows
功率密度 power density
功率调节 power regulation
功率系数 power coefficient
功率因数 power factor
功率因数表 power factor meter
功率因数校正 power factor correction (PFC)
功率因数校正电容器 power factor correction capacitor [kə'pæsɪtə]
共用接地系统 common earthing system
共振 resonance ['rezənəns]
共轴的，同轴的 coaxial ['kəʊ'æksəl]
供电电压 supply voltage [sə'plɑɪ]
供电干线 main supply line; supply main
供电局 power supply authority [ɔ:'θɒrətɪ]
供电网络 electrical supply network
固定串联电容补偿 fixed series capacitor compensation [kə'pæsɪtə]
固定式低压配电屏 fixed LT distribution panel
管式避雷 tubular arrester ['tju:bjʊlə]
硅控整流器；半导体闸流管 thyristor [θɑɪ'rɪstə]

H

合闸 switch on
合闸操作 closing operation
合闸回路 closing circuit
合闸继电器 closing relay
合闸脉冲 closing pulse [pʌls]
合闸时间 closing time
合闸速度 closing speed
合闸位置继电器 closing position relay
横差保护 transverse differential protection [ˌdɪfəˈrenʃəl]
后备保护 back-up protection
互感 mutual-inductor
互联 interconnection [ˌɪntəkəˈnekʃən]
户内交流金属铠装移动式开关柜 indoor AC armored movable switchgear [ˈɪndɔː] [ˈɑːməd] [ˈmuːvəbl] [ˈswɪtʃgɪə]
户内式 indoor (type)
户外式 outdoor (type)
护套电缆 sheathed cable [ʃiːət]
滑触线 trolley line [ˈtrɒlɪ]
换接 change-over circuit
换相开关 phase converter
换向 commutation
换向片 commutator segment
换向器 commutator

J

击穿电压 breakdown voltage
机端 generator terminal
机端电压控制 AVR
机组自用电 unit service power
基础接地体 foundation grounding
基准电压 reference voltage
极点 pole [pəʊl]
极化继电器 polarized relay
极数 number of poles
极限电流 limiting current
极限切除时间 critical clearing time
极限转矩 breakdown torque [tɔːk]
集成电路 IC (integrated circuit)
集电环 collector ring
集电线路 current collecting line/collection line
集肤效应 skin effect
计量屏 measurement panel [ˈmeʒəmənt] [ˈpænl]
计算容量 calculated capacity [kəˈpæsətɪ]
继电器 relay
架空电缆 overhead cable
架空电线 hook-up wire
架空干线 overhead main
架空线路 overhead line
架空引出 over-head leading out
架空引入 over-head leading-in
间接并网 indirect grid connection

[kə'nekʃən]

间接照明 indirect lighting

间隙 gap [gæp]

检修 overhauling [ˌəʊvə'hɔːlɪŋ]

检修接地 inspection earthing

减压起动器 voltage reducing starter

减压器 reducer [rɪ'djuːsə]

降压变压器 step-down transformer

交互作用（与电网） interaction (with grid)

交链聚乙烯电缆 XLPE cable (Cross Linked Polyethylene) [ˌpɒliː'eθəliːn]

交流 alternating current (AC)

交流操作 AC operation

交流电度表 AC kilowatt hour meter ['kɪləʊwɒt]

交流电机 alternating current machine

交流电流 alternating current ['ɔːltəˌneɪtɪŋ]

交流电压 alternating voltage

交流电源 AC power source

交流发电机 alternative machine/alternator

交流环电动机 ac motor

交流接触器 AC contactor

交流输电系统 AC transmission system [trænz'mɪʃən]

交流同步转换器 AC synchronous converter ['sɪŋkrənəs]

交流励磁机励磁系统 excitation system with alternate-current exciter [ɔːl'tɜːnət] [ɪk'saɪtə]

角式避雷器 horn arrester [ə'restə]

接插式母线 plug-in bus-bar

接触电压 touch voltage

接触器 contactor

接地 earthing

接地刀 earth switch

接地刀闸 earthing knife-switch

接地电路 earthed circuit [ɜːət]

接地电阻 earth resistance

接地电阻表 earth tester

接地干线 ground bus [graʊnd]

接地故障 earth fault

接地火花避雷器 earthing arrester

接地极 earth electrode

接地开关 earthing switch

接地螺栓 earthing bolt [bəʊlt]

接地母线 earth bus

接地系统 earthing system

接地线 earth conductor

接地线引入处 entrance of earth wire ['entrəns]

接地装置 earthing device

接零保护 neutral protection

接零干线 neutral main

接入电力系统 connecting wind farm to power network

接线板 terminal block ['tɜːmɪnəl]

接线端子 terminal

接线盒 terminal box; junction box ['dʒʌŋkʃən]

接线箱 connection box; junction box

接线柱 binding post

结合电容 coupling capacitor [kə'pæsɪtə]

解列 decouple
金属氧化物避雷器 metal-oxide arrester [ˈɒksaɪd]
进线 incoming line
近区供电 near region power supply [ˈriːdʒən]
晶闸管励磁系统（可控硅励磁系统） SCR excitation system（thyristor excitation)[ˌeksɪˈteɪʃən] [θaɪˈrɪstə]
静触头 fixed contact
静电感应 electrostatic induction [ɪˌlektrəˈstætɪk]
净电功率输出 net electric power output
静态开关 static switch
静态同步补偿器 static synchronous compensator
静止无功补偿 static var compensation (SVC)
绝缘 insulation [ˌɪnsjʊˈleɪʃən]
绝缘包布 insulating tape
绝缘比 insulation ratio
绝缘材料 insulating materials [məˈtɪərɪəlz]
绝缘电缆 insulated cable
绝缘电阻 insulation resistance
绝缘套管 insulating bushing
绝缘物 insulant [ˈɪnsjʊlənt]
绝缘子 insulator
均压环 grading ring [ˈgreɪdɪŋ]
均压网 voltage balancing net

K

卡口插座 bayonet socket [ˈbeɪənɪt]
开关 switch
开关操作 switching operations
开关电器 switching apparatus [ˌæpəˈreɪtəs]
开关熔丝 switch-fuse
开关设备 switchgear
开关站 switch station
开路特性 open-circuit characteristic [ˌkærəktəˈrɪstɪk]
开路运行 open-circuit operation
开启式开关 open switch
铠装电缆 armored cable [ˈɑːməd]
可变电阻 variable resistor [rɪˈzɪstə]
可调变压器 variable transformer
空气断路器 air(blast)circuit breaker
空载 no-load
空载电流 no-load current
空载电压 non load voltage
空载损耗 no-load loss
空载运行 no-load operation
控制角（移相角） control angle (delay angle)
控制器 controller
控制器增益 controller gain
控制设备 control gear
控制室 control building
控制台 console [kənˈsəʊl]
跨步电压 step voltage
跨度 span
跨接线；跨接；跨（短）接片

jumper
跨柱 across column [ˈkɒləm]
快速熔断器 quick fuse
框图 block diagram [ˈdɑɪəgræm]
馈电电缆 feed cable
馈线 feeder
扩大单元接线 multi-generator-transformer unit connection

L

雷电过电，雷涌 lightning surge [sɜːdʒ]
雷电过电压（大气过电压） lightning over-voltage
雷电或然率 lightning probability
雷电日 thunder day [ˈθʌndə]
雷电日数 number of lightning days
雷电闪络 lightning flash over
雷击 lightning stroke
离网 off-grid
励磁 excitation
励磁电压 exciting voltage
励磁机 exciter
励磁绕组 exciting winding
励磁系统 excitation system
励磁响应 excitation response
联合单元接线 united generator-transformer unit connection
联结 connection
联络变压器 interconnecting transformer
联络屏 connection panel
联络线 liaison line [lɪˈeɪzɒŋ]
链式 chain system [tʃeɪn]
两相短路 two-phase short circuit [feɪz]
两相短路电流 two-phase short-circuit current
临界电压 critical voltage[ˈkrɪtɪkəl]
灵敏性 sensitivity [ˌsensɪˈtɪvɪtɪ]
零线，中性线 neutral line(conductor)
零序阻抗 zero sequence impedance [ˈsiːkwəns] [ɪmˈpiːdəns]
六氟化硫断路器 sulfur hexafluoride breaker (SF_6 breaker)
六氟化硫封闭组合电器 SF_6 gas insulated switchgear(GIS)[ˈswɪtʃgɪə]
漏磁电抗 leakage reactance [rɪˈæktəns]
漏磁通 leakage flux
漏电 leakage
漏电保护开关 leakage protection switch
漏电断路器 residual current circuit-breaker[rɪˈzɪdjʊəl]
漏电继电器 leakage relay
漏电路径 leakage path
罗可夫斯基线圈 Rogowski coil
螺口插座 screw socket
螺旋式熔断器 screw fuse
裸导线 bare conductor
落地安装 installed on ground
落点 drop point
铝线 aluminum conductor [əˈljuːmɪnəm]
滤波器 filter [fɪltə]

M

马力 horsepower [ˈhɔːsˌpɑʊə]
马蹄形磁铁 horseshoe magnet [ˈhɔːsʃuː]
埋地敷设 led underground
脉动电流 pulsating current [pʌlˈseɪtɪŋ]
脉动电压 pulsating voltage
满载 full load
满载转矩 full-load torque [tɔːk]
灭磁 deexcitation (field-suppression, field-discharge) [ˌdiːeksɪˈteɪʃən]
灭磁过电压 deexcitation over voltage (field-discharge over voltage)
灭磁时间 deexcitation time (field-discharge time)
灭弧装置 arc-control device
明敷 exposed laying [ɪkˈspəʊzd]
明线 open wire
明装 surface mounted
模绕 form-wound
母差保护 bus differential protection
母线 bus, busbar
母线分段断路器 bus sectionalizing circuit breaker [ˈsekʃənəlaɪz]
母线间隔垫 bus-bar separator
母线联络断路器 bus interconnecting circuit breaker
母线伸缩节 bus-bar expansion

N

耐电压 proof voltage [pruːf]
内部过电压 internal over-voltage [ɪnˈtɜːnəl]
能量输送 power transfer
逆变器 inverter [ɪnˈvɜːtə]
逆流继电器 reverse-current relay

O

欧姆 ohm [əʊm]
耦合电容器 coupling capacitor
耦合器 electric coupling

P

盘面照明 dial lighting
旁路断路器 by-pass circuit breaker
配电电器 distributing apparatus [ˌæpəˈreɪtəs]
配电调度中心 distribution dispatch center

配电盘；开关屏 switch board

配电设备 power distribution equipment

配电所（变电所）distribution substation [ˈsʌbsteɪʃən]

配电网络 electrical distribution network

配电网自动化系统 distribution automation system [ˌɔːtəˈmeɪʃən]

配电系统 distribution system

配套电气设备 auxiliary electrical equipment

配电箱 distribution cabinet [ˈkæbɪnɪt]

配电站 distribution substation

频率 frequency [ˈfrɪkwənsɪ]

平衡负荷 balanced load

屏蔽电缆 shielded cable [ˈʃiːldɪd]

屏蔽接地 shielding earthing

屏幕显示 CRT display

Q

漆包电缆 lacquer-cover cable [ˈlækə]

起步阻力 breakaway force

起动电流 starting current

起动器 starter

起励 build-up excitation

汽轮发电机 turbogenerator [ˌtɜːbəʊˈdʒenəreɪtə]

气隙磁化线 air-gap line

气隙磁通 air-gap flux

铅包电缆 lead-covered cable [ˈkeɪbl]

铅避雷器 aluminum arrester [əˈljuːmɪnəm]

欠电压 under-voltage

欠频 under-frequency

欠压继电器 under-voltage relay

嵌装 flush mounted

强电 strong current

强行减磁 forced field discharge (forced decrease excitation)

强行励磁 reinforced excitation

强行励磁和强行灭磁 forced excitation and forced field discharge

强励倍数 forcing factor (forcing multiple)

墙式电缆槽 wall duct [dʌkt]

切断电流 cut-off current

侵入雷电波过电压 incoming lightning surge voltage [sɜːdʒ]

轻载开关 underload switch

球形避雷 spherical arrester [ˈsferɪkəl]

全厂公用电 common power demand of plant

全功率变流器 full-power converter (full converter)

全功率交流技术 full power variable flow mode

R

绕线转子 wound rotor
绕线转子感应电动机 wound-rotor induction motor
绕扎电缆 wrapped cable
绕组 winding
绕组系数 winding factor
热效应 thermal effect [ˈθɜːməl]
人工接地 artificial earthing [ˌɑːtɪˈfɪʃəl]
人工照明 artificial lighting
熔断器 cutout, fusible cutout [ˈfjuːzəbl]
熔丝电流 fuse current
柔性交流输电系统 Flexible AC Transmission System (FACTS) [ˈfleksɪbl]
入射的 incident
入网 grid access
软电缆 flexible cable [ˈfleksɪbl]
弱电 weak current
弱电控制 weak-current control
弱电网 weak grid

S

三角接法 delta connection
三类负荷 third-class load
三绕组变压器 three-column transformer (ThrClnTrans); three-winding transformer
三相变压器 three-phase transformer
三相插头 3-phase plug [feɪz]
三相电度表 three-phase kilowatt hour meter [ˈkɪləʊwɒt]
三相短路 three-phase short circuit
三相短路电流 three-phase short-circuit current
三相故障 three phase fault
三相四极插座 3-phase 4-pole socket [ˈsɒkɪt]
三相四线制 3-phase 4-wire system
三相五线制 3-phase 5-wire system
三向开关 three-way switch
三心电缆 three core cable
闪变 flicker [ˈflɪkə]
闪变阶跃系数 flicker step factor
闪络 flashover [ˈflæʃˌəʊvə]
上限 upper limit [ˈʌpə]
少油断路器 minimum oil circuit breaker
少油式断路器 low oil circuit breaker
设备线夹 terminal connector
设备装设容量 installed capacity
升压变压器 step-up transformer
升压去磁 boost-buck
失配 mismatch [ˌmɪsˈmætʃ]
失去同步 loss of synchronization [ˌsɪŋkrənaɪˈzeɪʃən]
施工用电 construction power sup-

ply
时不变的 time invariant [ɪn'veərɪənt]
时间相位 time-phase
时限 time lag
实心电缆 solid cable ['sɒlɪd]
事故照明 accident lighting
事故照明切换屏 accident lighting change-over panel ['pænl]
试验，维护 test，maintenance ['meɪntənəns]
视在功率 apparent power [ə'pærənt]
视在容量 apparent capacity
手车式高压开关柜 handcart type high voltage switch cabinet
手动（自动）准同期 manual（automatic） precise synchronization [ˌsɪŋkrənaɪ'zeɪʃən]
手动操动机构 hand control mechanism ['mekənɪzəm]
手动控制 manual control
输出功率 output power
输出连接 output coupling
输出特性（机组） output characteristic of WTGS
输出轴 output shaft
输电系统 transmission system
输电系统 power transmission system，utility grid [juː'tɪlətɪ]
输电线路 transmission line
鼠笼式电机 squirrel cage motor ['skwɜːrəl]
鼠笼转子 shrouded rotors
鼠笼型感应电动机 squirrel cage induction motor ['skwɜːrəl]
衰减 attenuation [əˌtenjʊ'eɪʃən]
双股电缆 paired cable [peəd]
双回路放射式 two-circuit radial system ['reɪdɪəl]
双回路供电 two-feeder supply
双回同杆并架 double-circuit lines on the same tower
双馈 double-feds
双绕组变压器 double-column transformer（DblClmnTrans）
水电站 hydro plant ['haɪdrəʊ]
水电阻 water-resistant [rɪ'zɪstənt]
水力发电站 hydro power station
水轮发电机 hydrogenerator
水轮机 hydraulic turbine [haɪ'drɔːlɪk]
水平接地极 horizontal electrode [ˌhɒrɪ'zɒntəl]
塑料外壳式断路器 moulded case circuit breaker（MCCB）

T

他励的 separately excited
调幅 Amplitude Modulation（AM） ['æmplɪtjuːd] [ˌmɒdjʊ'leɪʃən]
调谐电路 tuned circuit
调压变压器 variable transformer ['veərɪəbl]
铁损 iron-loss ['aɪən]
铁芯 iron core

停电 power failure [ˈfeɪljə]
同步 synchronism/synchronization [ˈsɪŋkrənɪzəm]
同步发电机 synchronous generator [ˈsɪŋkrənəs]
同步器 synchronizer [ˈsɪŋkrənɑɪzə]
同步调相机 synchronous condenser
同步系数 synchronous coefficient [ˌkəʊɪˈfɪʃənt]
同步转速 synchronous speed
同期（同步） synchronizing
同轴电缆 coaxial cable [ˈkəʊˈæksəl]
铜损 copper loss [ˈkɒpə]
铜芯电缆 copper core cable
凸极 salient-pole [ˈseɪljənt] [pəʊl]

W

外部过电压 external over-voltage [ɪkˈstɜːnəl]
万能转换开关 universal switch [ˌjuːnɪˈvɜːsəl]
万用电表 universal meter
稳定变压器 stabilizing transformer [ˈsteɪbəˌlɑɪzɪŋ]
稳定短路电流 steady state short-circuit current
稳压器 stabilizer [ˈsteɪbɪlɑɪzə]
涡流 eddy current [ˈedɪ]
无功补偿 reactive power compensation
无功补偿器 reactive power compensation device
无功补偿系统 D-VAR system
无功电度表 reactive kilovolt ampere-hour meter [ˈkɪləʊvəʊlt]
无功电流 reactive current
无功负载 reactive load
无功功率 reactive power
无功功率表 reactive power meter
无功功率补偿 reactive power compensation
无功功率收费 reactive power charges
无功损耗 active loss
无刷励磁系统（它励旋转硅二极管励磁系统） brushless excitation system
无刷直流电机 brushless DC motor
无载运行 non-load operation
误差检测器 error detector

X

下限 lower limit
限波器 line trap
限幅器 limiter
限流器 current limiter
线电压 line voltage
线间距离 distance between lines
线卡子 guy clip
线路补偿器 LDC (line drop com-

pensation)
线路及安装 line and installation [ˌɪnstəˈleɪʃən]
线圈绕组 coil winding [kɒɪl]
线性区 linear zone [ˈlɪnɪə]
线与中性点间的 line-to-neutral
限流电路 limited current circuit
相电压 phase voltage
相对电压变化 relative voltage change
相复励调节 phase compounding regulation [kəmˈpɑʊndɪŋ]
相位 phase
相位补偿 phase compensation [ˌkɒmpenˈseɪʃən]
相位超前(滞后) phase lead (lag)
相位裕度 phase margin [ˈmɑːdʒɪn]
相序 sequential order of the phase [sɪˈkwenʃəl]
相移 phase displacement (shift)
消弧接触器 arc extinction contactor [ɪkˈstɪŋkʃən]
消弧线圈 arc-suppressing coil
斜率 slope [sləʊp]
谐波 harmonics [hɑːˈmɒnɪks]
谐波励磁系统 harmonic excitation system
泄漏电流 leakage current
星形联结 star connection
蓄电池室 battery room
旋转变压器 rotary transformer
旋转磁场 rotating magnetic field [mægˈnetɪk]
旋转电机 electrical rotating machine
巡回检测 cyclic checking-measuring (cyclic detection) [ˈsɑɪklɪk]

Y

压力开关 pressure switch
沿……敷设 run along
氧化膜避雷器 oxide film arrester [ˈɒksɑɪd]
氧化锌 zinc oxide [ˈɒksɑɪd]
氧化锌避雷器 metal oxide arrester (MOA) [əˈrestə]
一次电流 primary current
一次电压 primary voltage
一次系统图 primary system diagram
一次主要设备 preliminary main equipment
一类防雷区 first class protection
一类负荷 first-class load
移频键控 frequency shift keying (FSK) [ˈfrɪkwənsɪ] [kiːɪŋ]
移相器 phase shifter [ˈʃɪftə]
乙烯绝缘软性电缆 vinyl cabtyre cable
异步电动机 asynchronous motor [eɪˈsɪŋkrənəs]
异步发电机 asynchronous generator [eɪˈsɪŋkrənəs]
引上 led-up
引下 led-down

应急电源 emergency source [ɪˈmɜːdʒənsɪ]
应急照明 emergency lighting
应急照明线 emergency lighting line
永磁发电机 permanent-magnet generator
永磁同步电机 permanent-magnet synchronism motor [ˈsɪŋkrənɪzəm]
涌流 inrush current [ˈɪnrʌʃ]
用电量 power consumption [kənˈsʌmpʃən]
油断路器 oil circuit breaker
油浸变压器 oil-immersed transformer
油浸电缆 oil-immersed cable
有功电流 active current
有功负载 active load
有功功率 active power
有功功率表 active power meter
有功损耗 active loss
有效值 virtual value，effective value
有载 on-load
有载分接 on-load tap changers
有载高压变压器 transformer fitted with on-load tap-changing [ˈfɪtɪd]
有载调压变压器 transformer fitted with OLTC
右手定则 right-hand rule
裕度 margin
原动机 prime mover
原理图 schematic diagram [skiːˈmætɪk]
云母 mica [ˈmɑɪkə]

Z

匝比；变比 turn ratio
杂散电感 stray inductance [ɪnˈdʌktəns]
杂散电容 stray capacitance [kəˈpæsɪtəns]
载波 carrier
暂态稳定 transient stability
增压 boost
闸刀开关 closing switch
斩波电路 chopper circuit
障碍照明 obstacle lighting [ˈɒbstəkl]
照明变压器 lighting transformer
照明电缆 lighting cable
照明电压 lighting voltage
照明方式 lighting pattern [ˈpætən]
照明负荷 lighting load
照明干线 main lighting line
照明馈电屏 lighting feeder panel [ˈpænl]
照明配电箱 lighting distributing box
照明支线 lighting branch line
真空断路器 vacuum circuit breaker [ˈvækjuəm]
真空开关 vacuum switch
整流器 rectifier [ˈrektɪfɑɪə]
正常照明 normal lighting

正常状态 normal condition
正序阻抗 positive sequence impedance [ˈsiːkwəns][ɪmˈpiːdəns]
支架 support
支线 branch line
直击雷 direct stroke
直击雷过电压 direct lightning stroke over-voltage
直接并网 direct grid connection
直接起动 direct starting
直接照明 direct lighting
直流 direct current (DC)
直流操作 DC operation
直流电 direct current
直流电动机 direct current motor
直流电机 direct current machine
直流电压 direct voltage
直流电源 DC power source
直流发电机 direct current generator
直流接触器 DC contactor
直流励磁机励磁系统 excitation system with direct-current exciter [ˌeksɪˈteɪʃən]
直埋 buried directly underground
直轴瞬变时间常数 direct axis transient time constant [ˈæksɪs]
制动线圈 restraint coil
中等电压 medium voltage
中性点 neutral point
中性点接地 neutral point earthing
中性点有效接地系统 system with effectively earthed neutral [ɪˈfektɪvlɪ]
中性分层 neutral stratification [ˌstrætɪfɪˈkeɪʃən]
中性线，零线 neutral
中压 middle voltage
中压输出 MV transmission
终端变电站 terminal substation [ˈsʌbsteɪʃən]
周期 period [ˈpɪərɪəd]
主变压器 main transformer
主触头 main contact
主电路 main circuit
主开关 main switch
转/分 revolutions per minute
转/秒 revolutions per second
转差率 slip
转子电阻 rotor resistance
转子铁芯 rotor core
装机容量 installed capacity
自动电压控制（自动电压调整，无功功率控制） automatic voltage control (AVC) (reactive power-voltage control) [ˌɔːtəˈmætɪk]
自动点火装置；自动启动器 self-starter
自动功率调整器 automatic power regulator
自动空气断路器 automatic air breaker
自动空气式星三角起动器 automatic air star-delta starter
自动励磁调节 automatic excitation control
自动励磁调节器（自动电压调节器） automatic excitation controller (automatic voltage regulator)

自动灭磁 automatic deexcitation, automatic field-suppression, automaticfield-discharge[ˌdiːeksɪ'teɪʃən]

自动频率控制（自动调频）automatic frequency control（AFC）

自动调压器 automatic regulator

自动重合闸 auto-reclosing（ARC）

自动自同期重合闸 automatic self-synchronization reclosing [ˌsɪŋkrənɑɪ'zeɪʃə]

自感 self-inductor

自励的 self-exciting

自耦变压器 auto-transformer

自适应控制系统 adaptive control system

自同期 self-synchronization [ˌsɪŋkrənɑɪ'zeɪʃən]

自动相关函数 auto correlation function

自感 self-induction

总配电箱 general distribution box [ˌdɪstrɪ'bjuːʃən]

阻波器 preventing reactor

阻抗 impedance

阻抗电压 impedance voltage

阻抗继电器 impedance relay

阻尼 damping

阻燃铜芯塑料绝缘电线 flame retardant copper core plastic insulated wire [rɪ'tɑːdənt] ['ɪnsəˌleɪtəd]

组合电器 assembled switch gear [ə'sembəld]

组合开关 combination switch [ˌkɒmbɪ'neɪʃən]

组合滤波器 combined filter

组合式互感器 combined transformer

左手定则 left-hand rule

自动电压控制器 automatic voltage controller（AVC）

自动调温器；恒温器 thermostat ['θɜːməustæt]

工 程 力 学

A

安全寿命 safe life

C

脆性材料 brittle material [ˈbrɪtl] [məˈtɪərɪəl]

脆性断裂 brittle fracture [ˈfræktʃə]

脆性损伤 brittle damage [ˈdæmɪdʒ]

D

单晶体材料 single crystalline materials [ˈkrɪstəlɑɪn] [məˈtɪərɪəlz]

低周疲劳 low cycle fatigue [fəˈtiːg]

抵抗力 resistance [rɪˈzɪstəns]

断裂韧性 fracture toughness [ˈtʌfnɪs]

多晶体材料 polycrystalline materials [ˌpɒlɪˈkrɪstəlɑɪn] [məˈtɪərɪəlz]

F

腐蚀疲劳 corrosion fatigue [kəˈrəʊʒən] [fəˈtiːg]

G

各向同性 isotropic [ˌɑɪsəʊˈtrɒpɪk]

过载效应 overloading effect

H

横向 transverse

宏观损伤 macroscopic damage [ˌmækrəʊˈskɒpɪk]

环境效应 environmental effect [ɪnˌvɑɪərənˈmentəl]

J

交变应力 alternating stress [ˈɔːltəˌneɪtɪŋ]

交变载荷 alternating load

金属的疲劳 fatigue in metals [fəˈtiːg]

晶体结构 crystal structure [ˈkrɪstəl]

均匀应变状态 homogeneous state of strain [ˌhɒməˈdʒiːnɪəs]

均匀应力状态 homogeneous state of stress

累积损伤 accumulated damage

N

能量释放率 energy release rate

P

疲劳 fatigue [fəˈtiːg]

疲劳断裂 fatigue fracture [ˈfræktʃə]

疲劳辉纹 fatigue striations

疲劳裂纹 fatigue crack

疲劳破坏 fatigue rupture [ˈrʌptʃə]

疲劳强度 fatigue strength [streŋθ]

疲劳失效 fatigue failure [ˈfeɪljə]

疲劳寿命 fatigue life

疲劳损伤 fatigue damage

疲劳阈值 fatigue threshold [ˈθreʃhəʊld]

Q

强度判据/强度准则 strength criterion [kraɪˈtɪərɪən]

R

韧性材料 ductile material [ˈdʌktaɪl] [məˈtɪərɪəl]

韧性断裂 ductile fracture

蠕变变形 creep deformation [kriːp] [ˌdiːfɔːˈmeɪʃən]

蠕变疲劳 creep fatigue [fəˈtiːg]

S

时间相关变形 time-dependent deformation [dɪˈpendənt]

塑性变形 plastic deformation

随机疲劳 random fatigue

['rændəm]

损伤变量 damage variable ['dæmɪdʒ] ['veərɪəbl]

损伤强化 damage strengthening ['streŋθənɪŋ]

损伤软化 damage softening ['sɒfnɪŋ]

损伤矢量 damage vector ['vektə]

损伤演化方程 damage evolution equation [ɪ'kweɪʒən]

损伤与断裂 damage and fracture ['fræktʃə]

损伤阈值 damage thresh old [θreʃ]

损伤张量 damage tensor ['tensə]

损伤准则 damage criterion [kraɪ'tɪərɪən]

T

弹性理论 theory of elasticity

W

微观损伤 microscopic damage [ˌmaɪkrə'skɒpɪk]

X

虚功原理 principle of virtual work

Y

应变能 strain energy ['enədʒɪ]

应变疲劳 strain fatigue [fə'ti:g]

应变协调方程 equation of strain compatibility [kəmˌpætə'bɪlətɪ]

应力比 stress ratio

应力不变量 stress invariant [ɪn'veərɪənt]

应力幅值 stress amplitude ['æmplɪtju:d]

应力集中 stress concentration [ˌkɒnsən'treɪʃən]

应力疲劳 stress fatigue [fə'ti:g]

应力强度因子 stress intensity factor [ɪn'tensətɪ]

应力循环 stress cycle

Z

纵向 longitudinal [ˌlɒndʒɪ'tju:dɪnəl]

机 械 工 程

A

安全带 safety belt

安全阀 safety valve [vælv]

安全方案 safety concept

安全联轴器 security coupling [sɪ'kjʊərɪtɪ]

安全帽 safety helmet ['helmɪt]

安装，组装 assemble [ə'sembl]

凹面铣刀 concave cutter [kɒn'keɪv]

奥氏体的 austenitic [ˌɒstɪ'nɪtɪk]

B

扳手 spanner

半径 radius

半透明的 translucent [trænz'ljuːsənt]

保持架 retainer [rɪ'teɪnə]

保持性 retention [rɪ'tenʃən]

变速箱 gearbox ['gɪəbɒks]

变位齿轮 gears with addendum modification [ə'dendəm]

变形 deformation

标准，原型 prototype ['prəutətaɪp]

表面淬火 case-hardening ['hɑːdənɪŋ]

表面加工 surface finish

剥线钳 wire-stripping pliers ['plaɪəz]

补偿轴对中偏差 compensate misalignment of shaft

不透明的 opaque [əʊ'peɪk]

C

槽 groove [gruːv]

槽钢 channels ['tʃænlz]

车床 lathe [leɪð]

车刀 lathe tool [tuːl]

车削 turning

衬垫 backing

衬套 bushing bolck ['bʊʃɪŋ]

尺寸稳定性 dimensional stability [dɪ'menʃənəl]

齿 tooth

齿槽 tooth space

齿顶圆 tip circle

齿高 tooth depth

齿根圆 root circle

齿厚 tooth thickness

齿距 pitch

齿宽 face width

齿廓修形 profile modification, pro-

file correction [ˈprəʊfaɪl]
齿轮 gear
齿轮泵 gear pump
齿轮齿条 pinion and rack
齿轮的变位 addendum modification on gears [əˈdendəm] [ˌmɒdɪfɪˈkeɪʃən]
齿轮副 gear pair
齿轮级 gear stage
齿轮架 wheel brace
齿轮加工 gear machining [məˈʃiːnɪŋ]
齿轮马达 gear motor [ˈməʊtə]
齿轮切削机 gear cutting machines
齿轮系 train of gears
齿面 tooth flank [tuːɵ]
齿啮式连接 dynamic coupling
齿数 number of teeth
冲击动载荷试验 impulse load tests [ˈɪmpʌls]
冲头 punch [pʌntʃ]
冲压机 stamping parts [ˈstæmpɪŋ]
冲子研磨器 punch formers
成套传动设备 complete drive systems
成型润滑剂 forming lubricant
传递比 transfer ratio
传动比 transmission ratio [trænzˈmɪʃən]
传动精度 transmission accuracy [ˈækjʊrəsɪ]
传动链 transmitted chains
传动误差 transmission error
传动系统 drive train
传送带，输送机 conveyor [kənˈveɪə]
纯度（润滑油） oil purity
从动齿轮 driven gear
淬火 quench [kwentʃ]
淬透性 hardenability [ˌhɑːdənəˈbɪlətɪ]
锉刀 file

D

大齿轮 wheel gear
带传动 belt drives
带锯 band saws
带轮 sheave
单级行星齿轮系 single planetary gear train [ˈplænɪtərɪ]
单卡头 single clamp
单螺旋齿轮 single-helical gear [ˈhelɪkəl]
单排滚珠轴承 single-row roller bearings
单向传输 simplex transmission
单向离合器 free wheeling [ˈhwiːlɪŋ]
单自由度系统 single degree of freedom systems [ˈfriːdəm]
挡板 orifice plate；stop plate [ˈɒrɪfɪs]
挡块 baffle plate [ˈbæfl]
刀尖 nose of tool
刀角 nose angle
刀具 cutter

刀片 blades
导管 conduit [ˈkɒndɪt]
低速轴 low-speed shaft
低碳铬镍钢 low carbon chrome-nickel steel
电池钻 battery power drill [drɪl]
电动刀具 electric power tools [ɪˈlektrɪk]
电镀 planting
电缆剪 cable cutter
电脑数控车床 CNC lathes
电脑数控机床配件 CNC machine tool fittings
电脑数控剪切机 CNC shearing machines
电脑数控磨床 CNC grinding machines
电脑数控镗床 CNC boring machines
电脑数控弯折机 CNC bending presses
电脑数控铣床 CNC milling machines
电脑数控线切削机 CNC wire-cutting machines
电脑数控钻床 CNC drilling machines [ˈdrɪlɪŋ]
电气钳 electric plier [ˈplɑɪə]
电子束焊接 electron beam welding
电子元件 electronic component [ˌɪlekˈtrɒnɪk]
电钻 electric screw driver
垫片；垫圈 washer [ˈwɒʃə]
垫片；填隙片 shim
吊架 hanger [ˈhæŋə]
吊装 lifting
顶料销 lifter pin
顶针板 ejector retainner plate [ɪˈdʒektə] [rɪˈteɪnə]
定位 orientation [ˌɔːrɪenˈteɪʃən]
定位板 guide plate [gɑɪd]
定位块 located block [ləʊˈkeɪtɪd]
定位驱动 orientation drive [ˌɔːrɪenˈteɪʃən]
定位圈 locating ring
定位销 located pin
定位支承块 supporting block for location
动力放大 dynamic magnification [ˌmægnɪfɪˈkeɪʃən]
动能 kinetic energy [kɪˈnetɪk]
抖动 whip
镀锌 zink spray [zɪŋk] [spreɪ]
镀锌层 zinc coating
断裂 fracture [ˈfræktʃə]
锻 forge [fɔːdʒ]
锻铝 forging，aluminium [ˌæljʊˈmɪnjəm]
锻模 forging dies
锻压机 presses，forging
锻造 forge
多级行星齿轮系 multiple stage planetary gear train [ˈmʌltɪpl] [ˈplænɪtərɪ]
多样性 versatility
多轴钻床 drilling machines，multi-spindle [ˈspɪndl]

F

阀门 valve
法兰连接 flange connection/joint [flændʒ]
法向齿距 normal pitch
反冲，无效行程 backlash ['bækˌlæʃ]
反馈 feedback
方栓；齿条；止转楔；花键 spline
方头螺栓 square-head bolt
方位角法兰 azimuth flange [flændʒ]
方位角分级 azimuthal binning
方位角位置 azimuthal position
方向齿 steering gear
防松螺母，对开螺母 locknut ['lɒknʌt]
防振锤 damper
飞溅润滑系统 splash-type lubricating system [splæʃ] ['luːbrɪkeɪtɪŋ]
非工作齿面 non-working flank
分度头 index head
分子链 molecular chain [məʊ'lekjʊlə]
粉末冶金 powder metallurgy [me'tælədʒɪ]
峰值 peak value
缝焊机 seam welding machine
幅值 amplitude ['æmplɪtjuːd]
俯仰力矩 pitch moment
复合材料 composite material ['kɒmpəzɪt] [mə'tɪərɪəl]
复合构造 composite structure ['kɒmpəzɪt]

G

盖板 cover plate
感应淬火 induction hardening
刚度 rigidity；stiffness [rɪ'dʒɪdətɪ] ['stɪfnɪs]
刚度中心 center of stiffness
刚度重量比 stiffness-to-weight ratio
刚体平面旋转 rigid body planar rotation
刚性 rigidity [rɪ'dʒɪdətɪ]
刚性齿轮 rigidity gear
刚性连接，刚性联轴器 rigid coupling
刚性轴 stiff shaft
高速车床 high-speed lathes
高速钻床 high-speed drilling machines
隔板 diaphragm ['dɑɪəfræm]
工件 workpiece ['wɜːkpiːs]
工具箱 kit
工作齿面 working flank
公差 tolerance ['tɒlərəns]
攻丝机 screw machine
共轭曲线 conjugate curves

[ˈkɒndʒʊgeɪt]
固定的 stationary
固定连接 integrated coupling
固定螺栓 fixing bolts
固定套 fixed set
固定销 dowel pin [ˈdaʊəl]
固有频率 natural frequency
[ˈfrɪkwənsɪ]
刮刀 scraper
挂板 clevis [ˈklevɪs]
挂钩 hook
挂环 link
管道 pipe
管钳子 box spanner
光电编码器 optical encoder
[ˈɒptɪkəl]
滚齿 hobbing [ˈhɒbɪŋ]
滚动接触轴承 rolling contact bearing
滚动轴承 rolling bearing
滚轴 roller
过渡配合 transition fit
过盈配合 interference fit
[ˌɪntəˈfɪərəns]

H

焊接 welded
焊接缺陷 welding imperfections
[ˌɪmpəˈfekʃəns]
合金 alloy
合力 resultant force
后轴承 rear bearing
虎钳 vises
护目镜 protection spectacles
[ˈspektəklz]
花键 spline
花键连接 splinted coupling
花键式连接 splined coupling
花键轴 splined shaft
花篮螺栓 turn buckle [ˈbʌkl]
划线 scribing [ˈskraɪbɪŋ]
划线盘 tosecan [təˈsekən]
滑动接触轴承 sliding contact bearing
滑动离合器 slipping clutch [klʌtʃ]
滑动制动器 sliding shoes
滑动轴承 sliding bearing
滑阀 slide valve
滑块 sliding block
滑块固定块 sliding dowel block
[ˈdaʊəl]
滑块连接 oldham coupling
滑轮 pulley [ˈpʊlɪ]
回火 temper
活动扳手 adjustable spanner
活动板 active plate
活塞 piston [ˈpɪstən]
火焰淬火 flame hardening

J

机床 machine tool [tuːl]
机床夹具 jig [dʒɪg]

机电一体化 mechanotronics；mechanical-electrical integration [mɪˈkænɪkəl]

机电的 electromechanical [ɪˌlektrəʊmɪˈkænɪkəl]

机械，力学 mechanism [ˈmekənɪzəm]

机械的，力学的 mechanical [mɪˈkænɪkəl]

机械加工余量 machining allowance [məˈʃiːnɪŋ] [əˈlɑʊəns]

机械零件 mechanical parts

激光钢板切割机 laser cutting for SMT stensil

激光切割 laser cutting

棘轮扳手 ratchet spanner [ˈrætʃɪt]

集成 integrate [ˈɪntɪgreɪt]

计算机辅助工程 computer-aided engineering (CAE) [ˌendʒɪˈnɪərɪŋ]

计算机辅助设计 computer-aided design (CAD)

计算机辅助制造 computer-aided manufacturing (CAM) [ˌmænjʊˈfæktʃərɪŋ]

计算机集成制造系统 computer-integrated manufacturing system (CIMS)

计算机数字控制 computerized numerical control (CNC) [kəmˈpjuːtərɑɪzd] [njuːˈmerɪkəl]

加工 machining

加工中心 machining center

加速度幅值 acceleration amplitude [əkˌseləˈreɪʃən] [ˈæmplɪtjuːd]

加速度踏板 acceleration pedal [ˈpedl]

夹具 fixture；clamping

夹盘 chucks

夹线器 conductor holder

间隔棒 spacer

间隙配合 clearance fit

减速箱 reduction gearbox [ˈgɪəbɒks]

减压阀 reducing valve

减震器 vibration isolator [ˈɑɪsəleɪtə]

剪切 shear

剪切机 shearing machines

渐开线 involute

键 key

键槽 keyseat

胶接 adhesive [ədˈhiːsɪv]

角锉 angle grinder [ˈgrɑɪndə]

角接触轴承 angular contact ball bearing [ˈkɒntækt]

角盘 angle plate

角速度 angular velocity

角铣刀 angle cutter

绞刀 reamer

绞孔 fraising

铰链接合 flapping hinge [hɪndʒ]

接合焊 seam welding

接口，界面 interface [ˈɪntəfeɪs]

节点 pitch point

节圆 pitch circle

结构动力学 structural dynamics [ˈstrʌktʃərəl] [dɑɪˈnæmɪks]

结构钢 structural steels [ˈstrʌktʃərəl]

截面 section [ˈsekʃən]

金工锯 metal saw

金属板成型机 sheet metal forming machines
金属板加工机 sheet metal working machines
金属腐蚀 corrosion of metals
金属工艺学 technology of metals
金属合金 metal alloy ['ælɒɪ]
金属疲劳 metal fatigue [fə'tiːg]
金属切削 metal cutting
金属网屏蔽 metal grating shield
金相分析仪 metallographic analyzer
金相显微仪 microstructure microscope
紧边 tight side [tɑɪt]
进给 fed
浸镀，热浸；热浸镀 hot-dip
径向销连接 radial pin coupling ['reɪdɪəl]
径向载荷 radial loads
镜头 lens
精细节距 fine pitch
径向速度 radial velocity [vɪ'lɒsətɪ]
径向销连接 radial pin coupling ['reɪdɪəl] ['kʌplɪŋ]
矩形的 rectangular ['rek'tæŋgjʊlə]
矩形螺纹 square thread
锯床 sawing machines
聚合体 polymer ['pɒlɪmə]
卷边工具 crimping tools [krɪmpɪŋ]
绝缘手套 insulating glove
绝缘靴 insulating boots
均压环 grading ring

K

卡口 bayonet ['beɪənɪt]
卡盘 chuck [tʃʌk]
卡钳 callipers
卡线钳 conductor clamp
开口扳手 open ring wrench
开环控制 open-loop control
开模槽 ply bar scot [skɒt]
抗拉强度 tensile strength ['tensɑɪl]
可拆卸的 demountable [diː'mɑʊntəbl]
可加工性 machinability [məʃiːnə'bɪlətɪ]
可塑性 moldability [ˌməʊldə'bɪlətɪ]
可调钳 adjustable pliers
孔加工 spot facing machining
扣链齿轮 sprocket

L

拉床 broaching machine
拉紧的 taut [tɔːt]
拉孔 broaching ['bəʊtʃɪŋ]
拉伸 pulling
老化试验 ageing tests
冷锻 cold forging
冷锻冲压机 presses, cold forging
冷挤压 cold extrusion [ek'struːʒən]
冷加工 cold machining

离合器 clutch [klʌtʃ]
离散悬臂梁 discretized cantilever beams [ˈkæntɪliːvə] [biːm]
离心的 centrifugal [senˈtrɪfjʊgəl]
离心压力机 presses eccentric [ɪkˈsentrɪk]
立式车床 vertical lathes
立式带锯 vertical band saws [ˈvɜːtɪkəl]
立式加工中心 vertical machining centers
立式刨床 vertical planing machines
立式铣床 vertical milling machines [ˈmɪlɪŋ]
立式钻床 vertical drilling machines
力臂 torque arm
力矩扳手 torque spanner/wrench [rentʃ]
力矩杆 torsion bar [ˈtɔːʃən]
力矩系数 moment/torque coefficients
连接法兰 connection flange
连接件 connection [kəˈnekʃən]
连续冲模 dies-progressive [prəʊˈgresɪv]
联板 yoke plate [jəʊk] [pleɪt]
联结 link
联轴器 coupling
链 chain
链传动 chain drives
梁 beam
临界阻尼 critical damping [ˈdæmpɪŋ]
零件 parts
六方扳手 allen wrench [ˈælən] [rentʃ]
六角螺帽 hexagon nut [ˈheksəgən]
六角头螺栓 hexagon headed bolt
轮齿弯曲应力 tooth bending stress
轮毂 hub
轮系；齿轮传动链 gear train
螺钉 screw
螺杆缸 screw cylinder
螺母 nut
螺栓，拧螺丝 bolt
螺栓法兰接头 bolted flange joint
螺栓紧固 bolt tightening
螺栓连接 bolted [bəʊtɪd]
螺栓疲劳应力 bolt fatigue stresses [fəˈtiːg] [stres]
螺栓载荷增量 bolt load increment [ˈɪnkrɪmənt]
螺栓组装 bolt assembly
螺丝 screw
螺丝攻 screw tap
螺丝起子 screw driver
螺纹 thread
螺纹插装阀 SICV
螺纹衬套 threaded bushes
螺纹管 solenoid [ˈsəʊlənɒɪd]
螺纹管状加热元件 ribbed tubular heating element
螺纹加工 thread processing
螺纹磨床 grinders thread
螺线管 solenoid
螺旋 helix [ˈhiːlɪks]
螺旋扳手 wrench [rentʃ]
螺旋测位器 micrometer screw gauge [mɑɪˈkrɒmɪtə] [geɪdʒ]
螺旋齿 spiral gear [ˈspɑɪərəl]

螺旋齿轮，斜齿轮 helical gearing
螺旋千斤顶 screw jack
螺旋涡 helical vortex [ˈvɔːteks]
螺旋涡片 helicoidal vortex sheet
螺旋形尾涡 helical wake
螺旋状的 helical

M

马氏体的 martensitic[ˌmɑːtɪnˈzɪtɪk]
埋弧焊 shielded metal arc welding
毛坯 rough
锚固点 rope suspensions
铆接 riveted
密封/封料 seal
密封垫圈 gasket
密封剂 sealants
密封系统 sealing system
迷宫式密封 labyrinth seal
模具 mold [məʊld]
模数 module [ˈmɒdjuːl]
模芯 mold core
摩擦 friction [ˈfrɪkʃən]
摩擦点 friction point
磨，碾 grind [graɪnd]
磨床 grinder
磨光器 polisher [ˈpɒlɪʃə]
磨合期磨损 running-in wear
磨轮 grinding wheels
磨损 wear；abrasion [əˈbreɪʒən]
磨削工具 grinding tools
母线间隔垫 bus bar separator [ˈsepəreɪtə]
母线伸缩节 bus-bar expansion [ɪkˈspænʃən]

N

耐腐试验 corrosion resistance tests [kəˈrəʊʒən]
耐用度 durability
耐张线夹 strain clamp
内齿轮 internal [ɪnˈtɜːnəl]
内齿轮副 internal gear pair
内齿圈 ring gear
内导柱 inner guiding post [ˈɪnə] [ˈgaɪdɪŋ]
内卡钳 internal callipers [ɪnˈtɜːnəl] [ˈkæləpəz]
内力 internal force
内六角螺钉 inner hexagon screw
内圈 outer race
内置锥 built-in coning
啮合 mesh
啮合部件 geared parts
啮合的 cogged [kɒgd]
啮合干涉 meshing interference
扭矩扳手 torqure spanner
扭力 torsion [ˈtɔːʃən]
扭转 twist
扭转刚度 torsional rigidity[ˈtɔːʃənəl]
扭转刚度系数 coefficient of torsional

P

抛光 buffing [ˈbʌfɪŋ]
刨 plane
刨床 planing machines
皮带传动 belt drive
匹配的 aligned
偏斜 deflection [dɪˈflekʃən]
平行的 parallel [ˈpærəlel]
平行轴齿轮副 gear pair with parallel axes
平面磨削 plane grinding
平速比 speed increasing ratio
评定标准 evaluation criteria [krɑɪˈtɪərɪə]
屏蔽环 shielding ring
剖面线 hatching [ˈhætʃɪŋ]

Q

起动力矩 starting torque
气垫板 air cushion plate [ˈkʊʃən]
气垫顶杆 air-cushion eject-rod
气动工具 pneumatic power tools [njuːˈmætɪk]
汽缸套 cylinder [ˈsɪlɪndə]
千斤顶 jack
牵引板 towing plate [ˈtəʊɪŋ]
铅锤测深 flumbing
钳工 locksmith [ˈlɒksmɪθ]
钳子 pinchers
嵌入 embed [ɪmˈbed]
强度 strength [streŋθ]
清理焊缝 trimming [ˈtrɪmɪŋ]
倾斜角 tilt angle
氰化 cyanide [ˈsɑɪəˌnɑɪd]
球面的 spherical [ˈsferɪkəl]
球墨铸铁材料 nodular cast iron structure
球面滚子轴承 spherical roller bearings [ˈsferɪkəl]
球头挂钩 ball-hook
球头挂环 ball-eye
球隙 sphere gap
球状石墨铸铁 spheroidal graphite iron [sfɪəˈrɔɪdəl] [ˈɡræfɑɪt]
球头挂钩 ball-hook
球头挂环 ball-eye
屈服点 yield point [jiːld]
曲尺；检验角尺 try square
全钢膜片 all-steel disc couplings

R

燃烧 combustion [kəmˈbʌstʃən]
热处理 heat treatment
热固性的 thermosetting [ˌθɜːməʊˈsetɪŋ]
热加工 hotwork
热塑性的 thermoplastic

[ˌθɜːməʊˈplæstɪk]
热轧 hot-rolled
人字齿轮 double-helical gear
韧性 toughness [ˈtʌfnɪs]
熔热处理炉 heating treatment funaces
柔性齿轮 flexible gear [ˈfleksɪbl]
柔性滚动轴承 flexible rolling bearing
柔性联轴器 flexible coupling
柔性套筒扳手 flexible spanner band [ˈspænə]
柔性支撑 resilient mountings
柔性轴 flexible shaft
柔性转子 flexible rotor [ˈfleksɪbl]
蠕变 creep [kriːp]
软管接头；软管总成 hose assembly
润滑 lubricate [ˈlʊbrɪkeɪt]
润滑系统 lubrication systems
润滑液 lubricants

S

三角带 V-belt
三维三分量模型 3-dimensional 3-component model
三爪、分割工具头 3-Jaws indexing spacers
三坐标测量仪 three-coordinates measuring machine
砂轮机 grinder [ˈgrɑɪndə]
砂轮修整器 wheel dressers
砂纸 sand paper
设定压力 setting pressure [ˈpreʃə]
深层淬火 through-hardening
深沟球轴承 deep-groove ball bearing [gruːv]
渗氮 nitride [ˈnɑɪtrɑɪd]
渗碳 carburization [ˌkɑːbjʊrɑɪˈzeɪʃən]
十字接头 cross joint
石墨 graphite [ˈgræfɑɪt]
石墨片 graphite flakes [ˈgræfɑɪt] [fleɪks]
实验设计 experimental design [ekˌsperɪˈmentəl]
手持式模具 handle mold
手动调整 manual adjustment
手提钻孔机 protable driller
手摇钻 hand brace
输出角 output shaft
输出连接 output coupling
输出轴 output shaft
输入角 input shaft
输送链 conveying chains [tʃeɪnz]
术语 terms [tɜːmz]
双卡头 double clamp
双列深沟圆柱滚珠轴承 double row deep groove ball bearing
双曲面齿轮 herring-bone gear [ˈherɪŋ] [bəʊn]
伺服电机 servomotor [ˈsɜːvəʊˌməʊtə]
松边 slack side
速度幅值 velocity amplitude

[vɪˈlɒsətɪ]
随机振动 random vibration [ˈrændəm]
缩回 retract [rɪˈtrækt]
锁扣 latch [lætʃ]

T

太阳轮 sun gear
碳氮共渗 carbonitride [ˌkɑːbəʊˈnaɪtraɪd]
碳化物 carbide [ˈkɑːbaɪd]
弹垫 spring washer
弹簧 spring
弹簧常数 spring constant
弹簧秤 spring balance
弹簧小圆规 spring-bow compass
弹簧箱 spring box
弹性 resilience [rɪˈzɪlɪəns]
弹性连接 elastic coupling [ɪˈlæstɪk]
弹性模量；弹性模数 elasticity modulus [ˌelæsˈtɪsətɪ] [ˈmɒdjʊləs]
弹性缩尺 elastic scaling
弹性阻力 elastic resistence
镗床 boring machine
陶瓷 ceramic [sɪˈræmɪk]
特征频率 eigen frequency [ˈfrɪkwənsɪ]
调节 modulate [ˈmɒdjʊleɪt]
调线线夹 jumper clamp
调心滚珠轴承 spherical roller bearing [ˈsferɪkəl]
调整板 adjusting plate [əˈdʒʌstɪŋ]
调整螺钉 adjusting screw
调质钢 hardened and tempered steel [ˈhɑːdənd] [ˈtempəd]
铁锤 hammer
铁合金 iron alloy [ˈaɪən]
铁素体的 ferritic [fəˈrɪtɪk]
凸轮 cam [kæm]
凸形铣刀 convex cutter [kɒnˈveks]
凸轮轴 camshaft [ˈkæmʃɑːft]
凸缘 flange [flændʒ]
凸缘接头 flange joint [dʒɒɪnt]
凸缘连接 flanged union
凸缘联轴器 flange coupling
凸缘螺母 flanged nut
推板 stripper plate [ˈstrɪpə]
退火 anneal [əˈniːl]
脱料板 striper plate
脱模钳 wiper
脱碳 decarburization [diːˌkɑːbjʊəraɪˈzeɪʃən]
陀螺负荷 gyroscopic loads [ˌdʒaɪərəsˈkɒpɪk]

W

外齿轮 external gear [ɪkˈstɜːnəl]
外观检查 visual inspection [ˈvɪʒjʊəl]
外圈 inner race

外形尺寸 overall dimension ['əʊvərɔ:l]
弯曲半径 bending radius
弯管机 tube bending machines [tju:b]
弯曲 flexure ['flekʃə]
弯曲刚度 flexural rigidity [rɪ'dʒɪdətɪ]
弯曲机 bending machines
弯曲应力 bending stress
万能磨床 universal grinding machines
万能手钳 universal pliers
万能铣床 universal milling machines
万向接头 knuckle joint ['nʌkl]
万向节 universal joint [ju:nɪ'vɜ:səl]
万向联轴器 universal coupling
万向套筒扳手 flexible spanner hand
万向轴 cardan shaft ['kɑ:dən]
位移幅值 displacement amplitude ['æmplɪtju:d]
涡轮 turbine
蜗杆 worm
蜗轮 worm wheel
蜗轮 worm gear
卧式带锯 horizontal band saws ['hɒrɪ'zɒntəl]
卧式加工中心 horizontal machining centers
卧式铣床 horizontal milling machines
无缝的 seamless
无外圈轴承 bearing without outer ring
无心磨床 centerless grinding machines
无增速齿轮 gearless
五金配件 hardware fitting ['hɑ:dweə]
误差 error; deviation

X

铣 mill
铣床 milling machines
铣刀 milling cutter
铣头 milling heads
系数 modulus ['mɒdjʊləs]
细碎物 chip
下滑块板 lower sliding plate
箱体 housing
镶嵌玻璃 glazing
项目管理 project management
销 pin
销轴连接 pin-connected
销子 stud
小齿轮 pinion ['pɪnjən]
校准 calibrate
楔子 wedge [wedʒ]
斜齿轮 helical gear
斜齿圆柱齿轮 helical gear single-helical gear
橡胶弹性元件 rubber elastic com-

ponent
橡胶构件 rubber profile
泄露 leakage [ˈliːkɪdʒ]
斜齿单螺旋齿轮 helical gear single-helical gear [ˈhelɪkəl]
斜齿轮 helical gear [ˈhelɪkəl]
芯串 strand
信号调理 signal conditioning and processing [prəʊˈsesɪŋ]
行星齿轮 planet gear
行星齿轮传动机构 planetary gear drive mechanism [ˈmekənɪzəm]
行星架 planet carrier
行星齿轮 planet gear
行星齿轮传动机构 planetary gear drive mechanism [ˈplænɪtərɪ] [ˈmekənɪzəm]
行星齿轮系 planetary gear train
行星架 planet carrier
行星轮轴承 planet wheel bearing
性能测试 performance text
虚拟样机 virtual prototyping [ˈvɜːtʃʊəl]
悬臂梁 cantilever beams [ˈkæntɪliːvə]
悬垂线夹 suspension clamp [səˈspenʃən]
旋转接头 rotating union
旋转弯曲 swivel bending [ˈswɪvəl]

Y

牙嵌式连接 castellated coupling
牙条 threaded rods
压模 pressing dies
压缩成型 compression molding
压纹效平 waffle die flattening [ˈwɒfəl] [ˈflætnɪŋ]
压铸冲模 die casting dies
压铸机 die casting machines
牙嵌式连接 castellated coupling
延展性 ductility
研磨 lapping
摇杆 racker
叶片 vane
液压 hydraulic pressure [haɪˈdrɔːlɪk]
液压泵 hydraulic pump
液压动力元件 hydraulic power units
液压缸 hydraulic cylinder
液压工具 hydraulic power tools
液压过滤器 hydraulic filter
液压回转缸 hydraulic rotary cylinders [ˈrəʊtərɪ]
液压马达 hydraulic motor
液压系统 hydraulic system
液压油 hydraulic fluid [ˈflu(ː)ɪd]
液压元件 hydraulic components [kəmˈpəʊnənt]
一致性 uniformity [ˌjuːnɪˈfɔːmətɪ]
仪表 meter
仪表盘 panel [ˈpænl]
溢流阀；安全阀 relief valve
应变 strain
应力 stress

应力分析 stress analysis [ə'næləsɪs]
硬度 rigidity [rɪ'dʒɪdətɪ]
硬度 hardness
硬母线固定金具 bus bar support
油封 oil seal
油冷却器 oil cooler
油漆 painting
游标卡尺 slide caliper ['kælɪpə]
有机无机材料 organic and inorganic material [ɔː'gænɪk] [ɪnɔː'gænɪk]
原材料 raw material
原理图 schematic diagram [skiː'mætɪk]
圆头手锤 ball peen hammer
圆周侧隙 circumferential backlash [səˌkʌmfə'renʃəl] ['bækˌlæʃ]
圆柱齿轮 cylindrical gear
圆柱状的 cylindrical [sɪ'lɪndrɪkəl]
圆锥滚子轴承 tapered roller bearing ['teɪpəd]
圆锥铰链 coning hinge
运动学 kinematics [ˌkɪnɪ'mætɪks]
运输装置 conveyor [kən'veɪə]

Z

增速比 speed increasing ratio
增速齿轮副 speed increasing gear
增速齿轮系 speed increasing gear train
振荡 oscillation [ˌɒsɪ'leɪʃən]
振动频率 vibration frequency
振动试验 vibration tests
正火 normalize ['nɔːməlaɪz]
正切 tangent ['tændʒənt]
支持系统 holding systems
直齿圆柱齿轮 spur gear [spɜː]
直角尺 square [skweə]
直角度 square trowel
直径 diameter
直径和半径 diameter and radius
直线运动 linear motion
止付螺丝 stop screw
制动器 brake
制作 fabrication [fæbrɪ'keɪʃən]
质量功能展开 quality function deployment [diː'plɒɪmənt]
质量体系要求 quality systems requirements [rɪ'kwaɪəmənt]
治具 jig
中空结构件 hollow structural sections（HSS）['sekʃənz]
中纹锉 second out file
中心距 center distance
中心轮 center gear
周期振动 periodic vibration [ˌpɪərɪ'ɒdɪk]
轴 shaft
轴衬 bushing
轴承 bearing
轴承配件 bearing fittings
轴承寿命 bearing life
轴承箱 bearing housings
轴承座 bearing seats

轴向齿距 axial pitch [ˈæksɪəl]
轴向分量 axial component
轴向推力 axial thrust load
轴向载荷 axial loads
肘接；弯头接合 toggle joint [ˈtɒgl]
主动齿轮 driving gear
主轴 spindle [ˈspɪndl]
主轴箱 headstock [ˈhedstɒk]
柱塞泵 piston pump
柱塞头 plug [plʌg]
柱销 pin
柱销套 roller
铸钢 casting steel
铸铁 cast iron [ˈɑɪən]
铸造 found；cast；mold [məʊld]
铸造设备 foundry equipment
专用扳手 special purpose spanner
转矩 torque
装配 assembly
装置 fixtures
锥齿轮 bevel gear [ˈbevəl]
自攻螺丝 self tapping screw
自锁螺母 locking nut
阻尼比 damping ratio
阻尼系数 damping coefficient [ˈdæmpɪŋ]
钻床 drill machine
钻床工作台 drilling machines bench [bentʃ]
钻模 jigs
钻石刀具 diamond cutters [ˈdɑɪəmənd]
钻台 drill stand
钻头 drills
左旋螺纹 left-hand thread
作动器 actuator [ˈæktjʊeɪtə]

监控与通信

B

半双工传输 half-duplex transmission [ˈdjuːpleks]
编码 encode [enˈkəʊd]
波特 baud [bɔːd]
波特速率 baud rate
波形采集 waveform acquisition

D

代码 code
单向传输 simplex transmission
单元控制 unit control
地址 address

F

返回信息 return information
分接头位置信息 tap position information
复位 reset

G

高增益 high-gain
告警 alarm [əˈlɑːm]
工作时间 operating time
公共供电点 point of common coupling
功率采集系统 power collection system [kəˈlekʃən]
共用接地系统 common earthing system
故障 fault
故障切除时间 fault clearing time [ˈklɪərɪŋ]

J

基准误差 basic error
集中控制 centralized control [ˈsentrəlaɪzd]
即插即用界面 plug-and-play system
记忆棒 USB memory stick
监控 supervisory control
监控系统 monitoring system [ˈmɒnɪtərɪŋ]
监视控制及数据采集 supervisory control and data acquisition

(SCADA) [ˈsjuːpəˌvaɪzərɪ] [ˌækwɪˈzɪʃən]

监视信息 monitored information

键盘操作 key-operate

接口 interface [ˈɪntəfeɪs]

就地控制 local control

K

可编程序控制 programmable control (PLC control) [ˌæpəˈreɪtəs]

可靠性 reliability [rɪˌlaɪəˈbɪlətɪ]

空指令 skip

控制电缆 control cable

控制电路 control circuit

控制电器 control apparatus

控制柜 control cabinet [ˈkæbɪnɪt]

控制器 controller

控制室 control room

控制台 control desk

控制装置 control device

L

累积值 integrated total integrated value [ˈɪntɪgreɪtɪd]

M

脉宽调制 pulse width modulation (PWM)

命令 command

模拟控制 analogue control [ˈænəlɒg]

模拟盘 analogue board [ˈænəlɒg]

模拟信号 analog signal [ˈænəlɒg]

P

批处理 batch processing

配电盘 switch board

屏蔽环 shielding ring [ˈʃiːldɪŋ]

屏幕显示 screen display

Q

起动信号 starting signal

前置机 front end processor [ˈprəʊsesə]

确认 acknowledgement [əkˈnɒlɪdʒmənt]

R

冗余技术 redundancy [riˈdʌndənsɪ]
软件平台 software platform [ˈplætfɔːm]
弱电控制 weak current control

S

设备故障信息 equipment failure information
设定值 set point value
实时 real time
实时控制 real-time control
事件信息 event information
试验数据 test data
视图信号处理 visualizing signal processing
数据电路 data circuit
数据库 data base
数据采集 data acquisition [ˌækwɪˈzɪʃən]
数据查询 data viewing
数据存储 data storage
数据电路 data circuit
数据分析 data analysis [əˈnæləsɪs]
数据库 data base
数据终端设备 data terminal equipment (DTE) [ˈtɜːmɪnəl]
数字控制 digital control
双工传输 duplex transmission

T

调制解调器 modulator-demodulator [diːˈmɒdjʊleɪtə]
调制器 modulator [ˈmɒdjʊleɪtə]
通信电缆 communication (telecommunication) cable [ˌtelɪkəˌmjuːnɪˈkeɪʃən]
通信和控制设备 communication and control equipment
通信线路 communication line
通信端口防雷 communication-port lightning protection

W

微机程控 minicomputer program
位 bit
温差电偶；热电偶 thermocouple [ˈθɜːməʊˌkʌpl]
温度传感器 temperature sensor
误差信号 error signal

无线电射频 RF（radio frequency）
无线干扰 radio interference [ˌɪntəˈfɪərəns]

X

系统软件 system software
现场可靠性试验 field reliability test
现场数据 field data
协议 protocol
信号地 signal earth
信号电缆 signal cable
信号电路 signal circuit
修复时间 repair time

Y

延时操作数 delay operator [ˈɒpəreɪtə]
遥测 telemetering；remote measuring [tɪˈlemɪtərɪŋ]
遥控 telecontrol；remote control [ˌtelɪkənˈtrəʊl]
遥调 teleregulation；remote regulation
遥信 telesignalisation
一对一控制方式 one-to-one control mode
以太网交换机 ethernet switch
译码 decode [ˌdiːˈkəʊd]
硬件平台 hardware platform [ˈhɑːdweə]
优先级（报警） fault warning priority
有效性 availability [əˌveɪləˈbɪlətɪ]
用户端口防雷 user-port lightning protection
用户应用程序 user application software
用户友好界面 user-friendliness
远程访问 remote access
远程监控系统 remote monitoring system（SCADA）
远程监视 tele-monitoring
远程显示服务 remote display service
远程终端机 remote terminal unit
远动 teleautomatics
远方控制 remote control
远方终端 remote-terminal unit（RTU）[ˈtɜːmɪnəl]
运输终端 remote terminal unit

Z

指示灯 display lamp
智能电网 smart grid
智能控制系统 intelligent control system
状态信息 state information
字节 byte

流 体 力 学

KDU 方程 KDU equation [ɪˈkweɪʒən]

U 形管 U-tube [tjuːb]

B

巴塞特力 Basset force

薄翼理论 thin-airfoil theory [ˈeəfɔɪl]

壁剪切速度 friction velocity [ˈfrɪkʃən] [vɪˈlɒsətɪ]

壁剪应力 skin friction, frictional drag [ˈfrɪkʃənl]

壁效应 wall effect

边界层 boundary layer [ˈbaʊndərɪ] [ˈleɪə]

边界层方程 boundary layer equation

边界层分离 boundary layer separation [ˌsepəˈreɪʃən]

边界层厚度 thickness of boundary layer

边界层理论 boundary later theory

边界元法 boundary element method [ˈmeθəd]

变分法 variational method [veərɪˈeɪʃənl]

表观粘度 apparent viscosity [əˈpærənt]

表面波 surface wave

表面力 surface force

表面张力 surface tension [ˈtenʃən]

伯格斯方程 Burgers equation [bɜːgəz]

伯努利定理 Bernonlli theorem [ˈθɪərəm]

伯努利方程 Bernoulli equation

不规则波 irregular wave [ɪˈregjʊlə]

不可压缩流 [动] incompressible flow [ˌɪnkəmˈpresəbl]

不可压缩流体 incompressible fluid [ˈflu(ː)ɪd]

不可压缩性 incompressibility

不稳定性 instability

C

糙率 roughness [ˈrʌfnɪs]

测速法 anemometry [ˌænɪˈmɒmɪtrɪ]

层流 laminar flow [ˈlæmɪnə]

层流边界层 laminar boundary layer [ˈlæmɪnə] [ˈbaʊndərɪ]

层流分离 laminar separation

掺气流 aerated flow

超空化流 supercavitating flow

超声速流 [动] supersonic flow [ˌsjuːpəˈsɒnɪk]

超压［强］over pressure
尺度效应 scale effect
冲击波 shock wave
出口 exit，outlet
出口压力 exit pressure
传播 propagation [ˌprɒpəˈgeɪʃən]
传导 conduction
传质系数 mass transfer coefficient
次层 sublayer [ˈsʌbˈleɪə]

D

单相流 single phase flow [feɪz]
当地马赫数 local Mach number
等熵流 isentropic flow [ˌɑɪsenˈtrɒpɪk]
等速流 homogeneous flow [ˌhɒməˈdʒiːnɪəs]
低速空气动力学 low-speed aerodynamics [ˌeərəʊdɑɪˈnæmɪks]
地面效应 ground effect
动力相似［性］dynamic similarity [dɑɪˈnæmɪk] [ˌsɪmɪˈlærɪtɪ]
动力粘性 dynamic viscosity [vɪˈskɒsətɪ]
动量方程 momentum equation
动量厚度 momentum thickness
动量交换 momentum transfer
动量守恒 conservation of momentum
动态响应 dynamic response
动态校准 dynamic calibration [ˌkælɪˈbreɪʃən]
对流 convection [kənˈvekʃən]
对流传热 convective heat transfer
对流扩散方程 convection diffusion equation

E

二次流 secondary flow [ˈsekəndərɪ]
二维流 two-dimensional flow [dɪˈmenʃənəl]

F

反流 reverse flow
反压 back pressure
非定常流 unsteady flow，non-steady flow [ˌʌnˈstedɪ]
非恒定流（不定常流）unsteady flow
非均匀流 non-uniform flow
非平衡流［动］non-equilibrium flow [ˌiːkwɪˈlɪbrɪəm]
非线性波 nonlinear wave [nɒnˈlɪnɪə]
非线性不稳定性 nonlinear instabil-

ity
分布 distribution
分层流 stratified flow [ˈstrætɪfɑɪd]
分离点 separation point
[ˌsepəˈreɪʃən]
分离流 separated flow
风洞 wind tunnel [tʌnəl]
风速管 pitot-static tube
冯·诺伊曼条件 von Neumann condition
弗劳德数 Froude number
浮力 buoyancy [ˈbɒɪənsɪ]
负压（负相对压强） negative pressure
附面层 boundary layer
附体激波 attached shock wave
附着点 attachment point
[əˈtætʃmənt]
附着涡 bound vortex [bɑʊnd]
[ˈvɔːteks]
复速度 complex velocity

G

高速空气动力学 high-speed aerodynamics [ˌeərəʊdɑɪˈnæmɪks]
管流 pipe flow, tube flow

H

含水层 aquifer [ˈækwɪfə]
焓厚度 enthalpy thickness
[enˈθælpɪ]
耗散 dissipation [dɪsɪˈpeɪʃən]
恒定流（定常流） steady flow
后缘 trailing edge [ˈtreɪlɪŋ]
滑移速度 slip velocity
环流 circulation
缓流 subcritical flow [ˌsʌbˈkrɪtɪkəl]
回流 back flow

J

积分方法 integral method
[ˈɪntɪgrəl]
激光多普勒测速计 laser Doppler anemometer, laser Doppler velocimeter [ˈdɒplə] [ˌænɪˈmɒmɪtə]
[ˌveləˈsɪmɪtə]
急流 supercritical flow
几何相似 geometric similarity
[dʒɪəʊˈmetrɪk]
计算流体力学 computational fluid mechanics [ˈflu(ː)ɪd]
计算区域 computational domain
[dəʊˈmeɪn]
迹线 path, path line
减阻 drag reduction
剪切层 shear layer

剪切流 shear flow
渐变流 gradually varied flow [ˈgrædjuəlɪ]
介质 medium [ˈmiːdɪəm]
进口 entrance，inlet [ˈentrəns]
近场流 near field flow
静压头 static head [ˈstætɪk]
局域相似 local similarity [ˌsɪmɪˈlærɪtɪ]
矩量法 moment method
绝对粗糙度 absolute roughness [ˈrʌfnɪs]
绝对压强 absolute pressure
绝热流 adiabatic flow [ˌædɪəˈbætɪk]
绝热边界层 adiabatic boundary layer [ˌeɪdɑɪəˈbætɪk]
绝热变化 adiabatic change
绝热变化图 adiabatic chart
绝热递增率 adiabatic lapse rate
绝热方程式 adiabatic equation
绝热放气 adiabatic degassing [diːˈgæsɪŋ]
绝热过程 adiabatic process
绝热冷却 adiabatic cooling
绝热流 adiabatic flow
绝热膨胀 adiabatic expansion [ɪkˈspænʃ(ə)]
绝热平衡 adiabatic equilibrium [ˌiːkwɪˈlɪbrɪəm]
绝热区 adiabatic region
绝热曲线 adiabatic curve
绝热梯度 adiabatic gradient [ˈgreɪdɪənt]
绝热线 adiabatic line
绝热压缩 adiabatic compression
绝热增温 adiabatic heating/warming
绝热蒸发 adiabatic evaporation [ɪˌvæpəˈreɪʃən]
绝热指数 adiabatic index
绝热贮能 adiabatic energy storage

K

均匀流 uniform flow [ˈjuːnɪfɔːm]
卡门涡街 Karman vortex street [ˈkaːmən]
开尔文定理 Kelvin theorem [ˈkelvɪn] [ˈθɪərəm]
可压缩流体 compressible fluid [kəmˈpresəbl] [ˈflu(ː)ɪd]
空化 cavitation [ˌkævɪˈteɪʃən]
空气动力学 aerodynamics [ˌeərəudɑɪˈnæmɪks]
空蚀 cavitation damage
跨声速流 [动] transonic flow [trænˈsɒnɪk]
扩散 diffusion [dɪˈfjuːʒən]
扩散段 diffuser
扩散率 diffusivity
扩散速度 diffusion velocity [dɪˈfjuːʒən]

L

来流 incoming flow [ˈɪnˌkʌmɪŋ]
雷诺比拟 Reynolds analogy
雷诺数 Reynolds number
离散涡 discrete vortex
黎曼解算子 Riemann solver
理想流动 ideal flow [ɑɪˈdɪəl]
理想流体 ideal fluid [ˈfluːɪd]
理想气体 ideal gas
连续介质 continuum [kənˈtɪnjʊəm]
连续介质假设 continuous medium hypothesis [hɑɪˈpɒθɪsɪs]
连续介质力学 mechanics of continuous media
连续介质流动 continuum flow
连续介质模型 continuum model
连续流 continuous flow [kənˈtɪnjʊəs]
涟漪 ripple [ˈrɪpl]
两相流 two-phase flow
临界雷诺数 critical Reynolds number
临界流 critical flow
临界热通量 critical heat flux
流场 flow field
流场校测 flow calibration [ˌkælɪˈbreɪʃən]
流出边界条件 outflow boundary condition [ˈbɑʊndərɪ]
流动参量 flow parameter
流动分离 flow separation
流动特性 flow characteristics [ˌkærəktəˈrɪstɪks]
流动阻塞 flow blockage [ˈblɒkɪdʒ]
流动稳定性 flow stability
流动显示 flow visualization [ˌvɪzjʊəlɑɪˈzeɪʃən]
流管 stream tube
流函数 stream function
流量 flow rate，flow discharge
流量计 flow meter
流面 stream surface
流入边界条件 inflow boundary condition
流沙，风沙 wind-drift sand
流速计 anemometer [ˌænɪˈmɒmɪtə]
流态 flow regime
流体动力学 fluid dynamics [dɑɪˈnæmɪks]
流体动力阻尼 fluid dynamic damping [dɑɪˈnæmɪk]
流体压力计 manometer [məˈnɒmɪtə]
流体黏性 fluid viscosity [vɪˈskɒsətɪ]
流体运动学 fluid kinematics [ˌkɪnɪˈmætɪks]
流体质点 fluid particle
流体阻尼 fluid damping
流线 stream/airflow line
流线汇集效应 streamline squeezing effect
流线弯曲效应 streamline curvature effect [ˈkɜːvətʃə]
流线型 streamline shape

流线型外廓 streamline contour [ˈkɒntʊə]
流线型罩 streamline fairing
流向分量 streamwise component [kəmˈpəʊnənt]
流向涡量 streamwise vorticity [vɔːˈtɪsətɪ]
流向压力梯度 streamwise pressure gradient [ˈgreɪdɪənt]
流注击穿 streamer breakdown

M

马赫波 Mach wave [mɑːk]
马赫角 Mach angle
马赫数 Mach number
马赫线 Mach line
马赫锥 Mach cone
马蹄涡 horseshoe vortex [ˈhɔːsʃuː] [ˈvɔːteks]
脉线 streak line
摩擦损失 friction loss
摩擦因子 friction factor

N

纳维-斯托克斯方程 Navier-Stokes equation
内流 internal flow
能量传递 energy transfer
能量法 energy method
能量方程 energy equation
能量厚度 energy thickness
能量守恒 conservation of energy [ˌkɒnsəˈveɪʃən]
能量输运 energy transport
牛顿流体 Newtonian fluid [njuːˈtəʊnɪən]
浓度 concentration [ˌkɒnsənˈtreɪʃən]
努塞特数 Nusselt number

O

欧拉方程 Euler equation
欧拉平衡方程 Euler's equilibrium equation [ˌiːkwɪˈlɪbrɪəm]
欧拉数 Euler number

P

喷管 nozzle [ˈnɒzl]
皮托管 pitot tube [ˈpiːtəʊ]
频率响应 frequency response
平面流 plane flow
普朗特数 Prandtl number
普雷斯顿管 Preston tube [ˈprestən]

Q

气动加热 aerodynamic heating [ˌeərəʊdaɪˈnæmɪk]
气动力 aerodynamic force
气动热力学 aerothermodynamics [ˈeərəʊˌθɜːməʊdaɪˈnæmɪks]
气动噪声 aerodynamic noise
气动中心 aerodynamic center
气化 gasification [ˌgæsɪfɪˈkeɪʃən]
气体动力学 gas dynamics
气体润滑 gas lubrication [ˌluːbrɪˈkeɪʃən]
前缘涡 leading edge vortex [ˈvɔːteks]
强迫对流 forced convection
区域分解 domain decomposition [ˌdiːkɒmpəˈzɪʃən]

R

扰动 disturbance，perturbation [dɪˈstɜːbəns] [ˌpɜːtɜːˈbeɪʃən]
绕射 diffraction [dɪˈfrækʃən]
热膜流速计 hot-film anemometer [ˌænɪˈmɒmɪtə]
热线流速计 hot-wire anemometer
人工压缩 artificial compression [ˌɑːtɪˈfɪʃəl]
人工黏性 artificial viscosity [vɪˈskɒsətɪ]
瑞利流 Rayleigh flow
瑞利数 Rayleigh number

S

三维流 three-dimensional flow [dɪˈmenʃənəl]
散度型 divergence form [daɪˈvɜːdʒəns]
散射 scattering
熵函数 entropy function [ˈentrəpɪ]
熵条件 entropy condition
熵通量 entropy flux [ˈentrəpɪ]
射流 jet [dʒet]
失速 stall
势 potential
势流 potential flow
数值边界条件 numerical boundary condition [njuːˈmerɪkəl] [ˈbaʊndərɪ]
数值耗散 numerical dissipation
数值扩散 numerical diffusion [dɪˈfjuːʒən]
数值模拟 numerical simulation
数值色散 numerical dispersion
数值通量 numerical flux
数值黏性 numerical viscosity [vɪˈskɒsətɪ]
水动［力］噪声 hydrodynamic noise

[ˌhaɪdrəʊdaɪˈnæmɪk]

水动力学 hydrodynamics

水静力学 hydrostatics

水翼 hydrofoil [ˈhaɪdrəʊfɔɪl]

速度环量 velocity circulation

速度亏损律 velocity defect law

速度剖面 velocity profile

速度势 velocity potential

T

泰勒不稳定性 Taylor instability

泰勒数 Taylor number

泰勒涡 Taylor vortex [ˈvɔːteks]

特征线法 method of characteristics [ˌkærəktəˈrɪstɪks]

投影法 projection method

湍流边界层 turbulent boundary layer

湍流分离 turbulent separation

W

外流 external flow [ɪkˈstɜːnəl]

完全气体 perfect gas

网格雷诺数 cell Reynolds number

微压计 micromanometer

尾流 wake flow

未扰动流 undisturbed flow [ˌʌndɪˈstɜːbd]

位移厚度 displacement thickness [dɪsˈpleɪsmənt]

温度边界层 thermal boundary layer

涡 eddy

涡层 vortex layer

涡动传递理论 eddy transfer theory

涡动黏性（系数） eddy viscosity [vɪˈskɒsɪtɪ]

涡动阻力 eddy resistance [rɪˈzɪst(ə)ns]

涡对 vortex pair

涡方法 vortex method

涡管 vortex tube

涡环 vortex ring

涡街 vortex street

涡量 vorticity

涡量方程 vorticity equation

涡量计 vorticity meter

涡量拟能 cnstrophy [ˈenstrəfɪ]

涡流 eddy current；vortex [ˈvɔːteks]

涡流，湍流 burble

涡流普朗特数 eddy Prandtl number

涡流损失 vortex losses [ˈvɔːteks]

涡流黏度模型 eddy viscosity model [vɪˈskɒsətɪ]

涡流柱 vortex cylinder [ˈsɪlɪndə]

涡流阻尼器 eddy current damper

涡面 vortex surface

涡片 vortex sheet

涡线 vortex line

涡旋 vortex

涡旋脱落 vortex shedding [ˈʃedɪŋ]

涡黏性 eddy viscosity
无反射边界条件 nonreflecting boundary condition
无滑移条件 non-slip condition
无量纲参数 dimensionless parameter [ˌɪrəʊˈteɪʃənəl]
无旋流 irrotational flow
无压流 free surface flow
无黏性流体 nonviscous fluid, inviscid fluid [nɒnˈvɪskəs] [ɪnˈvɪsɪd]
物理解 physical solution
物理区域 physical domain [dəʊˈmeɪn]

X

吸出 suction [ˈsʌkʃən]
细长度 slenderness [ˈslendənɪs]
细长体 slender body
相对粗糙度 relative roughness
相对压强 relative pressure
相似理论 similarity theory [ˌsɪmɪˈlærɪtɪ]
相似律 similarity law
相似性解 similar solution
响应频率 response frequency [ˈfrɪkwənsɪ]
消能 energy dissipation
斜迎风格式 skew-upstream scheme [skiːm]
修正微分方程 modified differential equation

Y

压［力］降 pressure drop
压［强水］头 pressure head
压差 differential pressure
压差阻力 pressure drag
压力能 pressure energy
压力梯度 pressure gradient
压力中心 pressure center
压强传感器 pressure transducer [trænzˈdjuːsə]
压强分布图 pressure distribution diagram
压强计 manometer [məˈnɒmɪtə]
压缩波 compression wave
亚声速流［动］subsonic flow [ˈsʌbˈsɒnɪk]
衍射 diffraction [dɪˈfrækʃən]
叶栅流 cascade flow
液体动力润滑 hydrodynamic lubrication
液体动力学 hydrodynamics
液体静力学 hydrostatics [ˌhaɪdrəʊˈstætɪks]
液体静压 hydrostatic pressure
依赖域 domain of dependence
翼弦 chord
翼型 airfoil
迎风格式 upstream scheme, upwind scheme

迎角 angle of attack
影响域 domain of influence
涌波 surge wave
有限体积法 finite volume method
有压流 pressure flow
诱导速度 induced velocity
  [vɪˈlɒsətɪ]
诱导阻力 induced drag
远场边界条件 far field boundary condition
远场流 far field flow
约束涡 confined vortex
匀熵流 homoentropic flow
运动相似 kinematic similarity
  [ˌkɑɪnɪˈmætɪk]
运动黏性 kinematic viscosity

Z

暂态流 transient flow
黏度测定法 viscosimetry
黏度计 vicosimeter] meter
黏性底层（黏性次层） viscous-sublayer
黏性流 [动] viscous flow
真空度 vacuity [væˈkjuːətɪ]
蒸发 evaporation [ɪˌvæpəˈreɪʃən]
正激波 normal shock wave
指数格式 exponential scheme
  [ˌekspəʊˈnenʃəl] [skiːm]
质量传递 mass transfer
质量力（体积力） mass force
质量守恒 conservation of mass
重力波 gravity wave [ˈgrævɪtɪ]
周期流 periodic flow [ˌpɪərɪˈɒdɪk]
轴对称流 axisymmetric flow
  [ˌæksɪsɪˈmetrɪk]
注入 injection
驻点 stagnation point [stægˈneɪʃən]
驻涡 standing vortex
状态方程 equation of state
  [ɪˈkweɪʒən]
锥形流 conical flow [ˈkɒnɪkəl]
准定常流 quasi-steady flow
准谱法 pseudo-spectral method
  [ˈpsjuːdəʊ]
子扩散 molecular diffusion
自由对流 natural convection，free convection [kənˈvekʃən]
自由流 free stream
自由流线 free stream line
自由面 free surface
自由射流 free jet
总压 [力] total pressure
总压头 total head [ˈtəʊtəl]

风能发电专业词汇释义*

* 中华人民共和国国家标准，电工术语，《风力发电机组》，IEC60050—415：1999.

风力机和风力发电机组

半直驱式风电机组 multibrid technology WTGS

风轮经过一级齿轮增速驱动多极中速发电机的风电机组。

保护系统（风力发电机组）protection system (for WTGS)

确保风力发电机组运行在设计范围内的系统。

变桨距 variable pitch

叶片与轮毂通过轴承连接，桨距角可改变。

变转速风轮 variable speed rotor

风电机组运行时转速可以在一定范围内变化的风轮。

垂直轴风力机 vertical axis wind turbine

风轮轴垂直的风力机。

定桨距 fixed pitch

叶片与轮毂固定连接，桨距角不可改变。

风电场 wind power station; wind farm

由一批风力发电机组或风力发电机组群组成的电站。

风力发电机组 wind turbine generator system ; WTGS

将风的动能转换为电能的系统。

风力机 wind turbine

将风的动能转换为另一种形式能的旋转机械。

风轮扫掠面积 swept area of rotor

与风向垂直的平面上，风轮旋转时叶尖运动所生成圆的投影面积。

风轮直径 rotor diameter

叶尖旋转圆的直径。

风轮转速 rotor speed

风力机风轮绕其轴的旋转速度。

风场 wind site

进行风能资源开发利用的场地、区域或范围。

关机 shutdown

从发电到静止或空转之间的风力机过渡状态。

桁架式塔架 lattice tower

由钢管或角钢等组成的框架式结构的塔架。

恒速风轮 constant speed rotor

风电机组运行时转速基本保持不变的风轮。

恒速恒频风电机组 constant speed-constant frequency WTGS

并网运行后风轮转速不能随风速改变，由电网频率决定的风轮转速和电能频率在运行时基本保持不变的风电机组。

机舱 nacelle

设在水平轴风力机顶部包容电机、传动系统和其他装置的部件。

近海风电场 offshore wind farm

距离海岸比较近，而且风电机组基础与海底连接的风电场。

空气动力刹车 aerodynamic brake

风轮旋转时增加叶片运动方向上的空气阻力，达到降低转速目的的制动装置。

桨距角 pitch angle

在指定的叶片径向位置（通常为100%叶片半径处）叶片弦线与风轮旋转面间的夹角。

桨距调节 pitch regulated

利用叶片桨距角的变化改变叶片升阻力，调节风电机组输出功率。

紧急关机 emergency shutdown

保护装置系统触发或人工干预下，使风力机迅速关机。

静止 standstill

风力发电机组的停止状态。

空转 idling

风力机缓慢旋转但不发电的状态。

控制系统（风力机）control system (for wind turbines)

接受风力机信息和/或环境信息，调节风力机，使其保持在工作要求范围内的系统。

轮毂 hub

将叶片或叶片组固定到转轴上的装置。

轮毂高度 hub height

对水平轴风力机是从地面到风轮扫掠面中心的高度，对垂直轴风力机是赤道平面高度。

偏航 yawing

风轮轴绕垂直轴的旋转（仅适用于水平轴风力机）。

偏航机构 yawing mechanism

使水平轴风电机组的风轮轴绕塔架垂直中心线旋转的机构。

上风向式风电机组 upwind WTGS

风先通过风轮再经过塔架的风电机组。

失速调节 stall regulated

在风轮转速基本不变的条件下，利用叶片的气动力失速原理改变叶片升阻力，调节风电机组输出功率。

水平轴风力机 horizontal axis wind turbine

风轮轴基本上平行于风向的风力机。

顺桨 feather

将桨距角变化到风轮叶片处于零升力或升力很小条件的状态。

锁定 blocking

利用机械销或其他装置，而不是通常的机械制动盘，防止风轮轴或偏航机构运动。

停机 parking

风力机关机后的状态。

停机制动 parking brake

能够防止风轮转动的制动。

推力系数 thrust coefficient

风作用在风轮上产生的轴向力与未扰动气流的动压及风轮扫掠面积的乘积之比。

尾流效应损失 wake effect losses

在风电场中，气流经过上游风电机组风轮后所形成的尾流对下游风电机组性能产生的影响和能量损失。

下风向式风电机组 downwind WTGS

风先通过塔架再经过风轮的风电机组。

叶根 root of blade

叶片上与轮毂连接的部位。

叶尖 tip of blade

叶片上在展向上与叶根距离最大的部位。

叶尖速比 tip speed ratio

叶尖速度与风速之比。

叶尖速度 tip speed

风轮旋转时叶尖的线速度。

叶片 blade

风力机的部件。叶片上的每个剖面设计成翼型，从叶根到叶尖，其厚度、扭角和弦长有一定分布规律，具有良好的空气动力外形。

叶片安装角 setting angle of blade

在叶片安装到轮毂上时设定的桨距角。

叶片长度 length of blade

叶片在展向上沿压力中心连线测得的最大长度。

叶片桨距角 pitch angle of blade

叶尖剖面的弦线与风轮旋转平面的夹角。

叶片扭角 twist angle of blade

叶片各剖面弦线和风轮旋转平面的夹角。

翼型 airfoil

风轮叶片横截面的轮廓。

圆筒式塔架 tubular tower

由钢板弯曲焊接成圆筒形结构的塔架。

正常关机 normal shutdown

全过程都是在控制系统控制下进行的关机。

整流罩 nose cone

装在轮毂上呈流线型的罩子。

支撑结构 support structure

由塔架和基础组成的风力机部分。

制动器 brake

能降低风轮转速或能停止风轮旋转的装置。

直驱式风电机组 direct drive WTGS, gearless WTGS

又称“无齿轮箱式风电机组”。风轮直接驱动多极低速发电机的风电机组。

设计和安全参数

安全寿命 safe life

严重失效前预期使用时间。

极限状态 limit state

构件的一种受力状态，如果作用其上的力超出这一状态，则构件不再满足设计要求。

设计工况 design situation

风力机运行中的各种可能的状态，例如发电、停车等。

设计极限 design limits

设计中采用的最大值或最小值。

使用极限状态 serviceability limit states

正常使用要求的边界条件。

潜伏故障 latent fault; dormant failure

正常工作中零部件或系统存在的未被发现的故障。

外部条件（风力机）external conditions (for wind turbines)

影响风力机工作的诸因素，包括风况、其他气候因素（雪、冰等）、地震和电网条件。

严重故障（风力机）catastrophic failure (for wind turbines)

零件或部件严重损坏，导致主要功能丧失，安全受损。

载荷状况 load case

设计状态与引起构件载荷的外部条件的组合。

最大极限状态 ultimate limit state

与损坏危险和可能造成损坏的错位或变形对应的极限状态。

风　特　性

安全风速 survival wind speed

结构所能承受的最大设计风速的俗称。一般不采用，设计时可参考极端风速。

标准风速 standardized wind speed

利用对数风廓线转换到标准状态(10m 高，粗糙长度 0.05m) 的风速。

参考风速 reference wind speed

用于确定风力机级别的基本极端风速参数。

注：

1. 与气候有关的其他设计参数均可以从参考风速和其他基本等级参数中得到。

2. 对应参考风速级别的风力机设计，它在轮毂高度承受的 50 年一遇 10 分钟平均最大风速，应小于或等于参考风速。

测风塔 wind measurement mast

安装风速、风向等传感器以及风数据记录器，用于测量风能参数的高耸结构。

粗糙长度 roughness length

在假定垂直风廓线随离地面高度按对数关系变化情况下，平均风速变为 0 时算出的高度。

对数风切变律 logarithmic wind shear law

表示风速随离地面高度以对数关系变化的数学式。

额定风速（风力机）rated wind speed (for wind turbines)

风力机达到额定功率输出时规定的风速。

风功率密度 wind power density

与风向垂直的单位面积中风所具有的功率。

风廓线 风切变律 wind profile; wind shear law

风速随离地面高度变化的数学表达式。

风能 wind energy

空气流动所具有的能量。

风能玫瑰图 wind energy rose

用极坐标来表示不同方位风能相对大小的图解。

风能密度 wind energy density

在设定时段与风向垂直的单位面

积中风所具有的能量。

风能资源 wind energy resources

大气沿地球表面流动而产生的动能资源。

[风能资源评估] 代表年 representative year for wind energy resource assessment

分析过去多年测风资料得到的一个典型年，其风能资源参数是未来风电场经营期内的预测平均值，作为估算风电机组年发电量的依据。

风切变 wind shear

风速在垂直于风向平面内的变化。

风切变幂律 power law for wind shear

表示风速随离地面高度以幂定律关系变化的数学式。

风切变指数 wind shear exponent

通常用于描述风速剖面线形状的幂定律指数。

风矢量 wind velocity

标有被研究点周围气体微团运动方向，其值等于该气体微团运动速度（即该点风速）的矢量。

注:空间任意一点的风矢量是气体微团通过该点位置的时间导数。

风数据记录器 wind data logger

记录并初步处理测风数据的电子装置。

风速 wind speed

空间特定点的风速为该点周围气体微团的移动速度。

注:风速为风矢量的数值。

风速分布 wind speed distribution

用于描述连续时限内风速概率分布的分布函数。

注：经常使用的分布函数是瑞利和威布尔分布函数。

风速廓线 wind speed profile, wind shear law

又称“风切变律”，风速随离地面高度变化的数学表达式。

[风速或风功率密度] 年变化 annual variation

以年为基数发生的变化。风速（或风功率变化）年变化是从1月到12月的月平均风速（或风功率密度）变化。

[风速或风功率密度] 日变化 diurnal variation

以日为基数发生的变化。月或年的风速（或风功率密度）日变化是求出一个月或一年内，每日同一钟点风速（或风功率密度）的月平均值或年平均值，得到0点到23点的风速（或风功率密度）变化。

风向 wind direction

风的流动方向（在风速超过2m/s时测量）。

风向玫瑰图 wind rose

用极坐标表示不同风向相对频率的图解。

极端风速 extreme wind speed

t秒内平均最高风速，它很可能是特定周期（重现周期）T年一遇。

注：参考重现周期T＝50年和T＝1年，平均时间t＝3秒和t＝

10秒。极端风速即为俗称的“安全风速”。

基准粗糙长度 reference roughness length

用于转换风速到标准状态的粗糙长度。

注：基准粗糙长度定为0.05m。

空气的标准状态 standard atmospheric state

空气的标准状态是指空气压力为101325Pa，温度为15℃（或绝对288.15K），空气密度1.225kg/m³时的空气状态。

年风速频率分布 annual wind speed frequency distribution

在观测点一年时间内，相同的风速发生小时数之和占全年总小时数的百分比与对应风速的概率分布函数。

年际变化 inter-annual variation

以30年为基数发生的变化。风速年际变化是从第1年到第30年的年平均风速变化。

年平均 annual average

数量和持续时间足够充分的一组测量数据的平均值，供作估计期望值用。

注：平均时间间隔应为整年，以便将不稳定因素如季节变化等平均在内。

年平均风速 annual average wind speed

按照年平均的定义确定的平均风速。

平均风速 mean wind speed

给定时间内瞬时风速的平均值，给定时间从几秒到数年不等。

气流畸变 flow distortion

由障碍物、地形变化或其他风力机引起的气流改变，其结果是相对自由流产生了偏离，造成一定程度的风速测量误差。

切入风速 cut-in wind speed

风力机开始发电时，轮毂高度处的最低风速。

切出风速 cut-out wind speed

风力机达到设计功率时，轮毂高度处的最高风速。

瑞利分布 RayLeigh distribution

经常用于风速的概率分布函数，分布函数取决于一个调节参数—尺度参数，它控制平均风速的分布。

上风向 upwind

主风方向的相反方向。

扫掠面积 swept area

垂直于风矢量平面上的，风轮旋转时叶尖运动所生成圆的投影面积。

湍流尺度参数 turbulence scale parameter

纵向功率谱密度等于0.05时的波长。

湍流惯性负区 inertial sub-range

风速湍流谱的频率区间，该区间内涡流经逐步破碎达到均质，能量损失乎略不计。

注：在典型的10m/s风速，惯性负区的频率范围大致在0.02Hz～2kHz间。

湍流强度 turbulence intensity

标准风速偏差与平均风速的比

率。用同一组测量数据和规定的周期进行计算。

威布尔分布 Weibull distribution

经常用于风速的概率分布函数，分布函数取决于两个参数，控制分布宽度的形状参数和控制平均风速分布的尺度参数。

注：瑞利分布与威布尔分布区别在于瑞利分布形状参数 2。

下风向 down wind

主风方向。

旋转采样风矢量 rotationally sampled wind velocity

旋转风轮上某固定点经受的风矢量。

注：旋转采样风矢量湍流谱与正常湍流谱明显不同。风轮旋转时，叶片切入气流，流谱产生空间变化。最终的湍流谱包括转动频率下的流谱变化和由此产生的谐量。

阵风 gust

超过平均风速的突然和短暂的风速变化。

注：阵风可用上升—时间，即幅度—持续时间表达。

主风向 prevailing wind direction

在风能玫瑰图中风能最大的方位。

由流风速 free stream wind speed

通常指轮毂高度处，未被扰动的自然空气流动速度。

最大风速 maximum wind speed

10 分钟平均风速的最大值。

电网连接

电力汇集系统（风力发电机组） power collection system (for WTGS)

汇集风电机组输出电能的电力连接系统。包括风电机组电能输出端与电网连接点之间的所有电气设备。

电网连接点（风力发电机组）network connection point (for WTGS)

对单台风力发电机组是输出电缆终端，而对风电场是与电力汇集系统总线的连接点。

额定功率（风力发电机组）rated power (for WTGS)

正常工作条件下，风力发电机组的设计要达到的最大连续输出电功率。

风场电气设备 site electrical facilities

风力发电机组电网连接点与电网间所有相关电气装备。

互联（风力发电机组）interconnection (for WTGS)

风力发电机组与电网之间的电力连接，从而电能可从风力机输送给电网，反之亦然。

输出功率（风力发电机组）output power (for WTGS)

风力发电机组随时输出的电功率。

最大功率（风力发电机组）maximum power（of a WTGS）

正常工作条件下，风力发电机组输出的最高净电功率。

功率特性测试技术

测量功率曲线 measured power curve

描绘用正确方法测得并经修正或标准化处理的风力发电机组净电功率输出的图和表。它是测量风速的函数。

测量扇区 measurement sector

测取测量功率曲线所需数据的风向扇区。

测量误差 uncertainty in measurement

关系到测量结果的，表征由测量造成的量值合理离散的参数。

测量周期 measurement period

收集功率特性试验中具有统计意义的基本数据的时段。

额定功率 rated power（for WTGS）

正常工作条件下，风力发电机组所能达到的设计最大连续输出电功率。

复杂地形带 complex terrain

风电场场地周围属地形显著变化的地带或有能引起气流畸变的障碍物地带。

分组方法 method of bins

将实验数据按风速间隔分组的数据处理方法。

注：在各组内，采样数与它们的和都被记录下来，并计算出组内平均参数值。

风障 wind break

相互距离小于3倍高度的一些高低不平的自然环境。

净电功率输出 net electric power output

风力发电机组输送给电网的电功率值。

功率曲线 power curve

描绘风电机组净电功率输出与风速的函数关系图和表。

功率特性 power performance

风力发电机组发电能力的表述。

功率系数 power coefficient

净电功率输出与风轮扫掠面上从自由流得到的功率之比。

精度（风力发电机组）accuracy（for WTGS）

描绘测量误差用的规定的参数值。

距离常数 distance constant

风速仪的时间响应指标。在阶梯变化的风速中，当风速仪的指示值达到稳定值的63%时，通过风速仪的气流行程长度。

可利用率（风力发电机组）availability（for WTGS）

在某一期间内，除去风力发电机组因维修或故障未工作的时数后余下的时数与这一期间内总时数的比值，用百分比表示。

年发电量 annual energy production

利用功率曲线和轮毂高不同风速频率分布估算得到的一台风力发电机组一年时间内生产的全部电能。计算中假设可利用率为100%。

年理论发电量 annual energy production

利用功率曲线和轮毂高度代表年的风速频率分布，估算得到的一台风电机组一年时间内生产的全部电能。计算中假设可利用率为100%。

日变化 diurnal variations

以日为基数发生的变化。

试验场地 test site

风力发电机组试验地点及周围环境。

数据组（功率特性测试）data set (for power performance measurement)

在规定的连续时段内采集的数据集合。

外推功率曲线 extrapolated power curve

用估计的方法对测量功率曲线从测量最大风速到切出风速的延伸。

障碍物 obstacles

邻近风力发电机组能引起气流畸变的固定物体，如建筑物、树林。

最大功率 maximum power (for WTGS)

正常工作条件下，风电机组输出的最高净电功率。

噪声测试技术

基准高度 reference height

用于转换风速到标准状态的约定高度。

注：参考高度定为10m。

基准距离 reference distance

从风力发电机组基础中心到指定的各麦克风位置中心的水平公称距离。

注：基准距离以米表示。

掠射角 grazing angle

麦克风盘面与麦克风到风轮中心连线间的夹角。

注：

1. 拒用“入射角”这一术语。
2. 掠射角以度表示。

声的基准风速 acoustic reference wind speed

标准状态下（10m高，粗糙长度等于0.05m）的8m/s风速。它为计算风力发电机组视在声功率级提供统一的根据。

注：测声参考风速以m/s表示。

声级 weighted sound pressure level; sound level

已知声压与20μPa基准声压比值的对数。声压是在标准计权频率和标准计权指数时获得。

注：声级单位为分贝，它等于上述比值以10为底对数的20倍。

声压级 sound pressure level

声压与基准声压之比的以10为底的对数乘以20，以分贝计。

注：对风力发电机组，基准声压为20μPa。

视在声功率级 apparent sound power level

在测声参考风速下，被测风力机风轮中心向下风向传播的大小为1pW点辐射源的A—计权声级功率级。

注:视在声功率级通常以分贝表示。

音值 tonality

音值与靠近该音值临界波段的遮蔽噪音级间的区别。

注:音值以分贝表示。

指向性（风力发电机组）directivity (for WTGS)

在风力机下风向与风轮中心等距离的各不同测量位置上测得的A—计权声压级间的不同。

注：

1. 指向性以分贝表示。
2. 测量位置由相关标准确定。

太阳能发电专业词汇释义

太阳电池相关词汇

太阳电池 solar cell

将太阳辐射能直接转换成电能的器件。

单晶硅太阳电池：single crystalline silicon solar cell

以单晶硅为基体材料的太阳电池。

多晶硅太阳电池 amulti crystalline silicon solar cell

以多晶硅为基体材料的太阳电池

非晶硅太阳电池 amorphous silicon solar cell

用非晶硅材料及其合金制造的太阳电池。

薄膜太能能电池 thin - film solar cell

用硅、硫化镉、砷化镓等薄膜为基体材料的太阳电池。这些薄膜通常用辉光放电、化学气相淀积、溅射、真空蒸镀等方法制得。

多结太阳电池 multijunction solar cell

由多个 p—n 结形成的太阳电池。

化合物半导体太阳电池 compound semiconductor solar cell

用化合物半导体材料制成的太阳电池。

带硅太阳电池 silicon ribbon solar cell

用带状硅制造的太阳电池。

光电子 photo - electron

由光电效应产生的电子。

太阳电池的伏安特性曲线 *I—V* characteristic curve of solar cell

受光照的太阳电池，在一定的辐照度和温度以及不同的外电路负载下，流入的电流 I 和电池端电压 V 的关系曲线。

短路电流 short - circuit current (I_{sc})

在一定的温度和辐照度条件下，光伏发电器在端电压为零时的输出电流。

开路电压 open - circuit voltage (V_{oc})

在一定的温度和辐照度条件下，光伏发电器在空载（开路）情况下的端电压。

最大功率 maximum power (P_m)

在太阳电池的伏安特性曲线上，电流电压乘积的最大值。

最大功率点 maximum power point

在太阳电池的伏安特性曲线上对应最大功率的点，亦称最佳工作点。

最佳工作点电压 optimum operating voltage (V_n)

太阳电池伏安特性曲线上最大功率点所对应的电压。

最佳工作点电流 optimum operating current (I_n)

太阳电池伏安特性曲线上最大功率点所对应的电流。

填充因子 fill factor (curve factor)

太阳电池略去串联电阻和并联电阻之后，最大功率与开路电压和短路电流乘积之比。评估太阳电池负载能力的重要参数。

$$FF=(I_m \times V_m)/(I_{sc} \times V_{oc})$$

式中：I_{sc}为短路电流；V_{oc}为开路电压；I_m为最佳工作电流；V_m为最佳工作电压。

曲线修正系数 curve correction coefficient

测试太阳电池时，由于温度的不同而引起伏安特性曲线的变化，修正此项变化的系数称曲线修正系数。

太阳电池温度 solar cell temperature

太阳电池中势垒区的温度。

串联电阻 series resistance

太阳电池内部的与 p—n 结或 MIS 结等串联的电阻，它是由半导体材料体电阻、薄层电阻、电极接触电阻等组成。

并联电阻 shunt resistance

太阳电池内部的、跨连在电池两端的等效电阻。

转换效率 cell efficiency

受光照太阳电池的最大功率与入射到该太阳电池上的全部辐射功率的百分比。

暗电流 dark current

在光照情况下，产生于太阳电池内部与光生电流方向相反相成的正向结电流。

暗特性曲线 dark characteristic curve

在无光照条件下给太阳电池施加外部偏压所得到的伏安特性曲线。

光谱响应 spectral response (spectral sensitivity)

在各个波长上，单位辐照度所产生的短路电流密度与波长的函数关系。

太阳电池组件相关词汇

太阳电池组件 module (solar cell module)

具有封闭及内部联结的、能单独提供直流电输出的，最小不可分割的太阳电池组合装置。

隔离二极管 blocking diode

与太阳电池组件或太阳电池板串联的二极管，用于防止反向电流流过它们。

旁路二极管 bypass (shunt) diode

与太阳电池组件或太阳电池板并联的二极管，当部分太阳电池、太阳电池组件或太阳电池板被遮或损坏时，方阵中的太阳电池可由旁路二极管形成通路，以防止热斑现象。

组件的电池额定工作温度 NOCT (nominal operating cell temperature)

在辐照度为 $800W/m^2$ 温度为 20℃、风速为 1m/s 状态，在中午时太阳光垂直照射到敞开安装于框架中的组件上，在这个标准参考环境中，测得的组件内太阳电池的平均平衡温度中组件的电池额定工作温度。

短路电流的温度系数 temperature coefficients of I_{sc}

在规定的试验条件下，被测太阳电池温度每变化1℃，太阳电池组件短路电流的变化值。通常用α来表示。

开路电压的温度系数 temperature coefficients of V_{oc}

在规定的试验条件下，被测太阳电池温度每变化1℃，太阳电池组件开路电压的变化值。常用β来表示。

峰值功率的温度系数 temperature coefficients of Pm

在规定的试验条件下，被测太阳电池温度每变化1℃，太阳电池组件峰值功率的变化值。

组件效率 module efficiency

按组件外形（尺寸）面积计算的转换效率。

峰瓦 watts peak

太阳电池组件方阵，在标准测试条件下的额定最大输出功率。

额定功率 rated power

在规定的工作条件下，光伏发电器在额定电压下所规定的输出功率。

额定电压 rated voltage

在规定的工作条件下，依据同一类型光伏发电器的特性选择确定其输出电压，使这一类光伏发电器的输出功率都接近最大功率，这个电压叫额定电压。

额定电流 rated current

在规定的工作条件下，光伏发电器在额定电压下所规定的电流。

光伏电站相关词汇

太阳能光伏系统 solar photovoltaic (PV) system

包含所有逆变器（单台或多台）和相关的BOS（平衡系统部件）以及具有一个公共连接点的太阳电池方阵在内的系统。

并网太阳能光伏发电系统 Grid-Connected PV system

与电网连接的太阳能光伏发电系统，太阳能光伏方阵输出的电经逆变器转换后输入电网。在有公用电网的地区，一般采用并网运行光伏发电系统。

独立太阳能光伏发电系统 Stand alone PV system

需要有蓄电池作为储能装置，主要用于无电网的边远地区。由于必须有蓄电池储能装置，所以整个系统的造价比较高。

太阳能控制器 solar controller

太阳能控制器的作用是控制整个系统的工作状态，并对蓄电池起到过充电保护、过放电保护的作用。在温差较大的地方，合格的控制器还应具备温度补偿的功能。

逆变器 inverter

静态功率变换器，将光伏系统的直流电变换成交流电的设备。用于将光电功率变换成适用于电网使用的一种或多种形式的电功率的电气设备。

孤岛效应 islanding

电网失压时，光伏系统仍保持对失压电网中的某一部分线路继续供电的状态。

逆变器变换效率 inverter efficiency

逆变器的有用输出功率与输入功率的比值。

方阵（太阳电池方阵）array（solar cell array）

由若干个太阳电池组件或太阳电池板在机械和电气上按一定方式组装在一起并且有固定的支撑结构而构成的直流发电单元。地基、太阳跟踪器、温度控制器等类似的部件不包括在方阵中。

子方阵 sub-array（solar cell sub-array）

如果一个方阵中有不同的组件或组件的连接方式不同，其中结构和连接方式相同部分称为子方阵。

充电控制器 charge controller

按预定方式给蓄电池组充电，并根据蓄电池的荷电程度及时改变充电速率，防止过充电的控制装置。

直流/直流电压变换器 DC/DC converter（inverter）

把直流电压升高或降低的设备。

直流/直流电压变换器 DC/AC converter（inverter）

把直流电变成交流电的设备。

电网 grid

输电、配电的各种装置和设备、变电站、电力线路或电缆的组合。它把分布在广阔地域内的发电厂和用户联接成一个整体，把集中生产的电能配送到众多个分散的电能用户。在本标准中特指供电区电力变压器次级输出到用户端的输电网络。

太阳跟踪控制器 sun-tracking controller

使太阳电池方阵或测试设备按规定要求对准太阳的一种装置。

电网保护装置 protection device for grid

监测光伏系统电力并网的技术状态，在指标越限情况下将光伏系统与电网安全解列的装置。

并网接口 utility interface

在光伏系统与电网配电系统之间的互相联接。泛指发电设备与电网之间的并解列点。

光伏系统有功功率 active power of PV power station

光伏系统输出的总有功功率。

光伏系统无功功率 reactive power of PV power station

光伏系统输出的总无功功率。

光伏系统功率因数 power factor of PV power station

由光伏系统输出的总有功功率与总无功功率计算而得到的功率因数。一段时间内的平均功率因数（PF）计算公式为

$$PF=\frac{P_{out}}{\sqrt{P_{out}^2+Q_{out}^2}}$$

式中：P_{out}为光伏系统输出的总有

功功率，kW·h；Q_{out}为光伏系统输出的总无功功率，kvar·h。

公共连接点 point of common coupling

光伏系统与所接入的电力系统的连接处

接线盒 junction box

太阳能电池组件与外电路之间的连接器，接线盒通过硅胶与组件的背板粘在一起，组件内的引出线通过接线盒内的内部线路连接在一起，内部线路与外部线缆连接在一起，使组件与外部线缆导通。

发电量 power generation

在确定的时段内，光伏电站从光伏逆变器送出的总电量。

汇流箱 combiner box

在太阳能光伏发电系统中，为了减少太阳能光伏电池阵列与逆变器之间的连线使用到汇流箱。

配电箱 distribution box

按电气连线要求将开关设备、测量仪表、保护电器和辅助设备组装在金属柜中，构成配电装置。

电能表 supply meter

用来测量电能的仪表，又称电度表，火表，电能表，千瓦小时表，指测量各种电学量的仪表。

变压器 transformer

利用电磁感应的原理来改变交流电压的装置。

太阳能光伏建筑一体化 Building-integrated PV (BIPV)

直接将太阳电池组件做建筑材料，这样既可以做建材又可以发电，使太阳能光伏发电与建筑结合。

光伏电站检测相关词汇

U 辐射 radiation

以电磁波的形式或粒子（光子）形式传播能量的过程。

太阳辐照度 solar radiation

太阳辐射经过大气层的吸收、散射、反射等作用后到达固体地球表面上单位面积单位时间内的辐射能量。其单位为：瓦/平方米（W/m^2）。

散射辐照（散射太阳辐照）量 diffuse irradiation (diffuse insolation)

在一段规定的时间内，照射到单位面积上的来自天空辐射能量。

直射辐照 direct irradiation (direct insolation)

在一段规定的时间内，照射到单位面积上的来自天空太阳圆盘及其周围对照射点所张的半圆锥角为8°的辐射能量。

总辐射度（太阳辐照度）global irradiance (solar global irradiance)

入射于水平表面单位面积上的全部的太阳辐射通量。

辐射计 radiometer

测量辐射能的仪器。

依顺时针方向到目标方向线之间的

水平夹角。

倾斜角 tilt angle

太阳电池组件与水平线之间的夹角。

太阳常数 solar constant

在地球的大气层外，太阳在单位时间内投射到距太阳平均日地距离处垂直于射线方向的单位面积上的全部辐射能。

大气质量（AM）air mass

大气质量是太阳光束穿过大气层的光学路径，以该光学路径与太阳在天顶时其光束到达海平面所通过的光学路径的比值来表示。

光热发电相关词汇

开口直径 aperture diameter width

指槽式聚光镜抛物槽的开口直径大小。

方位角 azimuth angle

方位角又称地平经度（缩写Az)，是在平面上量度物体之间的角度差的方法之一。是从某点的指北方向线起，依顺时针方向到目标方向线之间的水平夹角。

空气质量 AM，air mass 的缩写

直射阳光光束透过大气层所通过的路程，以直射太阳光束从天顶到达海平面所通过的路程的倍数来表示。

集热器 collector

直接将太阳能转化为热能，使用高储热的物质诸如水或油等，之后使用热交换器使用所搜集的热量。是聚光太阳能设备的总称，其中包括，concentrator（聚光镜）和 receiver（接收器）。

集热回路 collecting loop

槽式太阳能集热回路包括两种模式，一种是双回路系统，包括导热油（HTF 系统）和（水蒸汽系统），另一种叫 DSG 系统，直接产生蒸汽系统。两种系统的区别在于，第一种是由导热油做为热量转换的中间介质，而后者是太阳能直接转化为水蒸气的热能。第一种系统效率低于第二种，第二种技术对集热管要求较高。

聚光比 concentration ratio

包括几何聚光比和能量聚光比。

陶瓷 ceramics

在槽式太阳能热管中，关键的技术就是在集热管的高温选择性吸收涂层。该图层要求即要有较高的吸收率又要有较低的发射率。而目前研发的有效集热管涂层采用的是 Ni/SiO_2 金属陶瓷和 $Mo-Al_2O_3$ 金属陶瓷。

菲涅尔透镜 Fresnel lens

用微分切割原理制成的薄板式透镜。

几何聚光比 geometrical concentrator ratio

聚光器面积与接收器面积之比叫几何聚光率。

地面覆盖率 GCR(Group Cover Tatio)

GCR为抛物槽聚光镜开口直径/两排槽式集热器中心线间的距离。一般取值为0.33，槽式聚光镜不同公司其开口直径不同，当今常规的槽式聚光镜开口槽的直径为5.77，采用的是西班牙的DISS电站的聚光镜尺寸（Skyfuel在设计中开口槽直径取值为6）。

光照强度 intensity

物理测量的单位面积的光照功率，单位是瓦特每平方米。

熔融盐 molten-salt

槽式太阳能电站中，带储能的系统中用到的一种储能物质。

单膜式镜片 monolithic stretched-membrane

指蝶式太阳能聚光镜的一种聚光镜形式。

支架 pylons

用于支撑聚光镜和抵御风载的钢结构，槽式聚光镜支架主要有扭矩盒式支架和扭管式支架。

反射薄膜技术 ReflecTech

在玻璃镜面上贴服铝制活银制薄膜，用以提高反射率。

二次聚光器 secondary concentrator

将通过聚光器的会聚阳光再一次进行会聚的光学装置叫二次聚光器。

照射方向自动跟踪 setpoint tracing

通过校对太阳板，从而使太阳光一直是垂直照在太阳板上。

太阳照射常数 solar constant

在日地平均距离条件下，地球大气上界垂直于太阳光线的面点所接受的太阳辐射通量密度为1.395 W/m^2，瓦每平方米。

太阳小时 sun hours

是在一个地区每年一共的光照小时数，是衡量一个地区是否适合安装太阳能发电装置的最重要的因素。

光谱 sun spectrum

由发光物质直接产生的光谱称为发射光谱。

连续光谱：由连续分布的一切波长的光组成，是炽热的固体，液体及高压气体发光产生的光谱，诸如太阳的光谱，在地球上观察太阳光谱很大受大气的影响。

系统效率 system efficiency

指光热系统向电网传输的能量除以光照能量。

太阳能倍数 solar multiple

太阳能镜场的热输出/发电机组的热输入。

太阳能发电系统 SEGS (Solar Electric Generation Systems)

槽式太阳能发电系统中的一种系统，在美国，由美国LUZ公司投资成立的。除了SEGS外，还有西班牙的DISS系统，西班牙的Andasol系统，美国的Solar－one系统等，其区别在于，在系统建设过程中，采用的系统流程及设备的技术都有所不同。

热效率 thermal efficiency

热效率＝光学效率（optical efficiency）－热管热损失/DNI×聚光镜面积

附　　录

一、风力发电机结构图解

1. 风力发电机

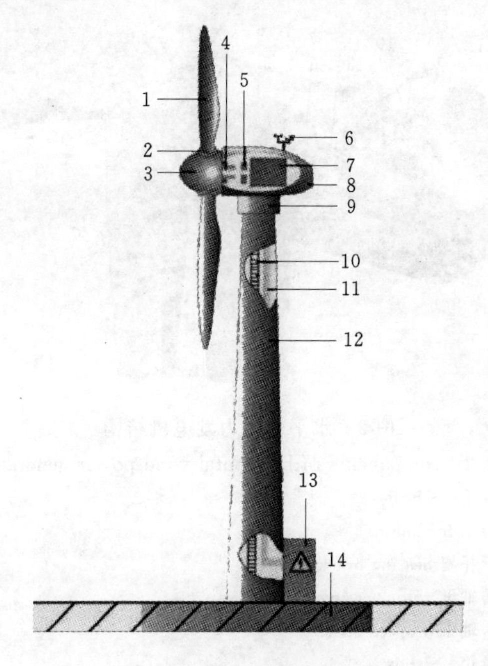

图 1 风力发电机
Fig. 1 wind power generator

1. 转轮叶片 rotor blades
2. 桨叶调整系统 pitch regulating system
3. 轮毂 rotor hub
4. 变速箱 gearbox
5. 刹车装置 brake unit
6. 测量装置 measuring instrument
7. 发电机 generator
8. 机舱 nacelle
9. 风向跟踪系统 yawing system
10. 升梯 lift
11. 电缆 power cable
12. 塔架 tower
13. 箱变 box-type transformer
14. 地基 foundation

2. 水平轴风力发电机结构

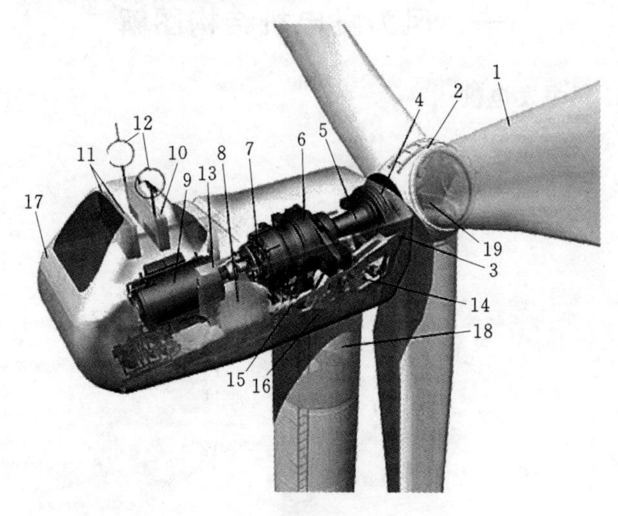

图 2 水平轴风力发电机结构

Fig. 2 components of horizontal wind power generator

1. 叶片 rotor blades
2. 轮毂 rotor hub
3. 轴承托架 bearing bracket
4. 回转轴承 rotor bearing
5. 低速轴 low speed shaft
6. 齿轮箱 gearbox
7. 联轴器 coupling
8. 高速轴 high speed shaft
9. 发电机 generator
10. 冷却系统 cooling system
11. 通风 ventilation
12. 风速风向仪 anemometer and wind vane
13. 控制柜 control panel
14. 偏航刹车 yaw brake
15. 偏航驱动 yawing driven
16. 偏航齿轮 yaw gear
17. 机舱罩 nacelle cover
18. 塔架 tower
19. 叶片轴承 blade bearing

3. 垂直轴风力发电机

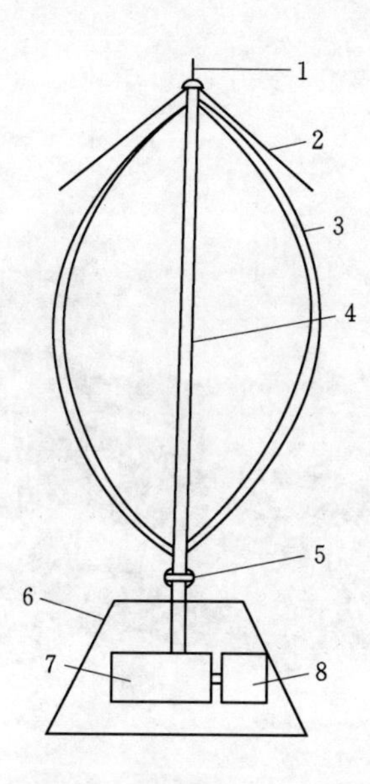

图 3 垂直轴风力发电机

Fig. 3 vertical wind power generator

1. 避雷针 lightning rod
2. 张线 bracing wire
3. 叶片 blades
4. 转轴 shaft
5. 抱闸刹车 band brake
6. 机架 main frame
7. 齿轮箱 gearbox
8. 发电机 generator

4. 直驱式风力发电机

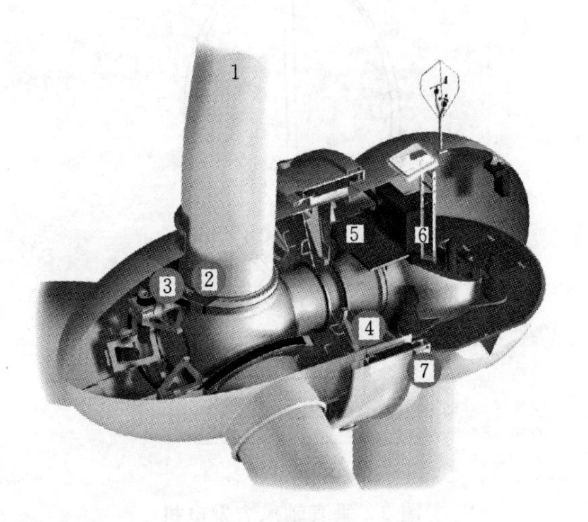

图 4 直驱式风力发电机

Fig. 4 directly drive wind power generator

1. 叶片 rotor blade
2. 铸造轮毂 cast hub
3. 变桨驱动 pitch drives
4. 发电机转子 generator rotor
5. 发电机定子 generator stator
6. 机座 base frame
7. 塔架 tower

二、风力等级划分标准（蒲福风级表）

风力等级	名称		相当于平地10m高处的风速		陆上地物征象	海面和渔船征象	海面大概的波高（m）	
	中文	英文	m/s	km/h			一般	最高
0	静风	Calm	0.0～0.2	小于1	静，烟直上	海面平静	—	—
1	软风	Light air	0.3～1.5	1～5	烟能表示风向，树叶略有摇动	微波如鱼鳞状，没有浪花。一艘渔船正好能使舵	0.1	0.1
2	轻风	Light breeze	1.6～3.3	6～11	人面感觉有风，树叶有微响旗子开始飘动，高的草开始摇动	小波，波长尚短，但波形显著，波峰光亮但不破裂。渔船张帆时，可随风移行每小时1～2海里	0.2	0.3
3	微风	Gentle breeze	3.4～5.4	12～19	树叶及小枝摇动不息，旌旗展开高的草，摇动不息	小波加大，波峰开始破裂，浪沫光亮，有时可有散见的白浪花。渔船开始簸动，张帆随风移行每小时3～4海里	0.6	1.0
4	和风	Moderate breeze	5.5～7.9	20～28	能吹起地面灰尘和纸张，树枝摇动高的草呈波浪起伏	小浪，波长变长，白浪成群出现。渔船满帆时，可使船身倾于一侧	1.0	1.5
5	清劲风	Fresh breeze	8.0～10.7	29～38	有叶的小树摇摆，内陆的水面有小波，高的草、波浪起伏明显	中浪，具有较明显的长波形状；许多白浪形成（偶有飞沫）。渔船需缩帆一部分	2.0	2.5
6	强风	Strong breeze	10.8～13.8	39～49	大树枝摇动，电线呼呼有声，撑伞困难高的草，不时倾伏于地	轻度大浪开始形成：到处都有更大的白沫峰（有时有些飞沫）。渔船缩帆大部分，并注意风险	3.0	4.0

续表

风力等级	名称		相当于平地10m高处的风速		陆上地物征象	海面和渔船征象	海面大概的波高（m）	
	中文	英文	m/s	km/h			一般	最高
7	疾风	Near gale	13.9～17.1	50～61	全树摇动，大树枝弯下来，迎风步行感觉不便	轻度大浪，碎浪而成白沫沿风向呈条状	4.0	5.5
8	大风	Gale	17.2～20.7	62～74	可折毁小树枝，人迎风前行感觉阻力甚大	有中度的大浪，波长较长，波峰边缘开始破碎成飞沫片；白沫沿风向呈明显的条带。所有近海渔船都要靠港，停留不出	5.5	7.5
9	烈风	Strong gale	20.8～24.4	75～88	草房遭受破坏，屋瓦被掀起，大树枝可折断	狂浪，沿风向白沫成浓密的条带状。波峰开始翻转，飞沫可影响能见度。机帆船航行困难	7.0	10.0
10	狂风	Storm	24.5～28.4	89～102	树木可被吹倒，一般建筑物遭破坏	狂涛，波峰长而翻卷；白沫成片出现，沿风向呈白色浓密条带；整个海面呈白色；海面颠簸加大有震动感，能见度受影响，机帆船航行颇危险	9.0	12.5
11	暴风	Violent storm	28.5～32.6	103～117	大树可被吹倒，一般建筑物遭严重破坏	异常狂涛（中小船只可一时隐没在浪后）；海面完全被沿风向吹出的白沫片所掩盖；波浪到处破成泡沫；能见度受影响，机帆船遇之极危险	11.5	16.0
12	飓风	Hurricane	＞32.6	＞118	陆上少见，其摧毁力极大	空中充满了白色的浪花和飞沫；海面完全变白，能见度严重地受影响	14.0	—

三、风电专业常用符号

A rotor area，scaling factor 风轮面积，比例因子

a_{AB} transformation matrix from system A to B 系统A到B的变换矩阵

a axial induction factor，tower radius 轴向感应系数，塔架半径

a' tangential induction factor 切向感应系数

a_n Fourier coefficient 傅立叶系数

B number of blades 叶片数

B_n Fourier coefficient 傅立叶系数

C damping matrix 阻尼矩阵

C_p power coefficient 功率系数

C_T thrust coefficient 推力系数

C_l lift coefficient 升力系数

C_d drag coefficient 阻力系数

C_m moment coefficient 力矩系数

C_n normal load coefficient 法向载荷系数

C_t tangential load coefficient 切向载荷系数

C_θ azimuthal component of axial velocity 轴向速度的方位分量

c chord 弦长

D rotor diameter，drag 风轮直径，阻力

dA normal vector to area 面积的法向矢量

E modulus of elasticity 弹性模量

ED moment of centrifugal stiffness 离心刚度矩

EI moment of stiffness inertia 惯性矩

ES moment of stiffness 刚性力矩

F force (vector) 力（矢量）

F_g generalized force (vector) 广义力（矢量）

F force，Prandtl's tip loss correction 力，普朗特叶尖损失修正

f external body force (vector) 外部质量力（矢量）

f force，probability，frequency 力，概率，频率

f_s separation function 分离函数

f_n frequency in discrete Fourier transformation 离散傅立叶变换频率

GI torsional stiffness 扭转刚度

H tower height，form factor 塔架高，形状系数

h　height above ground level　地面上高度

h_w　Weibull distribution　威布尔分布

I　moment of inertia，turbulence intensity　惯性矩，湍流强度

K　stiffness matrix　刚度矩阵

k　form factor　形状因子

k_t　terrain factor　地形因子

L　lift，distance between two points in space　升力，空间两点距离

l　length scale　湍流长度尺度

M　torque (vector)，mass matrix　力矩（矢量），质量矩阵

M　torque，aerodynamic moment　力矩，空气动力力矩

M_G　generator torque　发电机力矩

M_{flap}　flapwise bending moment　副翼方向弯曲力矩

M_a　Mach number　马赫数

$\dot{m}$　mass flow　质量流量

m　mass per length　单位长度质量

n　rotational speed of shaft　轴转速

P　momentum (vector)　动量（矢量）

P　power　功率

PSD　power spectral density function　功率谱密度函数

P　pressure，load　压强，载荷

p_N　load normal to rotor plane　风轮法向载荷

p_T　load tangential to rotor plane　风轮切向载荷

p_c　centrifugal load　离心载荷

Q　rate of heat transfer　传热量

Re　Reynolds number　雷诺数

R　rotor radius，resistance　风轮半径，电阻

r　radius (vector)　半径（矢量）

SL slip　转差率

S_{ij}　coherence function　相干函数

T　thrust，total time　推力，总时间

t　time　时间

U　boundary layer edge velocity　边界层外缘速度

u　x-component of velocity vector，axial velocity at rotor plane，deflection　x 轴速度分量，风轮轴向速度，挠度

u_1　velocity in wake　尾涡速度

u_i internal energy 内能

structural velocity 结构速度

$\ddot{u}$ structural acceleration 构造加速度

u^{1f} flapwise eigenmode deflection first 副翼方向固有模式第一偏转

u^{1e} edgewise eigenmode deflection first 沿边固有模式第一偏转

u^{2f} flapwise eigenmode deflection second 副翼方向固有模式第二偏转

V velocity（vector）速度（矢量）

V_b blade velocity（vector）叶片速度（矢量）

V_o wind speed 风速

V_∞ velocity at infinity 无穷远速度

V_{rel} relative velocity to aerofoil 相对机翼速度

V_θ tangential velocity component 切向速度分量

V_2 velocity in rotor plane for a shrouded rotor 遮挡风轮平面内速度

V_{10min} time averaged wind speed over a period of 10 minutes 10 分钟周期内的时均风速

V_{2s} time averaged wind speed over a period of 2 seconds 2 秒钟周期内的时均风速

v angle between chord line and first principal axis 弦长与主轴夹角

v y-component of velocity vector y 轴速度分量

w z-component of velocity vector z 轴速度分量

w induced velocity 诱导速度

W_y tangential component of induced velocity 诱导速度切向分量

W_z normal component of induced velocity 诱导速度法向分量

q_{2s} dynamic pressure based on V_{2s} 基于 V_{2s} 的动压

x local tip speed ratio 局部叶尖速率

z_o roughness length 粗糙长度

四、风电专业常用希腊字母

α angle of attack 冲角

β twist of blade 叶片扭角

Γ circulation 环量

Δt time increment 时间增量

δ boundary layer thickness 边界层厚度

δ^* displacement thickness 位移厚度

ε augmentation factor, strain 增大系数，张力

θ momentum thickness, local pitch 动量厚度，局部齿距

θ_{cone} cone angle 锥角

θ_{p} pitch angle 桨距角

θ_{o} azimuthal position where blade is deepest into the wake 尾流区内的叶片方位角

θ_{wing} azimuthal position of blade 叶片方位角

θ_{yaw} yaw angle 偏航角

κ curvature about the principal axis 主轴曲率

λ tip speed ratio 叶尖速比

υ kinematic viscosity, wind shear exponent 运动黏度系数，风切变指数

ρ density 密度

σ solidity, stress, standard deviation 体积，压力，标准偏差

σ_{r} stress range 应力范围

σ_{m} mean stress 平均应力

τ shear stress, time constant 剪切应力，时间常数

φ flow angle 流动角

χ wake skew angle 尾流倾斜角度

ω angular velocity of rotor, eigenfrequency 风轮角速度，本征频率

ω_{n} frequency in discrete Fourier transformation 离散傅立叶变换频率

五、常用数学式的读法

$\frac{1}{2}$ a half 或 one half

$\frac{1}{3}$ a third 或 one third

$\frac{2}{3}$ two thirds

$\frac{1}{4}$ a quarter 或 one quarter; a fourth 或 one fourth

$\frac{1}{10}$ a tenth 或 one tenth

$\frac{1}{100}$ a (one) hundredth

$\frac{1}{1000}$ a (one) thousandth

$\frac{1}{1234}$ one over a thousand two hundred and thirty four

$\frac{3}{4}$ three fourths 或 three quarters

$\frac{4}{5}$ four fifths 或 four over five

$\frac{113}{300}$ one hundred and thirteen over three hundred

$2\frac{1}{2}$ two and a half

$2\frac{7}{8}$ two and sever over eight 或 two and seven eighths

$3\frac{1}{8}$ three and one eighth

$4\frac{1}{3}$ four and a third

$125\frac{3}{4}$ a (one) hundred twenty-five and three fourths (quarters)

0.1（及.1） O point one 或 zero point one 或 nought point one

0.01（及.01） O point O one 或 zero point zero one 或 nought point nought one

0.250（及025） nought point two five 或 point two five

0.045 decimal (point) nought four five

235 two point three five

$4.\dot{9}$ four point nine recurring

$3.03\dot{2}\dot{6}$ three point nought three two six, two six recurring

45.67 four five (forth-five) point six seven

38.72 three eight point seven two 或 thirty-eight decimal seven two

0.001（及.001） O point O O one 或 nought point nought nought One 或 zero point zero zero one 或 point nought nought one

六、常用数学用语

Where alternative ways of saying the expressions are given, the first is generally more formal or technical.

在给出两种可供选的说法时，第一种一般是比较正式的或专业性的。

+	plus/and	加(上)
	positive	正的
—	minus/take away	减(去)
	negative	负的
±	plus or minus	加减
	positive or negative	正负
×	(is) multiplied by/times (or when giving dimensions) by	乘(或在表示面积、体积时说成),乘以
÷	(is) divided by	除以
=	is equal to/equals	等于
≠	is not equal to/does not equal	不等于
≈	is approximately equal to	约等于
≡	is equivalent to/is identical with	全等于
<	is less than	小于
≮	is not less than	不小于
⩽	is less than or equal to	小于或等于
>	is more than	大于
≯	is not more than	不大于
≧	is more than or equal to	大于或等于
%	percent	百分之……
∞	infinity	无限大
∝	(varies as/is proportional to 3 : 9 :: 4 : 12) three is to nine as four is to twelve	(随……而不同,与……成比例)3 比 9 等于 4 比 12
$\log_e$	natural logarithm or logarithm to the base e/:/	自然对数或以 e 为底数的对数
$\sqrt{\ }$	(square) root	平方根
$\sqrt[3]{\ }$	cube root	立方根
x^2	x/eks/squared	x 平方
x^3	x/eks/cubed	x 立方
x^4	x/eks/to the power of four/to the fourth	x 的四次方(或四次幂)
p	pi/pai/	圆周率
r	/a:(r)/=radius of a circle	半径
∫	the integral of	……的积分
°	dearee	度
′	minute (or an arc); foot or feet (unit of length)	(弧的)分,英尺(长度单位)
″	second (of an arc);inch or inches (unit of length)	(弧的)秒;英寸(长度单位)

七、单位制及换算
Units and their conversions

1. 国际单位制

SI Base Units

Quantity 量	Name of unit 单位名称	Symbol 单位符号
length 长度	meter 米	m
mss 质量	kilogram 千克	kg
time 时间	second 秒	s
electrical current 电流	ampere 安［培］	A
thermodynamic temperature 热力学温度	kelvin 开［尔文］	K
luminous intensity 发光强度	candela 坎［德拉］	cd
amount of a substance 物质的量	mole 摩［尔］	mol

* SI System International

2. 国际单位制导出单位

SI derived Units

Quantity 量	Name of unit 单位名称	Symbol 单位符号
acceleration 加速度	meter per second squared 米/秒^2	m/s^2
angular acceleration 加速度	radian per second square 弧度/秒^2	rad/s^2
angular velocity 角速度	radian per square 弧度/秒	rad/s
area 面积	square meter 平方米	m^2
(electrical) capacitance 电容	farad 法拉	F

续表

Quantity 量	Name of unit 单位名称	Symbol 单位符号
capacity rate 力率	Watts per degree 瓦特/度（开尔文）	W/K
density 密度	kilogram per cubic meter 千克/立方米	kg/m^3
dynamic viscosity 动力黏度	newton-second per square meter 牛顿·秒/平方米	$N \cdot s/m^2$
electrical conductivity 电导率	ampere per volt meter 安培/伏特·米	$A/(V \cdot m)$
electrical field strength 电场强度	volt per meter 伏特/米	V/m
electrical resistance 电阻	ohm 欧姆	Ω
electromotive force 电动势，电位	volt 伏特	V
force 力	newton 牛顿	N
frequency 频率	hertz 赫兹，周期/秒	Hz
heat flux density 热流密度	watts per square meter 瓦特/平方米	W/m^2
heat transfer coefficient 传热系数	watts per square meter degree 瓦特/平方米·度	W/mK^2
illumination 光照度	lux，lumen per square meter 勒克斯，流明/平方米	lx lm/m^2
inductance 电感	henry 亨利	H
kinematic viscosity 运动黏度	square meter per second 平方米/秒	m^2/s
luminous flux 光通量	lumen 流明	lm

续表

Quantity 量	Name of unit 单位名称	Symbol 单位符号
luminance 光亮度	candela per square meter 坎特拉/平方米	cd/m^2
magnetic 磁通量	weber 韦伯	Wb
magnetic flux density 磁感应强度	tesla，weber per square meter 特斯拉，韦伯/平方米	T，Wb/m^2
magnetic field strength 磁场强度	ampere per meter 安培/米	A/m
magnetic permeability 磁导率	henry per meter 亨利/米	H/m
magnetomotive force 磁动势	ampere 安培	A
plane angle 平面角	radian 弧度	rad
power，heat flux 功率，热流	watt 瓦特	W
quantity of electricity 电量	coulomb 库伦	C
pressure 压力，压强	pascal 帕斯卡	Pa
radiant intensity 辐射强度	watts per square meter，steradian 瓦特/平方米	W/m^2
stolid angle 立体角	sferadian 球面度	Sr
specific heat 比热	joules per kilogram kelvin 焦耳/千克·开尔文	J/kg·K
surface tension 表面张力	newton per meter 牛顿/米	N/m
thermal conductivity 热导率	watts per meter kelvin 瓦特/米·开尔文	W/(m·K)

续表

Quantity 量	Name of unit 单位名称	Symbol 单位符号
velocity 速度	meter per second 米/秒	m/s
volume 容积	cubic meter 立方米	m^3
volumetric flow rate 体积流量，容积流量	cubic meter per second 立方米/秒	m^3/s
work，energy，heat，torque 功、能、热、转矩	joule 焦耳	J

* 国际单位的辅助单位只有两个，即：平面角（Plane angle）和立体角（Solid angle）。
按规定，它们可以随意作为基本单位或导出单位。

八、度 量 衡 表
Tables of Measures and Weights

1. 公制 The Metric System

长度 Linear measure

Name of unit 单位名称	Symbol 符号	Ratio to the primary unit 与基本单位的比
micron 微米	μ	1/1000000 m
centimillimeter 忽米	cmm	1/100000 m
decimillimeter 丝米	dmm	1/10000 m
millimeter 毫米	mm	1/1000 m
centimeter 厘米	cm	1/100 m
decimeter 分米	dm	1/10 m
meter 米	m	1 m
decameter 十米	dam	10 m
hectometer 百米	hm	100 m
kilometer 公里	km	1000 m

2. 体积，容积 Capacity measure

Name of unit 单位名称	Symbol 符号	Ratio to the primary unit 与基本单位的比
milliliter 毫升	ml	1/1000 l
centiliter 厘升	cl	1/100 l
deciliter 分升	dl	1/10 l
litre 升	l	1
decalitre 十升	dal	10 l
hectolitre 百升	hl	100 l
kilolitre 千升	kl	1000 l

3. 质量 Mass

Name of unit 单位名称	Symbol 符号	Ratio to the primary unit 与基本单位的比
milligram 毫克	mg	1/1000000 kg
centigram 厘克	cg	1/100000 kg
decigram 分克	dg	1/10000 kg
gram 克	g	1/1000 kg
decagram 十克	dag	1/100 kg
hectogram 百克	hg	1/10 kg
kilogram 公斤	kg	1
quintal 公担	q	100 kg
ton 吨	t	1000 kg

4. 面积及地积 Area

Name of unit 单位名称	Symbol 符号	Metric value 折合公制
square mile 平方英里	sq. mi.	2.59 平方公里
acre 英亩	a.	4.047 平方米
square yard 平方码	sq. yd.	0.8361 平方米
square foot 平方英尺	sq. ft.	926 平方厘米
square inch 平方英亩	sq. in.	6.451 平方厘米

九、单 位 换 算
Conversion Factors

1. 长度单位换算 Length

Unit 单位	cm 厘米	m 米	in 英寸	ft 英尺	yd 码	mile 英里
1cm 厘米	1	0.01	0.3937	0.03281	0.01094	6.2137×10^{-6}
1m 米	100	1	39.37	3.281	1.0936	6.2137×10^{-4}
1in 英寸	2.54	0.0254	1	0.08333	0.02777	1.5783×10^{-5}
1ft 英尺	30.48	0.3048	12	1	0.3333	1.8939×10^{-4}
1yd 码	91.44	0.9144	36	3	1	5.6818×10^{-4}
1mile 英里	1.609×10^{5}	1609	6.336×10^{4}	5280	1760	1

2. 面积单位换算 Area

Unit 单位	cm^2 平方厘米	m^2 平方米	in^2 平方英寸	ft^2 平方英尺	yd^2 平方码	$mile^2$ 平方英里
$1cm^2$	1	10^{-4}	0.1550	1.0764×10^{-3}	1.1960×10^{-4}	3.8610×10^{-11}
$1m^2$	10^4	1	1550.003	10.7639	1.1960	3.8610×10^{-7}
$1in^2$	6.4516	6.4516×10^{-4}	1	6.9444×10^{-3}	7.7160×10^{-4}	2.4910×10^{-10}
$1ft^2$	929.0304	0.09290	144	1	0.1111	3.5870×10^{-8}
$1yd^2$	8361.273	0.8361	1296	9	1	3.2283×10^{-7}
$1mile^2$	2.5900×10^{10}	2.5900×10^{6}	4.0145×10^{9}	2.7878×10^{7}	3.0976×10^{6}	1